Small Area Estimation and Microsimulation Modeling

Small Area Estimation and Microsimulation Modeling

Azizur Rahman
Charles Sturt University
Australia

Ann Harding
University of Canberra
Australia

CRC Press
Taylor & Francis Group
Boca Raton London New York

CRC Press is an imprint of the
Taylor & Francis Group, an **informa** business
A CHAPMAN & HALL BOOK

CRC Press
Taylor & Francis Group
6000 Broken Sound Parkway NW, Suite 300
Boca Raton, FL 33487-2742

First issued in paperback 2019

© 2017 by Taylor & Francis Group, LLC
CRC Press is an imprint of Taylor & Francis Group, an Informa business

No claim to original U.S. Government works

ISBN-13: 978-1-4822-6072-4 (hbk)
ISBN-13: 978-0-367-26126-9 (pbk)

Visit the Taylor & Francis Web site at
http://www.taylorandfrancis.com

and the CRC Press Web site at
http://www.crcpress.com

To my parents, Gulam Rubbany Mondal and Halima Rubbany

and

my loved ones, Rani and Ayra

Azizur Rahman

To my children

Ann Harding

Contents

List of Figures xiii
List of Tables xvii
Preface xxi
Acknowledgments xxiii
List of Abbreviations xxv

1. Introduction 1
  1.1   Introduction 1
  1.2   Main Aims of the Book 5
  1.3   Guide for the Reader 6
  1.4   Concluding Remarks 10

2. Small Area Estimation 11
  2.1   Introduction 11
  2.2   Small Area Estimation 11
        2.2.1   Concept of Small Area 12
        2.2.2   Advantages of SAE 12
        2.2.3   Why SAE Techniques? 13
        2.2.4   Applications of SAE 13
  2.3   Approaches to SAE 15
  2.4   Direct Estimation 17
        2.4.1   H-T Estimator 17
        2.4.2   Generalized Regression Estimator 18
        2.4.3   Modified Direct Estimator 18
        2.4.4   Design-Based Model-Assisted Estimators 19
        2.4.5   A Comparison of Direct Estimators 22
  2.5   Concluding Remarks 24

3. Indirect Estimation: Statistical Approaches 27
  3.1   Introduction 27
  3.2   Implicit Models Approach 28
        3.2.1   Synthetic Estimation 28
        3.2.2   Composite Estimation 29
        3.2.3   Demographic Estimation 30
        3.2.4   Comparison of Various Implicit Models–Based
                Indirect Estimation 32
  3.3   Explicit Models Approach 33
        3.3.1   Basic Area Level Model 33
        3.3.2   Basic Unit Level Model 35

      3.3.3   Generalized Linear Mixed Model 36
      3.3.4   Comparison of Various Explicit Models–Based
               Indirect Estimation 41
  3.4   Methods for Estimating Explicit Models 42
      3.4.1   EBLUP Approach 42
      3.4.2   EB Approach 43
      3.4.3   HB Approach 45
  3.4   A Comparison of Three Methods 47
  3.5   Concluding Remarks 49

4. Indirect Estimation: Geographic Approaches 51
  4.1   Introduction 51
  4.2   Microsimulation Modeling 52
      4.2.1   Process of Microsimulation 52
      4.2.2   Types of Microsimulation Models 54
               4.2.2.1   Static Microsimulation 55
               4.2.2.2   Dynamic Microsimulation 57
               4.2.2.3   Spatial Microsimulation 59
      4.2.3   Advantages of Microsimulation Modeling 68
  4.3   Methodologies in Microsimulation Modeling Technology 69
      4.3.1   Techniques for Creating Spatial Microdata 69
      4.3.2   Statistical Data Matching or Fusion 70
      4.3.3   Iterative Proportional Fitting 72
      4.3.4   Repeated Weighting Method 74
      4.3.5   Reweighting 80
  4.4   CO Reweighting Approach 81
      4.4.1   Simulated Annealing Method in CO 83
      4.4.2   Illustration of CO Process for Hypothetical Data 85
  4.5   Reweighting: The GREGWT Approach 87
      4.5.1   Theoretical Setting 88
      4.5.2   How Does GREGWT Generate New Weights? 90
      4.5.3   Explicit Numerical Solution for Hypothetical Data 91
  4.6   Comparison between GREGWT and CO 97
  4.7   Concluding Remarks 99

5. Bayesian Prediction–Based Microdata Simulation 101
  5.1   Introduction 101
  5.2   Basic Steps 102
  5.3   Bayesian Prediction Theory 103
  5.4   Multivariate Model 103
  5.5   Prior and Posterior Distributions 106
  5.6   The Linkage Model 109
  5.7   Prediction for Modeling Unobserved Population Units 110
  5.8   Concluding Remarks 117

6. Microsimulation Modeling Technology for Small Area Estimation... 119
  6.1   Introduction 119
  6.2   Data Sources and Issues.. 120
          6.2.1   Census Data 120
          6.2.2   Survey Data Sets 122
  6.3   Microsimulation Modeling Technology–Based Model Specification.. 124
          6.3.1   Model Inputs.. 125
                  6.3.1.1   General Model File............................... 126
                  6.3.1.2   Unit Record Data Files..................... 126
                  6.3.1.3   Benchmark Files 128
                  6.3.1.4   Auxiliary Data Files 128
                  6.3.1.5   GREGWT File...................................... 132
          6.3.2   Generating Small Area Synthetic Weights...................... 132
          6.3.3   Model Outputs 134
  6.4   Housing Stress... 136
          6.4.1   Definition 136
          6.4.2   Measures of Housing Stress..................................... 136
          6.4.3   Comparison of Various Measures 138
  6.5   Small Area Estimation of Housing Stress 141
          6.5.1   Inputs at the Second-Stage Model 141
                  6.5.1.1   Consumer Price Index File................ 141
          6.5.2   Model Execution Process.. 142
          6.5.3   Final Model Outputs 143
  6.6   Concluding Remarks 144

7. Applications of the Methodologies... 147
  7.1   Introduction 147
  7.2   Results of the Model: A General View 147
          7.2.1   Model Accuracy Report 147
          7.2.2   Scenarios of Housing Stress under Various Measures..... 148
          7.2.3   Distribution of Housing Stress Estimation 150
          7.2.4   Lorenz Curve for Housing Stress Estimates................... 151
          7.2.5   Proportional Cumulative Frequency Graph and Index of Dissimilarity 152
          7.2.6   Scenarios of Households and Housing Stress by Tenures 154
  7.3   Estimation of Households in Housing Stress by Spatial Scales... 155
          7.3.1   Results for Different States 155
          7.3.2   Results for Various Statistical Divisions.......................... 157
          7.3.3   Results for Various Statistical Subdivisions.................... 159

7.4   Small Area Estimates: Number of Households in
       Housing Stress.. 162
       7.4.1   Estimated Numbers of Overall Households in
               Housing Stress.. 165
       7.4.2   Estimated Numbers of Buyer Households in
               Housing Stress.. 166
       7.4.3   Estimated Numbers of Public Renter Households
               in Housing Stress.. 167
       7.4.4   Estimated Numbers of Private Renter Households
               in Housing Stress.. 169
       7.4.5   Estimated Numbers of Total Renter Households in
               Housing Stress.. 170
7.5   Small Area Estimates: Percentage of Households in
       Housing Stress.. 171
       7.5.1   Percentage Estimates of Housing Stress for Overall
               Households 171
       7.5.2   Percentage Estimates of Housing Stress for Buyer
               Households 174
       7.5.3   Percentage Estimates of Housing Stress for Public
               Renter Households 175
       7.5.4   Percentage Estimates of Housing Stress for Private
               Renter Households 176
       7.5.5   Percentage Estimates of Housing Stress for Total
               Renter Households 177
7.6   Concluding Remarks.. 177

8. Analysis of Small Area Estimates in Capital Cities............ 181
8.1   Introduction 181
       8.1.1   Scenarios of the Results for Major Capital Cities 182
       8.1.2   Trends in Housing Stress for Some Major Cities............. 183
       8.1.3   Mapping the Estimates at SLA Levels within
               Major Cities.. 184
8.2   Sydney... 186
       8.2.1   Housing Stress Estimates for Overall Households......... 186
       8.2.2   Small Area Estimation by Households' Tenure Types... 188
               8.2.2.1   Estimates for Buyers 188
               8.2.2.2   Estimates for Public Renters 188
               8.2.2.3   Estimates for Private Renters.................. 189
               8.2.2.4   Estimates for the Total Renters.............. 189
8.3   Melbourne.. 190
       8.3.1   Housing Stress Estimates for Overall Households......... 191
       8.3.2   Small Area Estimation by Households' Tenure Types.... 192
               8.3.2.1   Estimates for Buyers 192
               8.3.2.2   Estimates for Public Renters 192

       8.3.2.3   Estimates for Private Renters.............................. 193

       8.3.2.4   Estimates for Total Renters 194

8.4   Brisbane... 195

    8.4.1   Housing Stress Estimates for Overall Households......... 196

    8.4.2   Small Area Estimation by Households' Tenure Types....197

       8.4.2.1   Estimates for Buyers 197

       8.4.2.2   Estimates for Public Renters 197

       8.4.2.3   Estimates for Private Renters.............................. 198

       8.4.2.4   Estimates for the Total Renters........................... 198

8.5   Perth... 199

    8.5.1   Housing Stress Estimates for Overall Households.........200

    8.5.2   Small Area Estimation by Households' Tenure Types....200

       8.5.2.1   Estimates for Buyers 200

       8.5.2.2   Estimates for Public Renters 201

       8.5.2.3   Estimates for Private Renters..............................202

       8.5.2.4   Estimates for the Total Renters...........................203

8.6   Adelaide...204

    8.6.1   Housing Stress Estimates for Overall Households.........204

    8.6.2   Small Area Estimation by Households' Tenure Types....205

       8.6.2.1   Estimates for Buyers 205

       8.6.2.2   Estimates for Public Renters 206

       8.6.2.3   Estimates for Private Renters..............................207

       8.6.2.4   Estimates for the Total Renters...........................207

8.7   Canberra...208

    8.7.1   Housing Stress Estimates for Overall Households.........209

    8.7.2   Small Area Estimation by Households' Tenure Types....210

       8.7.2.1   Estimates for Buyers 210

       8.7.2.2   Estimates for Public Renters 210

       8.7.2.3   Estimates for Private Renters.............................. 211

       8.7.2.4   Estimates for the Total Renters........................... 211

8.8   Hobart... 212

    8.8.1   Housing Stress Estimates for Overall Households......... 212

    8.8.2   Small Area Estimation by Households' Tenure Types....213

       8.8.2.1   Estimates for Buyers 213

       8.8.2.2   Estimates for Public Renters 213

       8.8.2.3   Estimates for Private Renters.............................. 214

       8.8.2.4   Estimates for the Total Renters........................... 214

8.9   Darwin.. 215

    8.9.1   Housing Stress Estimates for Overall Households......... 216

    8.9.2   Small Area Estimation by Households' Tenure Types....217

       8.9.2.1   Estimates for Buyers 217

       8.9.2.2   Estimates for Public Renters 217

       8.9.2.3   Estimates for Private Renters.............................. 218

       8.9.2.4   Estimates for the Total Renters........................... 218

8.10 Concluding Remarks 219

9. Validation and Measure of Statistical Reliability 221
    9.1   Introduction 221
    9.2   Some Validation Methods in the Literature 222
    9.3   New Approaches to Validating Housing Stress Estimation 225
           9.3.1   Statistical Significance Test of the MMT Estimates 225
           9.3.2   Results of the Statistical Significance Test 228
           9.3.3   Absolute Standardized Residual Estimate Analysis 235
           9.3.4   Results from the ASRE Analysis 236
    9.4   Measure of Statistical Reliability of the MMT Estimates............ 239
           9.4.1   Confidence Interval Estimation 240
           9.4.2   Results from the Estimates of Confidence Intervals....... 242
    9.5   Concluding Remarks 246

10. Conclusions and Computing Codes 249
    10.1   Introduction 249
    10.2   Summary of Major Findings 249
    10.3   Limitations 257
    10.4   Areas of Further Studies 259
    10.5   Computing Codes and Programming 260
           10.5.1 The General Model File Codes... 260
           10.5.2 SAS Programming for Reweighting Algorithms 269
           10.5.3 The Second-Stage Program File Codes 283
    10.6   Concluding Remarks 294

References 295

Appendix A: The Newton–Raphson Iteration Method 321

Appendix B: Topics Index of the 2005–2006 Survey of Income and Housing: CURFs 325

Appendix C: Tables of the Housing Stress for 50 SLAs with the Highest Numbers and Percentages Estimates 341

Appendix D: Distribution of SLAs, Households, and Housing Stress by SSDs in Eight Major Capital Cities 361

Appendix E: Spatial Analyses by Households Tenure Types for the Eight Capital Cities 365

Appendix F: SAS Programming for the Reweighting Algorithms from Parts 2 to 10 441

Index 485

List of Figures

Figure 2.1 A summary of different techniques and estimators for SAE.....16

Figure 4.1 Data matching. 71

Figure 4.2 A simple example of repeated weighting. 76

Figure 4.3 A flowchart of the simulated annealing algorithm. 85

Figure 4.4 A simplified combinatorial optimization process. 86

Figure 4.5 A comparison of absolute distance (AD) and chi-squared distance measures. 95

Figure 4.6 Plots of sampling design weights and new weights for specific cases. 96

Figure 5.1 A diagram of new system for generating spatial microdata. 102

Figure 6.1 Overall process for the creation of small area synthetic weights. 133

Figure 6.2 Distribution of housing stress for three variants in Australia 2006. 140

Figure 6.3 Computation process of microsimulation modeling technology to produce small area microdata and synthetic estimates. 143

Figure 7.1 A scenario of the number of households in housing stress by three different measures. 149

Figure 7.2 Distribution of housing stress at small area levels. 150

Figure 7.3 A Lorenz curve diagram for low-income households in housing stress. 152

Figure 7.4 A proportional cumulative frequency graph with the index of dissimilarity. 153

Figure 7.5 Percentage distribution of households by tenure types. 154

Figure 7.6 Percentage distribution of households in housing stress by tenure types. 155

Figure 7.7 A spatial distribution of the small area estimates of housing stress with an alternative cumulative frequency legend. 163

Figure 7.8 A spatial distribution of the percentage estimates of
housing stress with an alternative cumulative frequency
legend. 172

Figure 8.1 Distribution of housing stress by tenure in the
eight capital cities. 184

Figure 8.2 Trends in housing stress estimates for some major
cities of Australia (a) for estimated numbers and (b) for
percentage estimates. 185

Figure 9.1 The standard normal distribution of Z-statistic for a
two-sided statistical hypothesis test 227

Figure 9.2 Results of the hypothesis test at statistical local area
(SLA)–level housing stress estimates for overall
households. 228

Figure 9.3 Results of the hypothesis test for different households by
tenure types. 233

Figure 9.4 Absolute standardized residual estimate analysis for
MMT estimates of total households in housing stress. 236

Figure 9.5 Absolute standardized residual estimate (ASRE) analyses
for statistical local areas (SLAs) in eight major capital cities. 237

Figure 9.6 Measure of confidence intervals of the statistical
local area–level housing stress estimates for overall
households. 242

Figure 9.7 Confidence interval (CI) measures for estimates of
housing stress for major capital. 244

Figure A.1 Graphical representation of the Newton–Raphson
iteration process. 322

Figure E.1 Spatial distributions of the estimates of housing stress for
overall households in Sydney. 365

Figure E.2 Spatial distributions of the estimates of housing stress for
buyer households in Sydney. 367

Figure E.3 Spatial distributions of the estimates of housing stress for
public renter households in Sydney. 369

Figure E.4 Spatial distributions of the estimates of housing stress for
private renter households in Sydney. 371

Figure E.5 Spatial distributions of the estimates of housing stress for
total renter households in Sydney. 373

Figure E.6 Spatial distributions of the estimates of housing stress for overall households in Melbourne. 375

Figure E.7 Spatial distributions of the estimates of housing stress for buyer households in Melbourne. 377

Figure E.8 Spatial distributions of the estimates of housing stress for public renter households in Melbourne. 379

Figure E.9 Spatial distributions of the estimates of housing stress for private renter households in Melbourne. 381

Figure E.10 Spatial distributions of the estimates of housing stress for total renter households in Melbourne. 383

Figure E.11 Spatial distributions of the estimates of housing stress for overall households in Brisbane. 385

Figure E.12 Spatial distributions of the estimates of housing stress for buyer households in Brisbane. 387

Figure E.13 Spatial distributions of the estimates of housing stress for public renter households in Brisbane. 389

Figure E.14 Spatial distributions of the estimates of housing stress for private renter households in Brisbane. 391

Figure E.15 Spatial distributions of the estimates of housing stress for total renter households in Brisbane. 393

Figure E.16 Spatial distributions of the estimates of housing stress for overall households in Perth. 395

Figure E.17 Spatial distributions of the estimates of housing stress for buyer households in Perth. 397

Figure E.18 Spatial distributions of the estimates of housing stress for public renter households in Perth. 399

Figure E.19 Spatial distributions of the estimates of housing stress for private renter households in Perth. 401

Figure E.20 Spatial distributions of the estimates of housing stress for total renter households in Perth. 403

Figure E.21 Spatial distributions of the estimates of housing stress for overall households in Adelaide. 405

Figure E.22 Spatial distributions of the estimates of housing stress for buyer households in Adelaide. 407

Figure E.23 Spatial distributions of the estimates of housing stress for public renter households in Adelaide. 409

Figure E.24   Spatial distributions of the estimates of housing stress
for private renter households in Adelaide. 411

Figure E.25   Spatial distributions of the estimates of housing stress
for total renter households in Adelaide. 413

Figure E.26   Spatial distributions of the estimates of housing stress for
overall households in Canberra. 415

Figure E.27   Spatial distributions of the estimates of housing stress for
buyer households in Canberra. 417

Figure E.28   Spatial distributions of the estimates of housing stress
for public renter households in Canberra. 419

Figure E.29   Spatial distributions of the estimates of housing stress for
private renter households in Canberra. 421

Figure E.30   Spatial distributions of the estimates of housing stress
for total renter households in Canberra. 423

Figure E.31   Spatial distributions of the estimates of housing stress
for overall households in Hobart. 425

Figure E.32   Spatial distributions of the estimates of housing stress for
buyer households in Hobart. 427

Figure E.33   Spatial distributions of the estimates of housing stress
for public renter households in Hobart. 428

Figure E.34   Spatial distributions of the estimates of housing stress
for private renter households in Hobart. 429

Figure E.35   Spatial distributions of the estimates of housing stress
for total renter households in Hobart. 430

Figure E.36   Spatial distributions of the estimates of housing stress for
overall households in Darwin. 431

Figure E.37   Spatial distributions of the estimates of housing stress
for buyer households in Darwin. 433

Figure E.38   Spatial distributions of the estimates of housing stress
for public renter households in Darwin. 435

Figure E.39   Spatial distributions of the estimates of housing stress
for private renter households in Darwin. 437

Figure E.40   Spatial distributions of the estimates of housing stress
for total renter households in Darwin. 439

List of Tables

Table 2.1   Comparison of various direct estimators for SAE 23

Table 3.1   A comparative table for implicit models–based indirect
            estimators 33

Table 3.2   Comparative table for various explicit models 41

Table 3.3   Properties of design-based model-assisted and
            model-based estimators for small areas or domains................... 47

Table 3.4   A comparison of EBLUP, EB, and HB approaches....................... 48

Table 4.1   Characteristics of microsimulation models 54

Table 4.2   Synthetic reconstruction versus the reweighting technique...... 81

Table 4.3   New weights and its distance measures to sampling
            design weights.. 94

Table 4.4   Comparison of the GREGWT and CO reweighting
            methodologies 98

Table 6.1   Benchmarks and benchmark classes used in reweighting....... 129

Table 6.2   Outlook of the household *synthetic weights* produced by the
            GREGWT algorithm for small area microdata........................... 135

Table 6.3   Comparison of the different measures of housing stress 139

Table 7.1   Accuracy report of the reweighting algorithm for creating
            synthetic new weights 148

Table 7.2   Estimates of the numbers of households and the housing
            stress by tenures for states and territories................................ 156

Table 7.3   Number of households in housing stress by statistical
            divisions 158

Table 7.4   Thirty-five SSDs with the highest housing stress estimates......... 160

Table 8.1   Scenarios of households and housing stress for
            eight capital cities... 182

Table 9.1   Area property of the curve for some important values of
            the Z-statistic.. 227

Table 9.2   Statistical local areas show insignificant Z-statistic with
            p-value <0.05 230

Table 9.3 Report on statistical decision for inaccurate microsimulation modeling technology estimates by tenure 234

Table B.1 Topics in the 2006 census of population and housing 326

Table B.2 BCP tables for the 2006 census available at *census data online* 327

Table B.3 Selected topics in the SIH—CURFs 329

Table C.1 Fifty SLAs with the highest number of households in housing stress 341

Table C.2 Fifty SLAs with the highest number of buyer households in housing stress 343

Table C.3 Fifty SLAs with the highest number of public renter households in housing stress 345

Table C.4 Fifty SLAs with the highest number of private renter households in housing stress 347

Table C.5 Fifty SLAs with the highest number of total renter households in housing stress 349

Table C.6 Fifty SLAs with the highest percentage of overall households in housing stress 351

Table C.7 Fifty SLAs with the highest percentage of buyer households in housing stress 353

Table C.8 Fifty SLAs with the highest percentage of public renter households in housing stress 355

Table C.9 Fifty SLAs with the highest percentage of private renter households in housing stress 357

Table C.10 Fifty SLAs with the highest percentage of total renter households in housing stress 359

Table D.1 Distributions of SLAs, Households, and housing stress by SSDs in sydney 361

Table D.2 Distributions of SLAs, households, and housing stress by SSDs in Melbourne 362

Table D.3 Distributions of SLAs, households, and housing stress by SSDs in Brisbane 362

Table D.4 Distributions of SLAs, households, and housing stress by SSDs in Perth 363

Table D.5   Distributions of SLAs, households, and housing stress by
             SSDs in Adelaide..363

Table D.6   Distributions of SLAs, households, and housing stress
             by SSDs in Canberra..363

Table D.7   Distributions of SLAs, households, and housing stress by
             SSD in Hobart..364

Table D.8   Distributions of SLAs, households, and housing stress by
             SSDs in Darwin..364

Preface

Small area estimation and microsimulation modeling share significant common goals; yet, as academic disciplines they have different roots and have grown apart. Most of the review articles in small area estimation highlight only methodologies that are fully based on various statistical models and theories, while another type of methodology called "microsimulation modeling technology" has emerged as a very robust and useful means for small area estimation and socioeconomic policy analysis, typically using different techniques based on geographic and economic theories. Even a casual survey of leading journals shows that cross-references to such methods are rare. This is unfortunate, as many real-world problems seek for a multidisciplinary approach. Both small area estimation and microsimulation modeling are necessary ingredients in any serious analysis of the sustainability of many social systems, including housing, education, and health care; in the evaluation of potential inequalities of various estimation-based allocations for funding and social welfare to local and small area governments; in the estimation of the size of elusive individuals or households such as households experiencing housing stress or poverty, smokers, drug users, the unemployed, or those with certain chronic diseases; in the investigation of the consequences of any social policy changes, such as family benefits; and so forth.

Reliable small area statistics are essential in this data-centric world for various reasons, including their worldwide rapidly growing use in formulating policies and programs, allocating funds and social welfare, evaluating the effects of policy changes and forecasting, regional level planning, small business and marketing decisions, and similar applications. As a result, small area estimation methods are becoming very useful tools across the globe where students, academics, researchers, and practitioners deal with understanding problems at the small area level on vital issues in *social and behavioral sciences, government and social statistics, applied economics and social policy analysis, health sciences, computational statistics, data simulation, business and marketing decision analysis, spatial statistics and modeling,* and related disciplines.

This book provides an extensive and comprehensive state-of-the-art presentation of all of the existing methodologies in small area estimation, including *direct* and *indirect model-based* estimation, and then to develop a novel and more robust methodology, termed the microsimulation modeling technology (MMT), for small area estimation with real-world applications. This book serves as a single source of reference to the readers on valuable information about theories, applications, advantages, and limitations of the usefulness of all of the small area estimation methodologies.

These methodologies encompass the direct small area estimation methods, indirect statistical model–based approaches (i.e., implicit and explicit) with empirical best linear unbiased prediction (EBLUP), empirical Bayes (EB), and hierarchical Bayes (HB) estimation methods; indirect geographic estimation methods (i.e., various techniques for creating spatial microdata); a newly developed and sophisticated MMT; and the means of statistical reliability measures, including confidence interval estimations for small area statistics. The book also offers all invaluable computing codes and programming used to build the models and produce empirical results from Australia.

Azizur Rahman
Wagga Wagga, New South Wales, Australia
Ann Harding
Canberra, Australian Capital Territory, Australia

Acknowledgments

We are deeply grateful to our colleagues Professors Robert Tanton and Shuangzhe Liu (University of Canberra), Professors John Lynch (University of Adelaide), Robert Stimson (University of Queensland), Paul Williamson (University of Liverpool), Raymond Chambers and David Steel (University of Wollongong), and Murray Aitkin (University of Melbourne) for their valuable time, excellent stimulus, and very useful assistance throughout the development of this book.

We gratefully acknowledge a range of funding from an E-IPRS Award from the Commonwealth of Australia, the Australian Capital Territory—Land Development Agency Research Award from the ACT Government through the Australian Housing and Urban Research Institute (AHURI), and the NATSEM Top-Up Research Award from the National Centre for Social and Economic Modelling (NATSEM). We also acknowledge a number of additional grants for research supports, conference attendances, and data-related aids from the Australian Research Council (ARC) research grants, the Spatially Integrated Social Sciences research network, AHURI, the Australian Bureau of Statistics, NATSEM, the University of Canberra, Wharton School at the University of Pennsylvania, the University of Adelaide, and Charles Sturt University. We sincerely thank four anonymous reviewers and three anonymous editors for their valuable insightful comments on the first draft, which were used to improve the book manuscript. We also thank Professor Risto Lehtonen at the University of Helsinki for his stimulus and useful review comments on a revised version, which certainly improved the final version of this book. Finally, we acknowledge all of the excellent professional communications and services from the acquisitions editors team, especially, from David Grubbs, responsible for Statistics at Chapman & Hall/CRC Press, Boca Raton, Florida.

Dr Azizur Rahman
Wagga Wagga, New South Wales, Australia
Professor Ann Harding
Canberra, Australian Capital Territory, Australia

List of Abbreviations

ABS	Australian Bureau of Statistics
ACT	Australian Capital Territory
AEMSE	Average empirical mean square error
AHURI	Australian Housing and Urban Research Institute
ASRE	Absolute standardized residual estimate
BCP	Basic community profile
CBD	Central business district
CI	Confidence interval
CO	Combinatorial optimization
CPI	Consumer price index
CPU	Central processing unit
CRA	Commonwealth Rent Assistance
CURF	Confidentialized unit record file
CV	Coefficient of variation
EB	Empirical Bayes
E-BLUP	Empirical best linear unbiased prediction
GREGWT	Generalized Regression Weighting Tool
HB	Hierarchical Bayes
HH	Household
HS	Housing stress
IMA	International Microsimulation Association
IPF	Iterative proportional fitting
LAUS	Local area unemployment statistics
MCMC	Monte Carlo–Markov Chain
ME	Margin of error
ML	Maximum likelihood
MMT	Microsimulation Modeling Technology
MMT	Spatial microsimulation model
NATSEM	National Centre for Social and Economic Modelling
NT	Northern Territory
pdc	Probability density curve
REML	Restricted maximum likelihood
SAE	Small area estimation
SAIPE	Small area income and poverty estimate
SD	Statistical division
SIH	Survey of income and housing

SLA	Statistical local area
SNC	Standard normal curve
SSD	Statistical subdivision
TAD	Total absolute distance
XPC	Expanded community profile

1

Introduction

Confucius recognized the importance of education and research in the fifth century BC:

> Only when things are investigated is knowledge extended; only when knowledge is extended are intentions sincere; only when intentions are sincere are hearts and minds rectified and personal lives cultivated; only when personal lives are cultivated can families become harmonious; only when families are harmonious are states well governed; and only when states are well governed is there peace in the world.
>
> **From "My personal anthology on different great ideas" by Chan (1963, p. 87)**

To govern a state well, various types of social, demographic, and economic research are vital, not only for understanding the major sociodemographic and economic issues within the state but also for finding solutions using the most up-to-date knowledge available from socioeconomic modeling. Small area estimation of various business and policy issues is well recognized as one of the major interests in all societies today.

1.1 Introduction

In today's data-centric world, *small area statistics* has emerged as a hot topic among students, academics, practitioners, politicians, and policy makers in all societies. As a quick example, households experiencing poverty and/or housing stress or financial stress are not equally distributed across the geography of a country. Even within a city or region, different suburbs and/or local areas have distinct incomes, house prices, and rental rates. Also, social policies such as housing policy and rental systems vary from state to state (Rahman et al. 2013). Therefore, to develop effective policies on sociodemographic and economic issues such as housing stress issue at the regional or local level, policy makers need to obtain accurate data on housing and/or poverty issues from small areas. Small area estimation (SAE) techniques can produce such reliable data (Rao 2003, 2005; Zhang and Chambers 2004; Wang et al. 2008; Chandra and Chambers 2009; Rahman 2011; Rahman and Harding 2012; Pfeffermann 2013; Lehtonen and Veijanen 2015a; Rao and Molina 2015).

Substantial spatial differences in socioeconomic growth and well-being exist across all countries in the world (see Van Wissen 2000; Stimson and McCrea 2004; Chin et al. 2005; Harding et al. 2006; McNamara et al. 2006; McNelis 2006; Stimson et al. 2006, 2008; Zhou and Kockelman 2010; Rahman and Harding 2014). Housing policies such as housing costs, rent assistance, mortgage subsidies, and land development planning for housing are typical examples of policies that have had significant impacts on individuals' or households' living standards, experiences, choices, constraints, decisions, and lifestyle preferences (Melhuish et al. 2004; Kelly et al. 2006). In addition, housing acts as a proxy for a host of other factors relevant to economic disadvantage and social inequalities at small area levels. Within a situation of rapidly rising housing costs, a large number of households are potentially facing housing unaffordability and/or poverty. For example, recent reports of Australia show that more than one million households have been living with housing stress (Sandel and Wright 2006; Rudd, in reference of Grattan 2008). But who are these households? Where do they live at small area levels? What are their socioeconomic backgrounds? Are they receiving any kind of social support, and if so how much? To analyze such imperative questions, reliable small area statistics are essential for knowledgeable decision-making on various socioeconomic and housing debates.

Despite this quick example situation, there are also various vital issues in the areas of social and behavioral sciences, applied economics and social policy analysis, government and/or social statistics, health sciences, computational statistics and data simulation, and spatial statistics and modeling that demand reliable small area statistics (Brouwers et al. 2004; Brown et al. 2004; Bell and Huang 2006; Chen et al. 2006; Harding et al. 2006; Anderson 2007a, 2013; Hanaoka and Clarke 2007; Harding and Gupta 2007; Nakaya et al. 2007; Pfeffermann et al. 2008; Hynes et al. 2009; Molina and Rao 2010; Morrissey and O'Donoghue 2011; Farrell et al. 2013; Rahman and Harding 2014; Daalmans 2015; Tzavidis et al. 2015). A choice of these issues is discussed in the later chapters with their policy implications. In today's data-centric world, accurate SAEs are essential due to various reasons including their worldwide promptly growing use in formulating policies and programs, allocating funds, evaluating the effects of policy changes and forecasting, regional level planning, small business decisions, and similar applications (Veldhuisen et al. 2000; Rao 2003; Franklin 2005; Rephann et al. 2005; Kavroudakis et al. 2009; O'Donoghue et al. 2012; Wu and Birkin 2013; Rahman and Upadhyay 2015; Rahman n.d.). Thus, both of the public and private sectors' decision-makers at national and regional levels require information on small area estimates. However, in practice, it is not very easy to calculate reliable estimates at small area level for many reasons.

One of the key reasons to explain difficulties is that data required for this purpose have to be gathered at a local level and cannot be afforded by the established national- or state-level surveys. Sometimes, information about the location is not recorded, to conform by the confidentiality constraints. As well, conducting a comprehensive survey for local-level data in all small areas is not feasible due to basic constraints in time and the large costs associated with it.

A range of methodologies are available for SAE practices, which include the conventional direct estimations to various indirect model–based estimations (Rao 2003; Longford 2005; Estevao and Sarndal 2006; Rahman 2008a; Kott 2009; Lehtonen and Veijanen 2009; Rao et al. 2009; Sinha and Rao 2009; Rahman et al. 2010; Nandram and Sayit 2011; Pfeffermann 2013; Rao and Molina 2015) to the nonparametric and calibration-based estimations (Wu and Sitter 2001; Lehtonen et al. 2003; Wu 2003; Montanari and Ranalli 2005; Chambers and Tzavidis 2006; Sarndal 2007; Opsomer et al. 2008; Chandra and Chambers 2009; Kim and Park 2010; Salvati et al. 2010; Lehtonen and Veijanen 2012, 2015b; Chambers et al. 2014). In general, most of the researchers in SAE have frequently highlighted methodologies that are fully based on various statistical models and theories (e.g., see Ghosh and Rao 1994; Rao 1999, 2003a, 2008; Pfeffermann 2002, 2013; Pfeffermann and Correa 2012; Pfeffermann and Tiller 2006; Torabi and Rao 2008; Datta 2009; Tzavidis et al. 2010; Datta et al. 2011). For example, implicit model–based approaches consist of synthetic, composite, and demographic estimations. As well, explicit area level and unit level model–based approaches have been widely studied through various statistical tools and techniques including empirical best linear unbiased prediction (EBLUP), empirical Bayes (EB), and hierarchical Bayes (HB) methods. Moreover, the nonparametric and calibration approaches are design-based model-assisted SAE methods. These approaches are not restricted to direct estimation only (Lehtonen and Veijanen 2009, 2015a; Lehtonen et al. 2011). The family of model-assisted estimations is broad and also covers indirect methods as a key application area where the assisting model can be, for example, any member of the generalized linear mixed models family, for instance, a logistic mixed model (e.g., Lehtonen et al. 2003; Ugarte et al. 2009). The family of nonparametric regression–based estimators as well as model-assisted and model-free calibration estimators are also good examples of modern design–based SAE methodologies. Further discussions of these approaches are offered in the later chapters.

Another type of geographic techniques called microsimulation modeling technology (MMT) is also providing small area estimates, particularly during the last decade (e.g., see Williamson et al. 1998; Ballas et al. 2003; Chin and Harding 2006; Anderson 2007; Lehtonen and Veijanen 2009; Morrissey et al. 2010; Rahman et al. 2010b, 2013; Tanton et al. 2011; Williamson 2013; Rahman and Upadhyay 2015). This sort of modeling is typically based on spatial economic theory and uses quite different methodologies such as different reweighting tools for small area microdata generation. Therefore, a comprehensive appraisal of various SAE methodologies is essential toward finding the appropriate methods for SAE for different situations.

A relatively new method, MMT, is operated through different reweighting techniques such as generalised regression weighting tool (GREGWT), combinatorial optimization, and the Bayesian prediction–based microdata simulation. Characteristically, such a reweighting tool involves very complex methodical algorithms and huge computer coding and is used for generating a synthetic spatial micropopulation data set at a spatial or individual scale. As a result, the simulation process of synthetic microdata is a very challenging task, not only due to constraints in computing capacity and time but also they significantly depend on the methodology. To check the performance of current reweighting methodologies, and if possible to develop an alternative methodology for microdata simulation have been appeared an existing demand in microsimulation modeling (Voas and Williamson 2000; Huang and Williamson 2001; Tanton et al. 2007; Rahman 2009a). Nonetheless, like several other research institutions across Europe, Canada, and the United States, the National Centre for Social and Economic Modelling (NATSEM) in Australia has an international pioneering reputation and excellent supporting capacities to build up such a complex MMT.

Moreover, the validation of the MMT-based model outputs is also challenging as the model is entirely based on synthetic spatial microdata that perhaps did not exist previously (see Williamson et al. 1998; Ballas et al. 2001; Harding et al. 2003; Taylor et al. 2004; Hynes et al. 2006; Edwards and Clarke 2009; Morrissey and O'Donoghue 2009, among many others). Some researchers have considered that one of the major drawbacks of MMT is the difficulty in validating the model outputs (e.g., see Ballas et al. 2001). Although a range of validation techniques are used by modelers in different countries, there is still a lacuna in the literature of the standard validation methodology for MMT-based models. Besides, if measures of statistical reliability can be created and be attached to the estimates from the MMT, internationally, such research would also be at the leading edge. While there have been some earlier international efforts to establish the statistical reliability of the estimates of static microsimulation models (Pudney and Sutherland 1994) and dynamic microsimulation models (Creedy and Kalb 2006), no such work has been undertaken for spatial microsimulation models (unless recently by Rahman and his colleagues; see Rahman 2011; Rahman et al. 2013).

It is apparent from the previous discussion that there are some imperative scope to contribute new knowledge to the SAE and socioeconomic modeling arena by producing empirical reliable small area estimates and then developing new techniques to the SAE methodology, especially for spatial microsimulation modeling.

The remainder of this chapter is organized in the following order: Section 1.2 provides the aims, focus, and objectives of this book. Section 1.3 outlines the overall structural plan of this book. Finally, Section 1.4 gives the concluding summary of the introduction chapter.

1.2 Main Aims of the Book

The principal aims of this book are to provide an extensive and comprehensive state-of-the-art presentation of all of the existing methodologies in SAE such as *direct* and *indirect model–based* estimation and then to develop a novel and more robust methodology that is termed as the MMT for SAE with its real-world applications. The book focuses on methodological developments by bridging classic ideas and conventional statistical SAE concepts with the latest Bayesian prediction–based computational method for spatial microdata simulation and an ultramodern MMT–based geographical SAE theories and then on the role that SAE and microsimulation modeling can play in supporting rational decision-making process and policy analysis. Further, key focus of the book is to overcome the lacuna in the validation techniques and measures of statistical reliability of the MMT-based small area estimates. All methodologies are empirically applied using the Australian data.

This book particularly focuses on the following objectives to achieve the aims:

- To discuss the notion of SAE with its advantages and applicability in the real-world context
- To give a general depiction of the overall methodologies for SAE, with a review on the methods of the direct SAE
- To offer a comprehensive and significant appraisal of various implicit and explicit statistical models for indirect SAE and its applications
- To describe a range of statistical procedures such as the EBLUP, EB, and HB estimation methods used for studying the explicit models with their critical comparisons
- To present an extensive review of the methodologies in the geographic approaches of indirect SAE, with a vital assessment of the different reweighting techniques
- To evaluate the overall SAE methodologies including various reweighting techniques used in spatial microdata simulation for choosing the appropriate and more robust method for an MMT-based model and ascertaining the prospect of developing an alternative method
- To present an alternative methodology that is termed as the Bayesian prediction–based microdata simulation technique
- To describe the construction of an MMT-based spatial microsimulation model for the SAE
- To apply the methodologies on real-world data and generate the small area statistics and then discuss empirical results at the general and various spatial scales by exploring the variation in small area statistics

- To analyze the scenarios of small area estimates by the characteristics of the households in rural and urban major cities and then look into the variation in the estimates at small areas within an individual city as well as between the cities using the spatial analysis technique
- To review a range of available validation methods and propose the theories of two alternative validation techniques for the MMT-based small area estimates and then present its applied results
- To offer the method of the measure of statistical reliability such as confidence interval estimation for the spatial microsimulation model and then discuss the results of confidence interval measures for the small area estimates
- To provide the general conclusions of the book including possibilities for and directions to future work in conjunction with a supplementary of all computing codes or programming that has been used to produce empirical results.

1.3 Guide for the Reader

In order to accomplish the specific objectives set out earlier, this book is structured into 10 chapters. Each chapter relates to one and/or more objectives, and each specific objective may be covered in more than one chapter as knowledge builds from one chapter to the next. The general outline of this book is briefly presented in the following.

Chapter 2 provides the notion of SAE along with a depiction of various concepts and overall methodologies for SAE. The advantages and applicability of various SAE techniques are discussed in the contexts of practical illustrations. The chapter also offers a good description of various methods for direct SAE. Findings reveal that the overall methodology of SAE encompasses a series of simple direct estimation techniques to very complex indirect model–based approaches classified as the statistical and geographic model–based indirect approaches. The direct estimation techniques rely on the observed data, and it is usable only when the survey data are sufficiently large enough to cover all the small areas. The theories of the direct small area estimators are very straightforward, but it has some deficiency in terms of statistical inferences. In particular, the direct estimations can produce unreliable estimates with large standard errors due to lack in sample size, the model misspecification, and/or the asymptotic design inconsistency of model in regard to the sampling framework. Use of the indirect model–based estimation approaches for producing accurate and more reliable small area estimates with enough statistical precision is in crucial demand.

Chapter 3 presents a comprehensive appraisal of various statistical approaches to indirect model–based SAE. The methodologies of these approaches mainly include two types of statistical models widely known as *implicit* and *explicit* models and a range of statistical procedures for making inference about the estimates from those models. Details of each of these models are accounted for, and justification of its applicability to SAE into different contexts is addressed. The chapter then discusses three statistical procedures such as *empirical best linear unbiased prediction, empirical Bayes,* and *hierarchical Bayes* procedures that are typically used in explicit model–based SAE. These statistical procedures not only play a significant role in studying various explicit model–based SAE but also have particular characteristics and complexities in terms of the theories and practical applications. Findings show that the implicit model–based SAEs are relatively easier to calculate— but they are characteristically biased estimates and limited for applications with the nature and suitability of the applicable data. The explicit model–based approaches include a range of statistical models such as *area level, unit level,* and *generalized linear mixed models* that are widely used in various fields of social sciences, government statistics, economic, public health, and agricultural sciences. Although the statistical approaches use data from different sources to obtain the small area estimates, they do not involve generating microdata file for small area, which is a significant resource for various further analyses.

In Chapter 4, an extensive discussion of various methodologies in the geographic approaches to indirect SAE is presented with a particular focus on the reweighting techniques. It describes the systematic process in microsimulation modeling and the types of microsimulation models widely used across the world with an outline of the advantages and limitations of microsimulation models. The chapter then provides the assessment of methodologies in microsimulation modeling, along with a report on synthetic reconstruction techniques and details of reweighting methods to spatial level microdata simulation. The chapter also portrays a comparison between the two reweighting techniques to reveal a sense of which one is appropriate for using into different applied situations. Findings reveal that the microsimulation model–based geographic approaches are associated with very sophisticated state-of-the-art methodologies including the microdata simulating technology—combinatorial optimization and GREGWT reweighting. The combinatorial optimization technique uses an intelligent searching algorithm *simulated annealing* and the *Metropolis criterion*. On the other hand, the GREGWT reweighting technique utilizes a truncated chi-squared distance function and the Newton–Raphson algorithm through the Lagrange multipliers. These geographic approaches of SAEs are robust, and they can use the simulated microdata file for further analysis and updating. It is also possible to measure small area effects of policy changes by linking spatial and static microsimulation technologies. However, the traditional statistical model–based approaches do not have such robustness.

A new alternative methodology is proposed in Chapter 5 for generating synthetic spatial microdata at small area levels. The method is based on the Bayesian prediction theory, and it takes consideration of the complete scenarios of micropopulation data units such as observed and unobserved population units at small area level, which means that it can produce more reliable spatial microdata for statistically consistent small area estimates and their variance estimation. As a probabilistic approach, it is also able to create the statistical reliability measures such as the Bayes credible region or confidence interval of the small area estimates. The basic steps of the new approach and the Bayesian prediction theory described are introduced first. Then the description of a multivariate model is provided toward linking the data sets. Later on, the Bayesian prediction distribution for the model is determined toward finding the probabilities of unobserved units in small area population. The chapter then designs the joint posterior density function of parameters for the observed units and unobserved population units to simulate the complete scenarios of the small area level micropopulation.

A thorough description of the stepwise construction process of an MMT for SAE is offered in Chapter 6. These include the discussions of various data sources and significant issues related to the applicable data sets, the specification of the model including variable selection in the process of generating synthetic weights, the procedures for developing a spatially disaggregated synthetic microdata, and then the small area estimates. In general, the model comprises two main phases of computations that are generating small area synthetic weights by reweighting and then producing small area estimates by creating synthetic spatial microdata. Typically, five groups of files such as the *model file, unit record data files, benchmark files, auxiliary data files,* and *reweighting algorithm file* are required to run the first stage of the model. Additionally, three groups of files such as the *survey level unit record data files, synthetic weight file,* and *CPI file* are necessary for further computation of the model with another model-specific computing *program file*. The second-stage computational process of the MMT creates the invaluable synthetic spatial micropopulation data set for all small areas and then determines the reliable small area estimates of interest.

In Chapter 7, various empirical results of the SAE are discussed. At first, it provides a general view of the results and then highlights the housing stress estimates for different states, statistical divisions, and statistical subdivisions. Further, the chapter includes comprehensive reports on small area distributions of the estimated numbers and percentages of housing stress across Australia, using the spatial analysis techniques. Findings demonstrate that small area estimates of housing stress vary with geographic units and by tenure types of households. About 1 of each 10 households in Australia is experiencing housing stress, and there are several major geographic regional areas throughout Australia where housing stress is concentrated in capital cities and noncapital major coastal centers. At the statistical division level, most of the households in housing stress are residing in Sydney,

Melbourne, Brisbane, Perth, Adelaide, and some other statistical divisions mainly located across coastal centers of New South Wales and Queensland. The nation's capital Canberra and other capital city statistical divisions have shown relatively low estimates of housing stress. Newcastle, Wollongong, Richmond-Tweed, Gold Coast, Sunshine Coasts, Wide Bay–Burnett, and Cairns City have shown most of the peak estimates of housing stress at statistical subdivision level in Australia. At the statistical local area level, the highest estimates are mostly obtained in the New South Wales coastal cities and the rapidly growing mining areas around inland locations into different states. Moreover, the private renter households in Australia are more prevalent in housing stress compared to all other tenures.

Chapter 8 focuses on the spatial analysis reports from SAE of housing stress data to reveal significant local-level variations in the estimates by households' tenure types within eight major capital cities of Australia. Findings demonstrate that almost two-thirds of households with housing stress were resided in Sydney, Melbourne, Brisbane, Perth, Adelaide, Hobart, Canberra, and Darwin. Among these capital cities, Sydney and Melbourne have about 40% of all housing stress households in Australia, and Sydney alone has one-third households compared to the total estimates from the eight capital cities. Nevertheless, housing stress estimates show rather mixed increasing trends for some capital cities, and there are significant variations observed between the SLAs within a city as well as between these cities. For example, the statistical local areas in Canterbury-Bankstown, Fairfield-Liverpool, Blacktown, and Gosford–Wyong in Sydney are identified as the hotspots for housing stress due to a significant increase of housing costs with a very limited supply of housing for low- and middle-income households, whereas almost all small areas in Canberra are with very low level of housing stress estimates relative to other cities, since this city has much better socioeconomic conditions, with big proportion of households having higher income. The supply of housings including the commonwealth rent-assisted housing is also relatively better in this regional and low population growth area.

There are two key foci to Chapter 9: first, in relation to whether it is feasible to develop some sort of statistical tools to check the statistical significance of the small area estimates generated by the MMT, and second, the establishment of any possible measure of statistical reliability for the model estimates. By the way of these motivations, the chapter reviews different validation techniques available in the literature. Then it proposes the theories of two new validation techniques for testing the accuracy of the small area estimates and presents empirical results with discussions. The chapter also offers a methodology for the creation of statistical reliability: "confidence interval" estimation with applied results. Findings reveal that although there are few techniques available in the literature for validating MMT-based estimates, none of these methods directly looked at on the statistical significance test and confidence interval estimation of the model estimates. Theories of the proposed validations are coordinated with the typical statistical procedures including the Z-test and standardized residual analysis. These new validation methods have several key advantages over other techniques, such as they are fully scientific, straightforward in terms of computational requirements, have standard index toward making inference, be able to show an exact degree of statistical accuracy of the estimates, can identify and describe the possible features of small areas that particularly show inaccurate estimates, and are practicable to create the measures of statistical reliability, that is, confidence intervals of the MMT-based small area estimates.

Chapter 10 concludes the book by summarizing the key findings and methodological developments with its applications, considering whether the aims and objectives have been met, and discussing the limitations. It then looks into the directions for further works in this rapidly growing field of SAE and microsimulation modeling before finishing with the key messages of this book and supplies all invaluable computing codes and programming that have been utilized to build up the MMT-based small area model and produce empirical results.

1.4 Concluding Remarks

This introductory chapter has provided a general framework for the remainder of the book. The rationale and fundamental setting of the book has been described with the evidence of relevant literature. The scope of the book, aims, and objectives have been stated and justified. An outline of the text has also been provided with the main contents for each chapter.

2

Small Area Estimation

2.1 Introduction

In recent years, "small area statistics" has emerged as a widely interested hot topic among politicians, academics, and policy makers throughout the developed and developing world. To develop effective policy on various crucial social, economic, and health issues at the regional or local level, policy makers need to use accurate data on those respective issues for small area levels. The small area estimation (SAE) techniques can produce such reliable estimates for small areas.

Typically, there are two types of methods for SAE: *direct* and *indirect* SAEs. It is noted that each of these methods is associated with very simple to complex theories, algorithms, and models. For instance, the direct SAE method comprises a range of estimators that are based on very simple statistical theories and straightforward formulas. On the other hand, most of the indirect SAE approaches encompass complex methodologies from statistics, economics, and geography.

In this chapter, various definitions of housing stress measures used in the literature are described with an outline of the SAE concept and methodologies for SAE. This chapter also offers a good discussion on various techniques for direct SAE.

The plan of this chapter is as follows: Section 2.2 provides a general discussion about SAE concept, its advantages, and applicability in real-world contexts. Section 2.3 outlines the overall methodologies for SAE with a diagrammatic representation. Section 2.4 offers a detailed discussion on the direct SAE. Finally, Section 2.5 gives the concluding remarks of the chapter.

2.2 Small Area Estimation

SAE has received much attention in recent decades due to the increasing demand for reliable small area statistics (Pfeffermann 2013; Rahman et al. 2013; Rahman and Upadhyay 2015). In SAE, one uses data from similar domains to estimate the statistics in a particular small area of interest, and this "borrowing of strength" is justified by assuming a model that relates to the small area statistics (Meeden 2003). SAE is the process of using statistical models to link survey outcome or response variables to a set of predictor variables known for small areas in order to predict small area–level estimates. Traditional area-specific estimates may not provide enough statistical precision because of inadequate sample observations in small geographic areas. In such situation, it may be worth checking whether it is possible to use indirect estimation approaches based on the linking models. An objective Bayesian approach has been proposed by Meeden (2003) to SAE, and in the paper, the author demonstrates that using this Bayesian technique one can estimate the population parameters other than the means of a small area and find sensible estimates of their precision.

2.2.1 Concept of Small Area

The term *small area* typically refers to a small geographic area or a spatial population unit for which reliable statistics of interest cannot be produced due to certain limitations of the available data. For instance, small areas include a geographic region such as a Central Business District, suburb, and statistical local area (SLA) or a demographic domain such as a specific *age ×* *sex × education × income* unit or a demographic subdomain within a geographic region. Some of the other terms used as a synonym to small area are "small domain," "minor domain," "local area," and "small subdomain" (see, e.g., Rao 2003; Rao and Molina 2015). The history of using small area statistics goes back to the eleventh century in England and seventeenth century in Canada (Brackstone 1987). This book uses the SLA as the small geographic area or small area in Australia.

2.2.2 Advantages of SAE

SAE methodologies are beneficial for business organizations, policy makers, and researchers who are interested in estimates for regional small domains but who lack adequate funds for a large-scale survey that could produce precise, direct survey estimates for the small domains (Rahman and Harding 2014). For instance, population estimates of a small area may be used in a range of purposes, such as business organizations using them to develop profiles of customers, to identify market clusters, and to determine optimal site locations for their business. In addition, state and local governments use them to establish political boundaries to monitor the impact of public policies, and to estimate the need for schools, roads, parks, public transportation, and fire protection. Also researchers use them to study urban sprawl, environmental conditions, and social trends. Such estimates are used as denominators to calculate many types of rates and to determine the allocation of money in public funds each year (Smith et al. 2002; Lehtonen and

Veijanen 2015a; Rao and Molina 2015). Therefore, it is clear that the inference of precise small area estimates is of significance for many reasons.

2.2.3 Why SAE Techniques?

For any small area containing respondents to a sample survey, an estimator of the small area characteristic can be constructed from the sample data. This type of estimation relies on the estimates that are fully based on the domain-specific sample data and usually known as direct estimation. This means that a direct estimator can be calculated for a local area by using sample observations that only come from the sample units in that local area. To obtain the reliable direct estimators in local areas, a suitable sample (enough representative observations) must be available for all small areas. However, in the real-world practices, sufficiently representative samples are unavailable for all small areas. Depending on the type of the study and the time and money constraints, it can be also impossible to conduct a sufficiently comprehensive sample survey to obtain an adequate sample from every small area we are interested in. A basic problem with national- or state-level surveys is that they are not designed for the efficient estimation for small areas (Heady et al. 2003). The estimation techniques can overcome this limitation of survey data by producing accurate small area estimates.

2.2.4 Applications of SAE

Although SAE techniques have been frequently used in countries such as the United States, Canada, the United Kingdom, and some other countries in Europe for several decades (especially in the United States), they have been more recently used in Australia. The National Center for Health Statistics (1968) in the United States pioneered the use of the synthetic estimation for developing state estimates (since the sample sizes in most states were too small for providing reliable direct state estimates) of disability and other health characteristics from the National Health Interview Survey (e.g., see also Gonzalez et al. 1966). Other examples of major SAE programs in the United States include the following:

- The Census Bureau's Small Area Income and Poverty Estimates (SAIPE) program that regularly produces estimates of income and poverty measures for various population subgroups for states, counties, and school districts (Bell et al. 2007; also refer to the SAIPE website [http://www.census.gov/hhes/www/saipe/] for more information)
- The Bureau of Labor Statistics' Local Area Unemployment Statistics (LAUS) program that produces monthly and annual estimates of employment and unemployment for states, metropolitan areas,

counties, and certain subcounty areas (detailed information can be found on the LAUS website at http://www.bls.gov/lau/)

- The National Agricultural Statistics Service's County Estimates program that produces county estimates of crop yield (USDA 2007; see also http://www.nass.usda.gov/)

- The estimates of substance abuse in states and metropolitan areas that are produced by the Substance Abuse and Mental Health Services Administration (Hughes et al. 2008; also refer to http://www.samhsa.gov/ for details)

In Canada, SAE techniques are used to produce monthly estimates of unemployment rates at the national, provincial, and subprovincial levels (You et al. 2003), youth smoking behaviors at national and provincial levels (Pickett et al. 2000), and life expectancy at birth in small cities (Fines 2006).

The Office for National Statistics, United Kingdom, ran a large project named EURAREA, a project for enhancing SAE techniques to meet European needs with participating countries in the European Union. The EURAREA project was a research program to investigate methods for SAE and their application in some European countries. The EURAREA's contributions in SAE are under the following headings (EURAREA Consortium 2004a; Heady and Ralphs 2005):

- Empirical evaluation of SAE methods

- Making SAE to be national statistical institutes friendly in participating countries

- Creation of an environment for future empirical research in SAE

The EURAREA has succeeded in creating an environment for continuing research in spatial estimation in the participating countries. Within the framework of the EURAREA project, each of the participating countries had to obtain estimations for two geographic levels: the province and any other very small level below the province (see Heady and Ralphs 2005; the EURAREA project report 2005 at http://www.ine.es/en/docutrab/eurarea/eurarea_05_en.doc [EURAREA 2005]; and also http://www.statistics.gov.uk/eurarea/consortium.asp [EURAREA Consortium 2004a] for details about this project). Molina et al. (2007) reveal the estimates of unemployment or employment characteristics in small areas, and Longford (2004) estimates local area rates of unemployment and economic inactivity by using the data from the UK Labour Force Survey. Small area poverty rates in Italy have been studied by D'Alò et al. (2005) based on the EURAREA project findings. Moreover, in Poland, SAE techniques are used in different fields such as estimation of some employment and unemployment characteristics by region and county levels using the 2002 Population Census data, estimation of some characteristics of the smallest enterprises by region and county, and estimation of some agricultural characteristics by region and county using agricultural sample surveys and agricultural census data, respectively (Kordos 2005).

In recent years, the Australian Bureau of Statistics and some other Commonwealth Government agencies are adopting and using SAE methods to calculate estimates for small domains. For example, small area estimates available in Australia include estimates of disability (Elazar 2004; Elazar and Conn 2005), employment (Commonwealth Department of Employment and Workplace Relations 2007), and crime (Tanton et al. 2001). A comprehensive discussion about model-based SAE, different statistical approaches for SAE, and its applications is provided in Rao (2003).

2.3 Approaches to SAE

Most of the review articles in SAE have highlighted the methodologies that are fully based on various statistical models and theories (see, e.g., Ghosh and Rao 1994; Rao 1999, 2003a, 2005; Pfeffermann 2002, 2013; Longford 2005). However, another type of technique called "spatial microsimulations" has been used in providing small area estimates during the last decade (see, e.g., Ballas et al. 2003, 2006; Brown et al. 2004; Brown and Harding 2005; Chin and Harding 2006, 2007; Anderson 2007; King 2007; Tanton 2007; Rahman 2008a; Rahman et al. 2013). The spatial microsimulation models are based on the economic theory and its methodologies are quite different from others. Therefore, overall methodologies for obtaining small area estimates that have emerged over the last few decades will be reviewed in this book with particular emphasis given to the indirect model–based SAE.

A summary diagram of overall methodologies for SAE is depicted in Figure 2.1. Traditionally, there are two types of SAE—direct and indirect estimation. The direct SAE is based on the survey design and includes four estimators called the H-T estimator, generalized regression (GREG) estimator, modified direct estimator, and design-based model-assisted estimator. Indirect approaches of SAE can be divided into two approaches—statistical and geographic (Rahman 2008a). The statistical approach is mainly based on different statistical models and techniques. However, the geographic approach uses techniques such as microsimulation modeling.

It should be noted that implicit model–based approaches include three types of estimations—synthetic, composite, and demographic estimations—whereas explicit models are categorized as area level, unit level, and generalized linear mixed models. Based on the type of the study of interest, each of these models is widely studied to obtain small area indirect estimates using the empirical best linear unbiased prediction, empirical Bayes, and

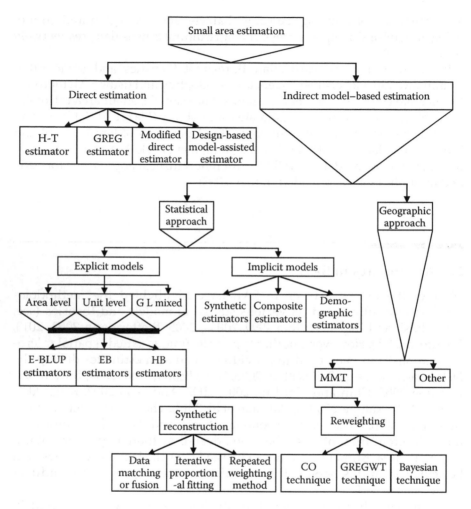

FIGURE 2.1
A summary of different techniques and estimators for SAE.

hierarchical Bayes methods. On the other hand, the geographic approach is based on microsimulation models, which essentially create synthetic/simulated micropopulation data to produce "simulated estimates." Obtaining reliable microdata at small area level is the key task for spatial microsimulation modeling. Synthetic reconstruction and reweighting are commonly used methods in microsimulation, and each of them is stimulated by different techniques to produce simulated estimators (see Figure 2.1). In particular, reweighting methodologies include CO, GREGWT, and Bayesian techniques. A description of different direct methods of SAE is addressed in Section 2.4. However, details of different indirect methods of SAE will be given in Chapters 3 through 5.

2.4 Direct Estimation

A *direct* estimation uses values of the variable of interest only from the time period of interest and only from units in the small domain or area of interest (Federal Committee on Statistical Methodology 1993). In this approach, all small areas to be sampled in order to produce a range of traditional direct and design-based indirect estimators. Although it is rare, when survey samples are large enough to cover all the study areas with sufficient data in each area, different direct estimators can be developed. The common direct estimations applied in SAE include the Horvitz–Thompson (H-T) estimator (Cochran 1977), GREG estimator (Sarndal et al. 1992), modified direct estimator (Rao 2003; Rao and Molina 2015) or survey regression estimator (Battese et al. 1988), and design-based indirect estimator (Lehtonen and Veijanen 2009). The following sections will cover discussion of each of these estimators in direct estimation.

2.4.1 H-T Estimator

Let a finite population $\Omega = \{1,2,\ldots,k,\ldots,N\}$ contain $E_i(E_i \subset \Omega)$ subpopulation for a small area $i(= 1,2,\ldots,P)$ with N_i elements and $\sum_i^P N_i = N$. Consider a sample $s(s \subseteq \Omega)$ is drawn from Ω with a given probability sampling design $p(\cdot)$, and $s_i(s_i \subset s)$ is the set of individuals that have been selected in the sample from the small area i. Suppose the inclusion probability $\pi_k = \Pr (k \in s)$ is a strictly positive and known quantity. For the elements $k \in s_i$, let (y_{ik}, x_{ik}) be a set of sample observations, where y_{ik} is the values of the variable of interest for the kth population unit in the small area i and $x_{ik} = (x_{ik1},\ldots,x_{ikj},\ldots,x_{ikJ})'$ is a vector of auxiliary information associated with y_{ik}.

Now, if Y_i and X_i represent the target variable and the available covariates for a small area i, then the H-T estimator (Cochran 1977) of the population total for ith small area can be defined as

$$\hat{Y}_{i[H-T]} = \sum_{k \in s_i} d_{ik} y_{ik} \tag{2.1}$$

where $d_{ik} = 1/\pi_{ik}(k \in s_i \subset s)$ are design weights depending on the given probability sampling design $p(\cdot)$.

It is worth noting that in principle the H-T estimator is not designed to use auxiliary information or covariates. However, it is possible to consider auxiliary information to evaluate this estimator (Sarndal et al. 1992). When $\pi_{ik} > 0$, for $\forall k \in E_i$, and there is sufficient sample observations available at ith small area, the H-T estimator is an unbiased estimator, that is, $E(\hat{Y}_{i[H-T]}) = Y_i = \sum_{k=1}^{N_i} y_{ik}$, but not efficient. In the context of SAE problems, with an inadequate sample, this estimator can be biased and more unreliable.

2.4.2 Generalized Regression Estimator

Suppose aggregated information about the auxiliary variables or covariates at the small area $i(= 1,2,...,P)$ is available and denoted by X_i. Then the GREG estimator should be obtained by combining this information X_i with the individual sample information from the survey data and can be defined as

$$\hat{Y}_{i[GREG]} = X_i'\hat{\beta} + \sum_{k \in s_i} d_{ik}(y_{ik} - x_{ik}'\hat{\beta}) \tag{2.2}$$

where
   $X_i = (X_{i1},...,X_{ij},...X_{iJ})'$ represents the vector of totals of the J covariates in the small area i

   $\hat{\beta} = \left(\sum_{k \in s_i} d_{ik}x_{ik}x_{ik}'\right)^{-1} \left(\sum_{k \in s_i} d_{ik}x_{ik}y_{ik}'\right)$ is the sample weighted least square (WLS) estimates of GREG (Rao 2003) and others are usual notation from previous section (refer to the H-T estimator part in Section 2.4.1)

Note that the GREG estimator could be negative in some small areas when the linear regression overestimates the variable of interest. In addition, when there is no sample observation in the small area i, the GREG estimator forms the model-based synthetic regression estimator that is fully based on the regression model and the small area covariates X_i.

A comprehensive discussion about the GREG estimator is given by Rao (2003). The GREG estimator is approximately design unbiased for SAE but not consistent because of high residuals; and when only area-specific auxiliary information is available, this estimator is also model unbiased under the assumption of the linear association between the response variable and the area-specific covariates (Rao 2003). This estimator can take a more general form (by choosing the "revised" weights, which is the product of the design weight and estimation weight, instead of design weights) to provide estimates of all target variables under different assumptions and at different domains, and hence the GREG estimator ensures consistency of results at different areas when aggregated over different variables of interest (Rao 2003, p. 13). In addition, this estimator also ensures consistency with the covariate totals but not with the estimation of target variable at the aggregated level. However, for a small area i, when the sample data provide consistent estimates of the covariates X_i, then it can also provide a good estimate of the target variable Y_i.

2.4.3 Modified Direct Estimator

The modified direct estimator, borrowing strength over small areas for estimating the regression coefficient, can be used to improve estimators' reliability. If auxiliary information or covariates X_i in the ith domain are available, then a modified direct estimator of population total is given by

$$\hat{Y}_{i[MDE]} = \hat{Y}_i + (X_i - \hat{X}_i)'\hat{\beta} \tag{2.3}$$

where

$\hat{Y}_i$ and $\hat{X}_i$ are the H-T estimators of the target variable Y_i and covariates X_i, respectively, for the small area i

$\hat{\beta} = \left(\sum_{k \in s} d_k x_k x_k'\right)^{-1} \left(\sum_{k \in s} d_k x_k y_k'\right)$ is the overall sample WLS estimates of regression coefficients

The modified direct estimator is approximately design unbiased as the overall sample size increases, even if the regional sample size is small. This estimator is also known as the modified GREG estimator or the survey estimator, and it uses the overall sample data at the aggregated level to calculate the overall regression coefficients. It is remarkable that although the modified direct estimator borrows strength for estimating the overall regression coefficients, it does not increase the effective sample size, unlike indirect small area estimators (e.g., refer to Rao 2003, p. 20). Nevertheless, the design-based model-assisted approach (that can also be defined as design-based *semidirect* or design-based *indirect* SAE) would increase the "effective" sample size with proper definition of variance estimators and residuals (Lehtonen and Veijanen 2009).

2.4.4 Design-Based Model-Assisted Estimators

There is a choice of design-based model-assisted SAE that borrows strength from other small areas and is capable of increasing "effective" sample size by using appropriate assisting models and model calibration or model-free calibration procedures. In such a case, borrowing strength is generally understood to mean that the estimator in use depends on data on the variable of interest from related areas, or more generally from a larger area, in an effort to improve the accuracy for the small area (Lehtonen and Veijanen 2009). Typically, dependencies arise through model-defined relationships between variable of interest and one or more auxiliary variables shared across two or more small areas. The assisting models used for borrowing strength can be parametric, semiparametric, or nonparametric (Saei and Chambers 2003; Opsomer et al. 2008; Goerndt et al. 2011; McRoberts 2011; Lehtonen and Veijanen 2015a,b). Models with a mixture of fixed and random effects are popular (Lohr and Prasad 2003; Breidt 2004), but estimates of variance components can be imprecise in SAE applications with few small areas of interests (e.g., <10) and when sample sizes vary greatly among those small areas (Datta and Lahiri 2000; Longford 2005, pp. 241–246; Fuller 2009, p. 314).

A range of statistical estimators applicable to the design-based model-assisted approach have been developed for SAE problems (Lehtonen and Veijanen 2009, 2012; Chambers and Clark 2012, p. 161). These estimators are also known as design-based *semidirect* or design-based *indirect* estimators for SAE. For example, an *extended family of GREG estimators* (i.e., model-assisted GREG estimators) with the use of generalized linear mixed models as the assisted models (Lehtonen and Veijanen 2009), model calibration estimators (Lehtonen et al. 2011; Lehtonen and Veijanen 2015a), and model-free calibration estimators (Särndal 2007; Lehtonen and Veijanen 2015b) for SAE are well known. This class of estimators is particularly useful in the situations where the response variables are binary or polytomous, or small areas of interests comprise the so-called unplanned domains, that is, the sample sizes of small areas are not controlled by stratification but are random in the sampling design.

Model-assisted GREG estimators are increasingly used for official statistics production in national statistical agencies and elsewhere. The ordinary GREG estimator in Equation 2.2 (i.e., within Section 2.4.2) incorporates a linear regression model to estimate the population total $\hat{Y}_i$ of a continuous study variable. However, for a binary or polytomous response variable, a linear model formulation will not necessarily fit the data well. A logistic model formulation will be a more realistic choice where the logistic GREG (LGREG) estimates the frequency f_i of a class C in ith small area. An LGREG model is fitted to the indicators $u_{ik} = I\{y_{ik} \in C\}$, $k \in s_i$ using the design weights. The fitted model yields estimated probabilities $\hat{p}_{ik} = P\{u_{ik} = 1; \mathbf{x}_{ik}, \hat{\beta}\}$. The LGREG estimator of the class frequency in E_i is defined as

$$\hat{f}_{i[LGREG]} = \sum_{k \in E_i} \hat{p}_{ik} + \sum_{k \in s_i} d_{ik}(u_{ik} - \hat{p}_{ik}). \qquad (2.4)$$

Here, $\sum_{k \in E_i} \hat{p}_{ik}$ is the sum of the predicted values in the population. Thus, it is necessary to have access to small area–level population information about the individuals' auxiliary variables. The last component of (2.4) is an H-T estimator of the residual total, which aims at correcting the possible bias of the first part. It is obvious that for certain model choices, notably for an area- or domain-specific model formulation, the last component vanishes.

Besides this, a small area or domain size correction (Lehtonen and Veijanen 2009) is incorporated into (2.4) to an estimator defined as

$$\hat{f}_{i[LGREG2]} = \sum_{k \in E_i} \hat{p}_{ik} + \frac{N_i}{\hat{N}_i} \sum_{k \in s_i} d_{ik}(u_{ik} - \hat{p}_{ik}); \hat{N}_i = \sum_{k \in s_i} d_{ik}, \forall i,k \qquad (2.5)$$

For the multinomial logistic GREG (i.e., MLGREG) estimator, an alternative logistic mixed model involving fitted values $\hat{p}_{ik} = P\{u_{ik} = 1; \mathbf{x}_{ik}, \hat{\beta}, \mathbf{z}_i\}$

(where z_i is a vector of domain- or area-specific random effects) is used instead of a fixed-effects logistic model (Lehtonen et al. 2005; Torabi and Rao 2008). The random effects are associated with domains or with regions. This model formulation may be a realistic option for many situations in practice. Geographically weighted regression (GWR) is a more recent example (Salvati et al. 2012). In GWR, borrowing strength relies on the presence of a positive distance-dependent spatial correlation.

Many agencies with the UN and EU that are interested in small area estimates on poverty, social exclusion, and Laeken indicators adhering to a design-based inference paradigm choose model-assisted estimators for SAE problems (Lehtonen et al. 2011; Lehtonen and Veijanen 2015b). A model-assisted estimator is design consistent and nearly design unbiased, but it can be imprecise. This type of estimator is design biased, but accuracy can be good if the model fits well in an area of interest. Bias will be serious if the selected model is inadequate. Sample sizes in a small area may, however, be too small to allow a meaningful assessment of model fit and precision (Rao 2003, p. 110; Longford 2005, p. 195).

Moreover, in classical *model-free calibration* (Särndal 2007; Lehtonen and Veijanen 2015b) and *model calibration* (Montanari and Ranalli 2005; Lehtonen and Veijanen 2015a), a calibration equation is imposed so that the weighted sample totals of auxiliary variables reproduce the known population totals. Calibration is typically used to construct an estimator as the sum of weighted sample, and the weights are chosen in such a way that the weighted sample sums of auxiliary variables are identical with known population totals (Lehtonen and Veijanen 2015b). In addition, model calibration represents a model-assisted technique, where an assisting model is explicitly stated (Montanari and Ranalli 2005). The weights are calibrated to reproduce the population total of the predictions derived via the specified model. In this approach, direct estimators are constructed by using models fitted separately in each small area or domain. So, a direct small area estimator can still incorporate auxiliary data outside the areas. This is relevant if accurate population data about the auxiliary variables are only available at a higher aggregate level. The later method consists of two phases: (1) the *modeling phase* and (2) *the calibration phase*.

There is much flexibility in both phases. At first, appropriate models for the study to be chosen, for example, a mixed model structure that accounts for spatial heterogeneity in the population, would be suitable for the heterogeneous data. The predictions calculated in the modeling phase are then used in the second phase when constructing calibration equation and a calibrated estimator. Calibration can be defined at the population level, at the area or domain level, or at an intermediate level, for example, at a regional level (neighborhood) that contains the domain of interest. Further, in the construction of the calibrated estimator, a "semidirect" approach involves using observations only from the area of interest, whereas in a "semi-indirect" approach, also observations outside the domain of interest are included.

In model calibration introduced by Wu and Sitter (2001) and Wu (2003), predictions are used instead of auxiliary variables. A model such as logistic regression or any other model is first fitted to the sample. Here, the estimator of the total frequency is a weighted sum of indicators over the whole sample, region, or domain. The weights are chosen so that the weighted sum of estimated probabilities over a subset of sample equals the sum of the predicted probabilities over a corresponding subset of population. The sum of weights over the sample subset must be equal to the size of the population subset. Further, the weights should be close to the design weights. The procedure of finding such weights is called calibration (Särndal 2007).

According to Wu and Sitter (2001), in small area population level calibration, the weights must satisfy a calibration equation

$$\sum_{k \in s} w_i = \sum_{k \in \Omega} z_i = \left[N, \sum_{k \in \Omega} \hat{p}_i \right]; \text{ where } z_i = (1, \hat{p}_i), \forall i, k \qquad (2.6)$$

This equation can be minimized using the Lagrange multiplier technique.

Details of the model calibration methods were studied for SAE by many researchers (e.g., see Särndal et al. 1992; Wu and Sitter 2001; Särndal 2007; Lehtonen et al. 2011; Lehtonen and Veijanen 2012, 2014, 2015a). A benefit of calibration is that models beyond the linear model can be specified, such as nonparametric and quantile models, logistic models, and other members of the generalized linear models family. This option offers a flexible treatment of different types of study variables including count variables and continuous variables whose relationship to the explanatory variables is nonlinear.

2.4.5 A Comparison of Direct Estimators

Table 2.1 shows a simple comparison of various direct estimators utilized in SAE in terms of methods, advantages, disadvantages, and their applications. Most of these estimators are essentially based on the real sample and applicable when there are adequately large sample data available at small area or domain levels. However, in terms of statistical inferences, the direct SAE can be considered as sampling design–based and statistical model–based estimators. It is noted that in a model-based approach, inferences are involved with statistical models, whereas in the design-based approach, inferences are fully based on the sampling design and/or making use of assisting models. For instance, the GREG estimator and the modified direct estimator are statistical model–based direct estimator, while the H-T estimator is a sampling design–based estimator, and the *extended family of GREG* (i.e., model-assisted GREG estimators) and model calibration estimators are design-based model-assisted estimators.

TABLE 2.1

Comparison of Various Direct Estimators for SAE

Direct SAE	Methods/ Comments	Advantages	Disadvantages	Applications
Various estimators	H-T estimator	Only based on real sample units	Easy to calculate and it is unbiased for large sample	It is unreliable and cannot use auxiliary data	Only if the sample size is large enough
	Generalized regression or GREG estimator	Based on real data and WLS estimates of regression	Can use auxiliary data at small area level, and approximately design and model unbiased	It could be negative in some cases and not a consistent estimator due to high residuals	When the sample size is large and reliable auxiliary data are available at local level
	Modified direct estimator	Based on real sample, auxiliary data and WLS estimate of regression	Design unbiased and uses overall aggregated data for coefficient estimation	Borrows strength from the overall data but cannot increase effective sample	When the overall sample size is large and reliable
	Design-based model-assisted estimator	Based on data from a related larger area, assisting models, and calibration procedures	Design unbiased and can increase effective sample size with proper variance estimators	Variance estimations can be vague with few areas and when sample sizes vary greatly among those areas	When response variables are categorical or counts and study involve unplanned area/domains

A brief description of direct SAE under the design-based methods and the model-based approaches has been given elsewhere (e.g., refer to Rao 2003; Lehtonen and Veijanen 2009; Rao and Molina 2015). The design-based direct estimation technique such as the H-T estimator can produce unreliable estimates due to lack of adequate sample size. While the design-based model-assisted or indirect techniques would improve such lack by borrowing strength from related small areas or more generally from a larger area, a proper definition of variance estimations and model-defined relationships is required for such effort to improve the accuracy for the small area. The variance estimations can be often imprecise because of small areas of interest and when sample sizes vary greatly among those areas. Nonetheless, the model-based direct estimation approaches can also perform poorly by model misspecification for increasing the sample size. This poor performance is mainly caused by asymptotic design inconsistency of the model-based estimator with respect to the stratified random sampling (Rao 2003). Therefore, the direct small area estimators have typically the following limitations:

- Estimates can only be computed for a subset of all areas that contain respondents to the sample survey.
- For those small sampled areas, the achieved sample size will usually be very small or not large enough and the estimator will thus have low precision, and this low precision will be reflected in rather wide confidence intervals for the direct estimates, thus making them statistically unreliable.
- Direct estimates can perform poorly by model misspecification and asymptotic design inconsistency of the sampling structure.

Moreover, in the context of SAE, the estimators based on direct estimation methodologies lead to unacceptably large standard errors because of disproportionately small samples from the small area of interest; in fact, no sample units may be selected from some small domains. This makes it necessary to find the indirect model–based estimators such as synthetic estimators that increase the sample size effectively and thus decrease the standard error for sufficient statistical precision. According to Gonzalez (1973), an estimator is called a "synthetic estimator" if a reliable direct estimator for a large area, covering several small domains, is used to drive an indirect estimate for a small domain, under the assumption that the small areas have the same characteristics as the large area. Due to the lack of enough sample information in small geographic areas, there is a lot of interest in creating indirect estimators for small areas in Australia (Tanton 2007).

2.5 Concluding Remarks

This chapter outlines the concept and methodologies of SAE which evidence that SAE plays vital roles in various real-world practices. SAE can produce reliable and accurate small area estimates to support the process of knowledgeable and effective decision-making and policy analysis at local or regional levels. Nowadays, most of the developed nations are using the SAE methodology. The overall methodology encompasses a series of simple direct estimation techniques to very complex indirect model–based approaches classified as the statistical and geographic model–based indirect approaches. Besides, a review of direct SAE has illustrated that the direct estimation techniques rely on the observed data and it is usable only when the survey data are sufficiently large enough to cover all the small areas or more generally from a larger geographic area. There are three common estimators for the direct SAE that include the H-T estimator, GREG estimator, modified direct estimator, or survey regression estimator. Also an additional class of estimators is defined as the design-based model-assisted estimators.

Although the theory and formulas of the direct small area estimators are very simple and straightforward, it has some deficiency in terms of statistical inferences. For instance, the sampling design–based *H-T* estimator can produce unreliable estimate due to lack of adequate sample size. In addition, the model-based GREG and *modified direct* estimators can also perform poorly because of the model misspecification for increasing sample size including the asymptotic design inconsistency of the model in regard to the sampling framework. Additionally, the model-assisted GREG estimators and model calibration estimators need appropriate auxiliary data from related small areas. They also require a proper definition of variance estimations and model-defined relationships between variables. Usually the variance estimations are imprecise due to variations in small areas and when sample sizes vary greatly among those areas. Furthermore, the direct small area estimates lead to unacceptably large standard errors as a result of disproportionately small samples from many small area of interest; in fact, no sample units may be selected from some small areas. Thus, it is essential to make use of the indirect model–based SAE approaches for generating accurate and more reliable estimates having sufficient statistical precision.

3

Indirect Estimation: Statistical Approaches

3.1 Introduction

An *indirect* small area or domain estimator uses values of the variable of interest from a domain and/or time period other than the domain and time period of interest (Federal Committee on Statistical Methodology 1993). The indirect small area estimations (SAEs) rely on different statistical or geographic models to produce estimates for all small areas. Once the model is chosen, whether it is statistical or geographic, its parameters are estimated using the data obtained from the sample survey. However, for the geographic models, microdata should be generated at small area levels to estimate the model parameters. An important issue in indirect SAE is that additional auxiliary information or covariates are needed and how to take full advantage of those covariates.

In the last chapter, a quick snapshot of the overall methodologies for SAE was presented, along with an account of various direct estimation methods. This chapter offers a significant appraisal on various statistical approaches to indirect SAE. The methodologies include various statistical models and a range of procedures for estimation.

The *statistical* approach to indirect SAE mainly uses two types of statistical models, which are commonly known as implicit and explicit models. The implicit models provide a link to related small areas through supplementary data from census and/or administrative records, whereas the explicit models account for small area level variations through supplementary data and they are termed "small area models" in the literature.

Furthermore, making inference from the explicit models or small area models, there are three statistical procedures that have been widely used in SAE. In the statistical science literature, these three methods are well known as the (empirical) best linear unbiased prediction (EBLUP), empirical Bayes (EB), and hierarchical Bayes (HB) procedures. Each of these processes has its own characteristics and complexities in terms of the theories and practical applications, but they have played a significant role in studying various small area models for indirect SAE.

The plan of this chapter is as follows. Various implicit statistical models for SAE are highlighted in Section 3.2. The explicit models are reviewed in Section 3.3. Different statistical procedures for studying the explicit models are discussed in Section 3.4. Comparisons of three statistical processes used in explicit models–based SAE are presented in Section 3.5. The concluding remarks of this chapter are given in Section 3.6.

3.2 Implicit Models Approach

This approach includes three statistical techniques of indirect estimation— which are synthetic, composite, and demographic estimations.

3.2.1 Synthetic Estimation

The idea of synthetic estimation and its application was first introduced in the United States by the National Center for Health Statistics (1968). They had used this indirect estimation technique to calculate state-level disability estimates. Gonzalez (1973) provides an excellent definition of a synthetic estimator—"an estimator should be synthetic when a reliable direct estimator for a large area is used to derive an indirect estimator for a small area belonging to the large area under the assumption that all small areas have the same characteristics as the large area." In addition, Levy (1979) and Rao (2003) provide extensive overviews of various synthetic estimation approaches and its applications in SAE.

The synthetic estimators can be derived by partitioning the whole population (e.g., national or state-wide data) into a series of mutually exclusive and exhaustive cells (e.g., age, sex, ethnicity, income) and deriving the estimate as a sum of products. For instance, suppose the whole sample domain Y is partitioned into $Y_{.j}$'s $(j = 1,..., J)$ large domains and a reliable direct estimate $\hat{Y}_{.j}$ of the jth domain total can be obtained from the survey data; the small area, i, may cut across j so that $Y_{.j} = \sum_i Y_{ij}$, where Y_{ij} is the total for cell (i,j). Let X_{ij} be the auxiliary information total available for the ith small area within jth large domain. Then a synthetic estimator of small area total $Y_i = \sum_j Y_{ij}$ can be defined as

$$\hat{Y}_{i[S]} = \sum_j \left(\frac{X_{ij}}{X_{.j}} \right) \hat{Y}_{.j},$$

where $X_{.j} = \sum_i X_{ij}$ (Ghosh and Rao 1994). This estimator is also known as the ratio-synthetic estimator.

Moreover, the commonly used regression-synthetic estimator of the ith area population total can be defined as

$$\hat{Y}_{i[S]} = X_i'\hat{\beta}$$

where

X_i is the known total of available auxiliary information in a small area i
$\hat{\beta}$ is the estimate of the population regression coefficients

It is noted that the synthetic estimator is essentially a biased estimator, and the bias is so large for some areas that the mean squared error for them is excessive. Even so, when the small area does not exhibit strong individual effects with respect to the regression coefficients, the synthetic estimator will be efficient, with small mean square error (MSE) (Rahman et al. 2010a). Moreover, the synthetic estimates are generally easy and inexpensive to obtain, since the independent variables are easily available from census or other administrative data and the regression coefficients are obtainable from national-level surveys.

3.2.2 Composite Estimation

Composite estimation is a kind of balancing approach between the synthetic and direct estimators. It is rational in that, as the sample size in a small area increases, a direct estimator becomes more desirable than a synthetic estimator. This is true whether or not the surveys are designed to produce estimates for small areas. In other words, when area-level sample sizes are relatively small, the synthetic estimator outperforms the traditional simple direct estimator, whereas when the sample sizes are large enough, the direct estimator outperforms the synthetic estimator. A weighted sum of these two estimators would be an alternative to choose one over the other to balance their degree of bias—and this type of estimator is commonly known as a "composite estimator."

According to Ghosh and Rao (1994), composite estimation is a natural way to balance the potential bias of a synthetic estimator against the instability of a direct estimator by choosing an appropriate weight. The composite estimator of population total Y_i for a small area i can be defined as

$$\hat{Y}_{i[C]} = \phi_i \hat{Y}_{i[D]} + (1 - \phi_i)\hat{Y}_{i[S]}$$

where

$\hat{Y}_{i[D]}$ and $\hat{Y}_{i[S]}$ are, respectively, the direct and synthetic estimators of Y_i
ϕ_i is a suitably chosen weight that lies between 0 and 1

The choice of weight ranges from simple to optimal weights (Ghosh and Rao 1994; Rao 2003; Rahman 2009a). For instance, one may get the optimal weight by minimizing the MSE of the composite estimator, $\hat{Y}_{i[C]}$, with respect to ϕ_i under the assumption that the covariance factor of $\hat{Y}_{i[D]}$ and $\hat{Y}_{i[S]}$ is too small relative to the MSE of $\hat{Y}_{i[S]}$, and then it should be negligible. Now if ϕ_i^{opt} represents the optimal weight, then it can be defined as

$$\phi_i^{opt} = \frac{\text{MSE}(\hat{Y}_{i[S]})}{\{\text{MSE}(\hat{Y}_{i[D]}) + \text{MSE}(\hat{Y}_{i[S]})\}}.$$

A number of the estimators proposed in the literature have the form of composite estimators—for example, the James–Stein (J-S) estimator (proposed by James and Stein 1961) that considers a common weight ϕ. Efron and Morris (1975) have derived the J-S estimator from other principles with a generalization and, in some cases, the J-S estimator is known as a "shrinkage estimator." Rao (2003) provides an account of composite estimation (with a review of the J-S method) and examples of its practical applications in the context of SAE. Composite estimators are biased and they may have improved precision, but this depends on the selection of the weight.

3.2.3 Demographic Estimation

Demographic estimation is another way to obtain indirect estimators based on implicit models. This approach mainly uses data from a recent census in conjunction with demographic information derived from various administrative record files. Three approaches are very popular in demographic estimation: the vital rates technique, the component method, and the sample regression method. The general description about these techniques is given by Rao (2003) and also more references therein. A summary of the different demographic techniques in SAE is given in the following.

The *vital rates approach* uses only the birth and death data for a certain year or the current year, say, t. Also, a population estimate at year t for a large area (e.g., a state or large domain) containing the small areas of interest should be known from the administrative records.

For a small area i, suppose b_{i0} and d_{i0} denote the births and deaths, respectively, for the census year and b_{it} and d_{it} are the births and deaths for the current year t that are known from administrative records. Then the crude birth and death rates r_{i0}^b and r_{i0}^d for the census year and r_{it}^b and r_{it}^d for the year t can be obtained from the relation $r_{it}^x = (x_{it}/p_{it})$ (where p_{it} is the population at small area i for the year t) using appropriate values of x (as b and d) and t (as 0). In particular, as $x = b$, then $x_{it} = b_{it}$ represents the births at small area i for the year t. Assume estimates $\hat{r}_{it}^b$ and $\hat{r}_{it}^d$ can be obtained from the relations $r_{it}^b = \phi^b r_{i0}^b$

and $r_{it}^d = \phi^d r_{i0}^d$, respectively, by obtaining the updating factors estimates $\hat{\phi}^b$ and $\hat{\phi}^d$ from the large domain data (for more information, refer to Rao 2003, p. 29). Hence, the estimated population of small area i for the year t is given as

$$\hat{p}_{it} = \frac{1}{2}\left(\frac{b_{it}}{\hat{r}_{it}^b} + \frac{d_{it}}{\hat{r}_{it}^d}\right).$$

It is worth mentioning that the success of the vital rates technique depends heavily on the validity of the implicit model that the updating factors φb and φd for the small area remain valid for the large domain containing the small area (Rao 2003; Rahman 2009a). However, such an assumption is often questionable in practice.

The *component method* takes account of net migration in the model. During the period between the census and the current year (t), if the net migration in the ith small area is denoted by m_{it}, then a population estimate is given by

$$\hat{p}_{it} = p_{i0} + b_{it} - d_{it} + m_{it}$$

where p_{i0} is the population of the small area i in the census year. Immigration statistics are not always available at small area levels but can be obtained from other administrative records such as from income tax returns information.

To define the *sample regression method*, let us assume that the sample estimates $\hat{Y}_i$ $(i = 1,\ldots, k \subset n)$ are available for k areas out of n small areas. So, it is possible to fit the regression equations

$$\hat{Y}_{i \in k} = \hat{\beta} X_i'; \quad \text{for } i = 1,\ldots,k$$

where
$\hat{\beta} = (\hat{\beta}_0,\ldots,\hat{\beta}_p)'$ is the vector of $(p + 1)$ regression coefficients
$X_i = (1, x_{i1},\ldots, x_{ip})'$ is the vector of covariates containing p characteristics

Now using these fitted regression coefficients and overall auxiliary information for all small areas $i(= 1,\ldots, n)$, it is then possible to predict on Y_i from the following fitted regression equations:

$$\hat{Y}_{i \in n} = \hat{\beta}_0 + \hat{\beta}_1 x_{i1} + \cdots + \hat{\beta}_p x_{ip}; \quad \text{for } i = 1,\ldots,n.$$

The sample regression estimator of the population at small area i for the current year t is then obtained as

$$\hat{p}_{it} = \hat{Y}_{i \in n}\left(\frac{p_{i1}}{P_1}\right)\hat{P}_t$$

where

p_{i1} is the population of the ith small area in the recent census from two consecutive censuses, $P_1 = \sum_i p_{i1}$

$\hat{P}_t$ is the population estimate at the current year t in a large area containing the small areas

It is worth noting that the demographic approach is only used for population estimates. The Australian Bureau of Statistics has used this method for their small area population estimates (e.g., refer to ABS 1999). As the size and composition of the residents in a geographic area may change over time, postcensal or noncensal estimates of population are essential for a variety of purposes, such as the determination of fund allocations, calculation of social and economic indicators (e.g., vital rates, unemployment rates, and poverty rates in which the population count serves as the denominator), and calculation of survey weights. In demographic methods of SAE, several regression symptomatic procedures, their properties, and significant applications are discussed by Rao (2003). The advantages of demographic estimation are that it is an easy estimation and the theory behind this method is very simple and easily understandable. However, the underestimate of population count due to omission, duplication, and misclassification in census is a big concern.

3.2.4 Comparison of Various Implicit Models–Based Indirect Estimation

As discussed in the previous sections, the implicit models–based statistical approaches to indirect SAE can produce generally three types of estimators that are *synthetic, composite,* and *demographic estimators.* A comparative summary for these indirect estimators is presented in Table 3.1. It is noted that although the implicit models–based indirect estimators have better precision and reliability compared to the direct small area estimators, each of these three estimators has distinct features in terms of their methods, advantages, disadvantages, and applications.

The synthetic estimator is characteristically biased, easy to calculate, and applicable to various fields in government and social statistics, whereas the composite estimator is a mixture of a direct small area estimator and an indirect synthetic estimator that fairly depends on a blanching weight chosen by the researchers. Like the synthetic estimator, a composite estimator is also biased and used in many areas in social and business statistics. Moreover, the demographic estimator is rooted in census data and time-dependent variables. The theory of demographic estimation is simple and straightforward. But this technique is only usable for estimating various rates in population such as birth and death rates, and the estimator is affected by any miscounts in the census data.

TABLE 3.1

A Comparative Table for Implicit Models–Based Indirect Estimators

Small Area Estimation		Methods/ Comments	Advantages	Disadvantages	Applications
Implicit models	Synthetic estimator	Need actual sample and auxiliary data for a large-scale domain	Very easy to formulate and inexpensive to calculate	Assumption that all small areas are similar to large area is not tenable and the estimate is biased	Used in various areas in government and social statistics
	Composite estimator	Based on direct and synthetic estimators	Have choice of balancing weight at small areas	Biased estimator and depends on the weight	Used in social and business statistics
	Demographic estimator	Rooted in data from census and with time-dependent variables	Easy to estimate and the underlying theories are simple and clear-cut	Only used for population estimates and affected by miscounts in census data	Used to find birth and death rates and various population estimates

3.3 Explicit Models Approach

This class of SAE approaches is mainly using different explicit models and in the literature it is termed as "small area models." Available small area models can be classified as the basic area level models and the basic unit level models (Rao 1999; Rahman 2009a). In the first type of models, information on the response variable is available only at the small area level, and in the second type of models, data are available at the unit or respondent level. In addition, a generalization of all of these area and unit level models is known as the generalized linear mixed model. A brief summary of these small area models is given in the following.

3.3.1 Basic Area Level Model

At first, to estimate the per capita income of small areas with a population size of less than 1000, Fay and Herriot (1979) used a two-level Bayesian model, which is currently well known as the Fay–Herriot model or basic area level mixed model. The Fay–Herriot model can be expressed as

$$\text{Linking model: } \theta_i \stackrel{iid}{\sim} N(x_i'\beta, \sigma_\varepsilon^2), \quad \text{where } i = 1, 2, \ldots, n;$$

$$\text{Matching sampling model: } \hat{\theta}_i \mid \theta_i \stackrel{iid}{\sim} N(\theta_i, \omega_i^2), \quad \text{where } i = 1, 2, \ldots, n.$$

The first type of model is derived from area-specific auxiliary data that are related to some suitable functions of the small area total to develop a linking model under normality assumptions with mean zero and variance σ_ε^2. Then the linking model is combined with a matching sampling model to generate finally a linear mixed model. In this case, the matching sampling model uses a direct estimator of the corresponding suitable function of the small area total and assumed normally distributed errors with mean zero and a known sampling variance ω_i^2. But these two assumptions of the matching model have been considered as the limitations of the basic area level model (Rao 2003, 2003a). The author argues that the assumption of known sampling variances for the matching sampling model is restrictive and the assumption of a zero mean may not be tenable if the small area sample size is very small and the relevant functional relationship is a nonlinear function of the small area total.

Let $x_i = (x_{1i}, x_{2i},..., x_{mi})$ represent an area-specific auxiliary data for the ith area ($i = 1, 2,..., n$) and Y_i represent the population total of small area. The basic area level model can be expressed by the following mathematical way:

$$\theta_i = x_i'\beta + \varepsilon_i \tag{3.1}$$

where
   $\theta_i = g(\cdot)$, some suitable function of Y_i is assumed to be related to x_i through the aforementioned linear model (3.1)
   β is the regression parameters' vector
   ε_i's are uncorrelated errors of the random small area effects that are assumed to be normally distributed with mean zero and variance σ_ε^2

The parameters of this model, β and σ_ε^2, are generally unknown and are estimated from the available data.

In the basic area level models, direct survey estimators of the small area population total, that is, $\hat{Y}_i$ (or mean), are available whenever the sample sizes of ith area are greater and/or equal to one (Rao 1999). Then it can be assumed that

$$\hat{\theta}_i = \theta_i + e_i \tag{3.2}$$

where
   $\hat{\theta}_i = \hat{g}(\cdot)$ is the function of direct estimators of the population of interest at ith small area
   e_i's are the sampling error terms—assumed to be independently and normally distributed with mean zero and known sampling variance ω_i^2

To estimate the sampling variance ω_i^2, Fay and Herriot (1979) utilized the generalized variance function method that uses some external information in addition to the survey data. Now by combining the matching sampling model (3.2) with linking model (3.1), we get the following form of the standard mixed linear model:

$$\hat{\theta}_i = x_i'\beta + \varepsilon_i + e_i \qquad (3.3)$$

Note that model (3.3) involves both model-based random errors ε_i and design-based random errors e_i. Models of this form, with $\theta_i = \log(Y_i)$, have recently been used to produce model-based county estimates of poor school-age children in the United States, and by using those estimates, the U.S. Department of Education allocated large-scale federal funds to counties, and the states distributed these funds among school districts in each county (Rao 1999). The author also indicates that the success of SAE through a model-based approach largely depends on getting good auxiliary information (x_i) that leads to a small model variance σ_ε^2 relative to known or estimated ω_i^2.

3.3.2 Basic Unit Level Model

On the other hand, the basic unit level model is based on unit level auxiliary variables. These are related to the unit level values of response through a nested error linear regression model, under the assumption that the nested error and the model error are independent to each other and normally distributed with common mean zero and common or different variances. This type of model can be represented by the following mathematical equation:

$$y_{ij} = x_{ij}'\beta + \varepsilon_i + e_{ij}, \qquad (3.4)$$

where
   x_{ij} represents unit-specific auxiliary data, which are available for areas
      $i = 1, 2,..., n$
   $j = 1, 2,..., N_i$ as N_i is the number of population units in the ith area
   β represents the vector of regression parameters

The unit responses y_{ij} are assumed to be related to the auxiliary values x_{ij} through the nested error regression equation (3.4). The ε_i's are normal, independent, and identically distributed with mean zero and variance σ_ε^2. The e_{ij}'s are independent of ε_i's and follow independent and identical normal distribution with mean zero and variance σ_e^2.

The nested error unit level regression model was first used to model county crop areas in the United States (Battese et al. 1988). The authors have used the normally distributed common error variance assumption and revealed that based on the fitting-of-constants method, the estimates of error variances are slightly different from each other. They also demonstrated some techniques for validating their model on the basis of unit level auxiliary variables. This type of model is appropriate for continuous value response variables. Besides, various extensions of this type of model have been proposed to handle binary responses, two-stage sampling within areas, multivariate responses, and others (Rao 2003a,b). For instance, if y_{ij} (where j represents small area in a county i) has binary responses, it may assume that y_{ij} follows independent Bernoulli distribution with parameter p_{ij} and that the p_{ij}'s are linked by assuming a logistic regression model $\log\{p_{ij}/(1-p_{ij})\} = x'_{ij}\beta + \varepsilon_i$, where ε_i has identical and normal distribution with mean zero and variance σ^2_ε. This is a special case of generalized linear mixed models. Moreover, a two-level logistic regression model on the p_{ij}'s with random slopes β_i has been used by Malec et al. (1999).

3.3.3 Generalized Linear Mixed Model

Most of the small area models are practically special cases of the generalized linear mixed models—that is, each type of the small area models can be extended to the corresponding standard form of the generalized linear mixed model, depending on how the process has an influence on the response variable. By usual notation, the generalized linear mixed model can be defined as

$$y = X\beta + Z\varepsilon + e \qquad (3.5)$$

where

y is a vector response
X is a known covariate matrix
β is the regression coefficient vector of X (usually termed the "fixed effects")
Z is a known structure matrix of the area random effects
ε is the random effects vector due to small area (e.g., SLA, postcode)
e is the random errors vector associated with sampling error or the variation of individual or unit level

The key assumptions for (3.5) are that the random effects (ε) and random errors (e) have mean zero and finite variances; that is, $\varepsilon \sim N(0, \sigma^2_\varepsilon \Phi)$ and $e \sim N(0, \sigma^2_e \Omega)$, where Φ and Ω are positive definite matrices, and the elements of ε and e are uncorrelated.

Although to allow correlation between small domains, the matrix of area random effects, Z, may have a complex structural form, in theory, researchers may consider Z as an identity matrix. In fact, the structure of Z depends on how spatial and temporal random effects are included in the model. Moreover, for an ith (i = 1, 2,..., n) area-specific auxiliary data $x'_i = (x_{1i}, x_{2i}, ..., x_{mi})$, the generalized linear mixed model in (3.5) is of the form of a basic area level model in (3.3). Whereas for a unit-specific auxiliary data x_{ij} that represents unit j ($j = 1, 2, \ldots, N_i$, where N_i is the number of population units in the ith area) in area i, the model (3.5) is of the form of a basic unit level model in (3.4). It is important to note that area level models have extensive scope in comparison to unit level models, because area level auxiliary data are more readily available than unit-specific auxiliary data (Rao 2003a; Rahman et al. 2010).

Rao and Yu (1994) extended the linear mixed model with a first-order autoregressive, that is, $AR(1)$ model, to combine cross-sectional data with time series data—that is, information observed in preceding periods. Also refer to the EURAREA Consortium (2004a) report for linear mixed models that allow for spatial and temporal autocorrelation in the random terms, with the purpose of improving the precision of small area estimators with sample information obtained in other periods and domains. Another model approach is considered by Pfeffermann and Burck (1990), where they combine time series data with cross-sectional data by modeling the correlation between the parameters of the time series models of the separate domains in a multivariate structural time series model.

The generalized linear mixed models (GLMMs) are also frequently used in SAE for binary, polytomous, and count variables (e.g., see Ghosh et al. 1998, 2009; Lehtonen and Veijanen 1999; Noble et al. 2002; Malec 2005; Jiang 2007; Lu and Larsen 2007; Molina et al. 2007; Torelli and Trevisani 2008; Montanari et al. 2010; Tzavidis et al. 2010, 2015; Lehtonen et al. 2011; Nandram and Sayit 2011; Berg and Fuller 2012, 2014; Erciulescu and Fuller 2013; Vizcaino et al. 2013; Rao and Molina 2015). Many authors study on various model-based and design-based indirect estimations that employ GLMMs in SAE with their applications. For example, Ghosh and Maiti (2004) study a model-based indirect approach linked to exponential quadratic variance function–based mixed models families for SAE and use the beta-binomial model to estimate poverty rates in counties in the United States. The normal linear mixed model is a special case of such families. Other examples include the beta-binomial, gamma-Poisson, and multinomial-Dirichlet distributions, which may be more appropriate than the normal assumptions when the true quantity to be predicted is a single proportion, a total, or a vector of proportions, respectively. They also compare their method to a similar procedure called lognormal mixed model that the U.S. Census Bureau uses. In contrast, Lehtonen and Veijanen (1999) study a design-based indirect approach by treating a logistic mixed model with domain-specific random effects as an assisting model and then doing design-based bias correction under the assumption of simple random sampling within small areas. They utilize this approach to small area poverty estimation in EU.

Montanari et al. (2010) utilize unit level linear mixed models and logistic mixed models for binary response variable and completely known auxiliary data. They also impose the benchmarking restriction by adding the constraint to a penalized likelihood for a logistic mixed model. Malec (2005) proposes Bayesian small area estimates for means of binary responses using a multivariate binomial and multinomial model. Nandram and Sayit (2011) develop a Bayesian procedure to incorporate a constraint in the context of a beta-binomial model.

A typical example of a GLMM design is a binomial logistic mixed model for a binary y-variable. To estimate the small area totals $\hat{t}_{i.} = \sum_{j \in E_i}^{N_i} y_{ik}$ for all areas E_i, the logistic mixed model can be defined as

$$\text{LMM}(y_{ik} \mid \varepsilon_i) = P\{y_{ik} = 1 \mid \varepsilon_i\} = \frac{\exp[x'_{ik}(\beta + \varepsilon_i)]}{1 + \exp[x'_{ik}(\beta + \varepsilon_i)]}; \ k \in E_i, \forall i \qquad (3.6)$$

where
   x_{ik} is a known vector value for ith small area
   β is a vector of fixed effects common for all areas
   ε_i is a vector of area-specific random effects
Hence, the small area estimates are calculated as

$$\hat{y}_{ik} = \frac{\exp[x'_{ik}(\beta + \varepsilon_i)]}{1 + \exp[x'_{ik}(\beta + \varepsilon_i)]}; \ \forall k \in E_i. \qquad (3.7)$$

An indirect small area estimator for domains is calculated with this mixed model with few special cases being studied by Lehtonen et al. (2005, 2011). However, a basic limitation is that if the sample size at domains is not large enough, the bias-corrected estimator may have large coefficient of variation because the bias correction is a direct estimator based only on the sample s_i in area i. Maiti (2001) describes a hierarchical Bayesian approach to fit a GLMM based on an outlier robust normal mixture prior for the random effects and uses of this model for SAE. Sinha (2004) proposes robust estimation of the fixed effects and the variance components of a GLMM, using a Metropolis algorithm to approximate the posterior distribution of the random effects.

Jiang (2007) reviews the classical inferential approach for linear and GLMMs and discusses the prediction for a function of fixed and random effects. Ghosh et al. (2009) consider a small area model where covariates have unknown distribution and estimate small area proportions using both HB and EB estimators. Vizcaino et al. (2013) derive small area estimators for labor force indicators in Galicia using a multinomial logit mixed model. In a Bayesian framework, Ghosh et al. (1998) demonstrate that the multinomial distribution for a multicategory response is in the exponential family and give a condition sufficient to guarantee that the posterior distribution is proper. In an application to exposure to hazards in the workplace, these authors model observed counts as conditionally independent Poisson random variables, given the small area totals and normally distributed random effects. Additionally, Lu and Larsen (2007) conduct a Bayesian analysis of a hierarchical Poisson log-linear model for enrollment in employment preparation courses in Iowa schools stratified by district size and area education agencies. Molina et al. (2007) predict employment, unemployment, and inactivity rates in the UK labor force, under the assumption that the observed totals in small areas follow a multinomial logit model with a random area effect. This study also compares a bootstrap estimator to an analytic estimator of the frequentist MSE.

Noble et al. (2002) demonstrate that the structure preserving estimation (i.e., SPREE that combines an auxiliary table, often from a previous census, to improve the estimators of the cell totals in a multiway contingency table) model is a special case of a GLMM. The estimators of the cell totals obtained from SPREE are the maximum likelihood (ML) estimators of the expected counts under a GLMM with a Poisson random component and a log link. Main effects for rows and columns are estimated with the direct estimators. Interactions are set equal to the interactions in a saturated log-linear model fit to the census two-way table. The Poisson model underlying SPREE is also extended to the larger family of GLMMs where the parameters of the linear predictor are partitioned into two sets: one set (e.g., the main effects in the case of SPREE) is estimated from the direct estimators and the second set (e.g., the interactions) from the auxiliary data. The authors also illustrate the generalization of SPREE through an application to estimation of unemployment rates from the Household Labour Force Survey conducted by Statistics New Zealand. In the application, main effects for age and sex are estimated from the Labour Force Survey, while fixed effects associated with nine regions and interactions between age and sex are estimated from a previous census.

Zhang and Chambers (2004) develop two extensions of the log-linear model underlying SPREE. Both are models for the true proportions of interest. The first one, called the generalized linear structural model (GLSM), is a log-linear model in which the interactions are assumed to be proportional to the interactions in a census. If the coefficient on the census interactions is assumed to equal one, then the GLSM simplifies to the log-linear model underlying SPREE. Predictors based on the GLSM are synthetic estimators. To obtain predictors with a smaller bias than predictors based on the GLSM, Zhang and Chambers (2004) extend the GLSM to a random effects model called the generalized linear structural mixed model (GLSMM). In the GLSMM, the vector of interactions for a single area is assumed to have a multivariate normal distribution with a singular covariance matrix (see Zhang 2009). Further, GLSMM could be extended to a more robust area level model having a multivariate Student-t distribution with appropriate location, scale, and shape parameters.

You et al. (2002) suggest an unmatched model as an alternative to a standard approach that applies nonlinear transformations to the direct estimators to justify the use of the normal linear mixed model. The unmatched model specifies a normal linear model for the conditional distribution of the direct estimator given the true value and a nonlinear model with normally distributed random effects for the true value. You et al. (2002) argue that unmatched models have several appealing features. For one, the normal linear model captures design properties of the direct estimators, such as design unbiasedness or known features of the design variance. Simultaneously, the nonlinear link function restricts the parameter space for the true value appropriately. Use of an unmatched model can avoid problems associated with nonlinear transformations of direct estimates. For instance, in small areas with small sample sizes, a nonlinear function of an unbiased direct estimator of a total may have a nonnegligible bias for the corresponding function of the expected value of the direct estimator of the total. Also, some nonlinear functions such as the log and the logit are undefined when the direct estimate is zero. You et al. (2002) apply unmatched models to estimate the net weight undercoverage in small domains from the Canadian Census and conduct a Bayesian analysis.

Several studies compare and contrast the types of models described earlier and also propose extensions. Mohadjer et al. (2007) extend the use of unmatched models to obtain Bayes estimators of the proportions of individuals with a low literacy level in states and counties in the United States based on the National and State Assessments of Adult Literacy. Fabrizi et al. (2008) extend it to estimate proportions of households in three ordered poverty classes (severe poverty, poverty, and at risk of poverty) in domains defined by cross-classifications of 20 administrative regions in Italy and nine household types. Moreover, Torelli and Trevisani (2008) review models for discrete data, arguing that the assumptions of linear mixed models for continuous responses are often unrealistic. They propose nonnormal hierarchical models to describe overdispersion in the sampling distributions of the direct estimators, conditional on the true underlying quantities of interest. In particular, they suggest Poisson-lognormal and gamma-Poisson mixture models for the direct survey estimators, conditional on the true values to be predicted. Also, specific issues from the Italian Labour Force Survey motivate these authors to develop adaptations for situations in which the geographic locations associated with auxiliary data differ from the small areas of interest.

Furthermore, Liu et al. (2007) evaluate the design properties of Bayes credible intervals for proportions under several hierarchical models. The models include a traditional normal linear mixed model and an unmatched model with normally distributed sampling errors and a logit link function. Liu et al. (2007) also propose a model in which the direct survey estimators are conditionally independent beta random variables, and the logits of the true proportions have normal distributions. The reviews by Rao (2003), Jiang and Lahiri (2006), and Rao and Molina (2015) include discussions of applications in which the quantities to predict are counts and proportions. More recently, Tzavidis et al. (2015) propose a method to SAE for counts based on M-quantile modeling that neither depends on strong distributional assumptions nor on a predefined hierarchical structure, and outlier robust inference is automatically allowed for. Torabi and Shokoohi (2015) also offer a new approach under GLMM design using P-spline regression models to cover normal and nonnormal responses for estimating counts or proportions in small areas, with applications to the number of asthma physician visits in small areas in Manitoba using a Canadian data set.

3.3.4 Comparison of Various Explicit Models–Based Indirect Estimation

The explicit models–based approaches to indirect SAE include a range of statistical models that are typically known as *area level, unit level,* and *generalized linear mixed models*. Although the descriptions of each of these explicit models are given in Sections 3.3.1 through 3.3.3, a comparison of them may give a clearer perception. A comparative summary of explicit models is presented in Table 3.2 according to their methods, advantages, disadvantages, and applications.

It is apparent from the table that the area level model is characteristically two levels, can utilize the direct small area estimator and the area level auxiliary data, and is widely used in various areas in statistics where the model assumptions (such as normal error model with known variance) are fitting. Besides, the unit level model is a nested error, useful for studying continuous variables and bivariate to multivariate data, and successfully used in various areas in agricultural statistics. Moreover, the generalized linear mixed model is a general form of all types of small area models, and it is practically useful in many situations where the specific area and/or unit level models cannot be used. For example, the generalized linear mixed

TABLE 3.2

Comparative Table for Various Explicit Models

Small Area Estimation		Methods/ Comments	Advantages	Disadvantages	Applications
Explicit models	Area level model	Based on two-level model as the Fay–Herriot model	Able to use direct estimator and area-specific auxiliary data	Assumptions of normality with known variance may be untenable at small sample	Various areas in statistics fitting with assumptions of the model
	Unit level model	Based on unit level auxiliary data and a nested error model	Useful for continuous variables, bivariate and multivariate data	Validating such a model estimate is quite complex and unreliable	Used successfully in many areas of agricultural statistics
	Generalized linear mixed model	A general model that encompasses all other small area models	Can allow correlation between small areas, the AR(1),[a] and time series data	Calculation and structure of matrix for area random effects are very complex	In all areas of statistics where data are useful for the general model

[a] The first-order autoregressive model.

models are applicable to the first-order autoregressive, that is, *AR(1)* model, time series data, and even when correlation exists between small areas.

A range of statistical procedures are commonly used in the explicit models–based SAE. For example, the EBLUP, EB, and HB methods have been utilized for statistical inferences on various explicit models, where the techniques of ML, restricted or residual maximum likelihood (REML), penalized quasi-likelihood, etc., have a significant role in variance estimates. Details of theoretical techniques for the estimation of the parameters for different small area models are discussed by Rao (2003) and references therein. In addition, an account of the algorithms and computational tools and techniques for the estimation of parameters of various small area models is provided in the EURAREA Consortium (2004) report. Further significant review on various methods utilized in the explicit models–based indirect estimation approach is offered in the next section of this chapter.

3.4 Methods for Estimating Explicit Models

This section provides a concise summary of the EBLUP, EB, and HB methodologies for estimating explicit models–based SAE. These three approaches have played a vital role in SAE in the context of different small area models. All of these methods are extensively discussed in the small area literature (e.g., refer to Ghosh and Rao 1994; Pfeffermann 2002; Rao 2003; Rahman et al. 2010).

3.4.1 EBLUP Approach

To predict the random effects or mixed effects for a small area model, the best linear unbiased prediction (BLUP) approach is widely used. The BLUP method was originated from Henderson (1950), and many authors used it in different ways and in various fields. A comprehensive overview of the derivations of the BLUP estimator with useful examples and applications is provided in Robinson (1991), You and Rao (2002a) and Rao (2003).

Consider the generalized linear mixed model in (3.5) with the following variance components for the random effects (ε) and random errors (e):

$$\text{Var}\begin{pmatrix} \varepsilon \\ e \end{pmatrix} = \begin{pmatrix} \Phi & 0 \\ 0 & \Omega \end{pmatrix} \sigma^2$$

where
   Φ and Ω are known positive definite matrices
   σ^2 is a positive constant

Let $\hat{\beta}$ and $\hat{\varepsilon}$ be the BLUP estimates of the regression coefficient vector β and the random effects vector ε, respectively. Then the BLUP estimates can be obtained by solving the following mixed model equations (Rao 2003):

$$\left. \begin{array}{c} X'\Omega^{-1}X\hat{\beta} + X'\Omega^{-1}Z\hat{\varepsilon} = X'\Omega^{-1}y \\ Z'\Omega^{-1}X\hat{\beta} + (Z'\Omega^{-1}Z + \Phi^{-1})\hat{\varepsilon} = Z'\Omega^{-1}y \end{array} \right\} \tag{3.8}$$

Hence, the BLUP estimate of random effects is $\hat{\varepsilon} = (Z'\Omega^{-1}Z)^{-1}Z'\Omega^{-1}(y - X\hat{\beta})$ with the BLUP of the regression coefficients $\hat{\beta} = \left\{ X'(\Omega + Z\Phi Z')^{-1}X \right\}^{-1} X'(\Omega + Z\Phi Z')^{-1}y$. It is worth noting that the BLUP estimate $\hat{\beta}$ is identical with the generalized least squares estimate of β (Robinson 1991).

Particularly for the basic area level model defined in (3.3), the BLUP estimator of $\hat{\theta}_i$ ($= \hat{g}(\cdot)$, a function of ith small area population estimates) will be

$$\tilde{\theta}_i = \gamma_i \hat{\theta}_i + (1 - \gamma_i)x_i'\hat{\beta}$$

where $\gamma_i = (\sigma_\varepsilon^2)/(\sigma_\varepsilon^2 + \omega_i^2)$ and $\hat{\beta}$ is the weighted least square estimator of β with weight $(\sigma_\varepsilon^2 + \omega_i^2)^{-1}$. It is worth mentioning that the BLUP estimator is the weighted combination of the direct estimator $\hat{\theta}_i$ and the regression synthetic estimator $x_i'\hat{\beta}$ (see Pfeffermann 2002). It takes proper account of "between area" variations relative to the precision of the direct estimator (Ghosh and Rao 1994). Besides, BLUP estimators minimize the MSE among the class of linear unbiased estimators and do not depend on a normality assumption of the random effects (Rao 2003, p. 95). However, they depend on the variance components of random effects that can be estimated by the method of moments.

In the usual BLUP approach, the variance components (σ_ε^2's) of the random effects are assumed to be known. But, in practice, it could be unknown, and they are estimated from the sample data (You and Rao 2002b). In those situations, two-step BLUP estimators, which are currently well known as EBLUP, can be developed. The variance components should be estimated using the ML or the REML method. The EBLUP approach is extensively described by many authors (Ghosh and Rao 1994; Pfeffermann 2002; Rao 2002, 2003; Rahman 2009). Besides, Rao (2003a) discusses the elaborate use of BLUP and EBLUP estimators for different models in the context of SAE.

3.4.2 EB Approach

Let us assume that the small area parameter of interest is μ and it can be defined as $\mu = l'\beta + q'\varepsilon$, a linear combination of regression parameters β and the vector of random effects ε, with known constants vectors l and q. Now if

$f(\cdot)$ represents the probability density function, then the EB approach can be described by the following points:

1. At first, find the posterior density of μ, $f(\mu|y,\kappa)$ given the data y, using the conditional density $f(y|\mu,\kappa_1)$ of, y given μ and the (prior) density $f(\mu|\kappa_2)$ of μ, where $\kappa' = (\kappa_1', \kappa_2')$ represents the vector of model parameters.
2. Estimate the model parameters κ (say, $\hat{\kappa}$) from the marginal probability density of observed data y, that is, $f(y|\kappa)$.
3. Use the estimated model parameters to obtain the estimated posterior density $f(\mu|y,\hat{\kappa})$.
4. Use the estimated posterior density for statistical inferences about the small area parameters of interest, μ.

Rao (2003) provides an excellent account of the EB method and its significant applications in SAE.

In particular, under the basic area level model in (3.3), the posterior density, $f(\theta_i|\hat{\theta}_i,\beta,\sigma_i^2)$, of θ_i given $\hat{\theta}_i,\beta$ and σ_ε^2 follows an identically independent normal probability distribution with mean $\hat{\theta}_i^B$ and variance $\gamma_i\omega_i^2$, that is, $\theta_i \overset{iid}{\sim} N(\theta_i^B, \gamma_i\omega_i^2)$, for $i = 1, 2,..., n$. Here, $\hat{\theta}_i^B = E(\theta_i|\hat{\theta}_i,\beta,\sigma_\varepsilon^2) = \gamma_i\hat{\theta}_i + (1-\gamma_i)x_i'\beta$ and $\gamma_i = \sigma_\varepsilon^2/(\sigma_\varepsilon^2 + \omega_i^2)$, and $E(\cdot)$ represents the expectation of θ_i with the posterior density $f(\theta_i|\hat{\theta}_i,\beta,\sigma_i^2)$.

Under quadratic loss function, $\hat{\theta}_i^B$ is the Bayes estimator of θ_i and it is optimal, which means the "mean square error" of $\hat{\theta}_i^B$, that is, $E(\hat{\theta}_i^B - \theta_i)^2$, is always smaller than the MES of any other kind of estimator $\tilde{\theta}_i$ of θ_i. Note that $\hat{\theta}_i^B$ depends on the model parameters β and σ_ε^2. From the model assumptions, it is clear that the direct estimate $\hat{\theta}_i$ of θ_i follows a normal distribution with appropriate parameters (i.e., $\hat{\theta}_i \overset{iid}{\sim} N(x_i'\beta, \sigma_i^2 + \omega_i^2)$), and we can find the estimates of the model parameters, $\hat{\beta}$ and $\hat{\sigma}_\varepsilon^2$, from the marginal distribution by the ML or REML method. Now using these estimators $\hat{\beta}$ and $\hat{\sigma}_\varepsilon^2$ into $\hat{\theta}_i^B$, the EB estimator of θ_i can be obtained as

$$\hat{\theta}_i^{EB} = \hat{\gamma}_i\hat{\theta}_i + (1-\hat{\gamma}_i)x_i'\hat{\beta}$$

where $\hat{\gamma}_i = (\hat{\sigma}_i^2)/(\hat{\sigma}_i^2 + \omega_i^2)$.

Note that the EB estimator $\hat{\theta}_i^{EB}$ is identical to the EBLUP estimator $\tilde{\theta}_i$ provided in the previous section and it is also identical to the J-S estimator (Rao 2003, p. 63) for the case of equal sampling variances $\omega_i^2 = \omega^2$ for all (small areas) i. Moreover, different properties and applications of EB estimator are extensively described elsewhere (e.g., Ghosh and Rao 1994; Rao 1999, 2003; Rahman 2009).

3.4.3 HB Approach

The HB approach is somewhat linked with the EB approach. In this case, a subjective prior density (say, $f(\kappa)$) of the model parameters vector $\kappa' = (\kappa_1', \kappa_2')$ is essential. A prior distribution is the expert knowledge on parameters, and $f(\kappa)$ may be informative (diffuse) or noninformative (uniform). When the prior density is specified, like the EB method, we need to obtain the posterior density, $f(\mu|y)$, of μ given the data y, using the Bayes theorem over the combined information from the prior density of the model parameters $f(\kappa)$ and conditional densities $f(y|\mu,\kappa_1)$ and $f(\mu|\kappa_2)$. Then inferences on a parameter of interest in small areas are fully based on the posterior density $f(\mu|y)$. For instance, if we are interested to estimate a small area parameter defined as $\phi = h(\mu)$, then the HB estimator of ϕ will be the posterior mean $\hat{\phi}^{HB} = E\{h(\mu|y)\}$. Hence, the posterior variance $\text{Var}\{h(\mu|y)\}$ will be used as a measure of precision of the HB estimator $\hat{\phi}^{HB}$.

The HB approach is very straightforward, inferences are clear-cut and exact, and it can handle high-dimensional complex problems using Monte Carlo Markov chain (MCMC) techniques such as Gibbs sampler (Rao 2003). Although theories behind the MCMC methods are complex, put simply they can generate samples from the posterior density and then employ the simulated samples to estimate the expected posterior quantities, for instance, $\hat{\phi}^{HB} = E\{h(\mu|y)\}$. Different software packages have been developed to run MCMC methods (Rao 2003 and also more references therein).

Rao (2003) gives a detailed description of the HB approach to the various small area models with different properties and significant applications of HB estimators. A brief summary is depicted here based on the basic area level model as defined in (3.3). In the HB approach, we require to select a joint prior density $g(\beta, \sigma_\varepsilon^2)$ of the model parameters β and σ_ε^2. Let us assume that the model parameters are independent to each other, and σ_ε^2 is known (initially, we may consider) with a uniform prior density, $g(\beta) \propto 1$, of β. Then, on the basis of this prior information, we can obtain the HB estimator of θ_i for the ith small area as

$$\hat{\theta}_i^{HB} = E(\theta_i|\hat{\theta}, \sigma_\varepsilon^2)$$

for $\hat{\theta} = (\hat{\theta}_1, \ldots, \hat{\theta}_i, \ldots, \hat{\theta}_n)'$ and with the posterior variance $\text{Var}(\theta_i|\hat{\theta}, \sigma_\varepsilon^2) = D_{1i}(\sigma_\varepsilon^2)$.

It is worth mentioning that in this prior selection, that is, considering σ_ε^2 is known and $f(\beta) \propto 1$, the HB estimator $\hat{\theta}_i^{HB}$ is identical to the BLUP estimator $\tilde{\theta}_i$ with the same variability. However, in practice, when σ_ε^2 is unknown, a prior distribution of σ_ε^2 is necessary in the HB method to take an account of the uncertainty about σ_ε^2. Let the prior density of σ_ε^2 be $g(\sigma_\varepsilon^2)$, and in that case the joint prior density of the model parameters is $g(\beta, \sigma_\varepsilon^2) = g(\beta).g(\sigma_\varepsilon^2) \propto g(\sigma_\varepsilon^2)$.

Hence, it is possible to obtain the posterior distribution $f(\sigma_\varepsilon^2|\hat{\theta})$ of σ_ε^2 utilizing the joint prior $g(\beta, \sigma_\varepsilon^2)$ and then the posterior mean or the HB estimator of θ_i as

$$\hat{\theta}_i^{HB} = E(\theta_i|\hat{\theta}) = E_{[\sigma_\varepsilon^2|\hat{\theta}]}(\hat{\theta}_i^{HB})$$

where $E_{[\sigma_\varepsilon^2|\hat{\theta}]}$ represents the expectation with respect to the posterior density $f(\sigma_\varepsilon^2|\hat{\theta})$. Here, the posterior variance of θ_i is given by

$$\text{Var}(\theta_i|\hat{\theta}) = E_{[\sigma_\varepsilon^2|\hat{\theta}]}\{D_{1i}(\sigma_\varepsilon^2)\} + \text{Var}_{[\sigma_\varepsilon^2|\hat{\theta}]}(\hat{\theta}_i^{HB}).$$

Note that no closed-form expressions exist for the evaluation of $\hat{\theta}_i^{HB}$ and $\text{Var}(\theta_i|\hat{\theta})$, and in this simple case, they can be approximated numerically using only one-dimensional integrations. But for complex models, multidimensional integration is often involved for these evaluations (Ybarra and Lohr 2008; Yan and Sedransk 2010), and MCMC techniques are useful to overcome the computational difficulties of high dimensions (Rao 2002; Rahman et al. 2010a).

Additionally, Lehtonen and Veijanen (2009) study on common design-based properties related to bias and accuracy of design-based model-assisted estimators (which are discussed in Section 2.4.4) and model-based indirect estimators (discussed in this chapter) for small areas or domains. An outline of those properties is presented in Table 3.3. Findings reveal that model-assisted estimators such as model-assisted generalised regression (GREG) and calibration are design consistent or nearly design unbiased by definition, but their variance can become large in domains where the sample size is small. Model-based small area estimators such as synthetic and EBLUP estimators are design biased: the bias can be large for domains where the model does not fit well. The variance of a model-based estimator can be small even for too small domains or areas, but the accuracy can be poor if the squared bias dominates the MSE, as shown, for example, by Lehtonen et al. (2005).

For a model-based estimator, the dominance of the bias component together with a small variance can cause poor coverage rates and invalid design-based confidence intervals. For design-based model-assisted estimators, on the other hand, valid confidence intervals can be constructed. Typically, model-assisted estimators are used for major or not very small domains/areas, and model-based estimators are used for minor or small domains/areas where design-based model-assisted estimators can fail.

Table 3.3 indicates that very small areas or domains present problems in the design-based model-assisted approach. Such a very small area is known as a mini small area when its share of population is less than 1% of the

TABLE 3.3

Properties of Design-Based Model-Assisted and Model-Based Estimators for Small Areas or Domains

Properties	Design-Based Model-Assisted Methods (e.g., Model-Assisted GREG and Calibration Estimators)	Model-Based Indirect Methods (Synthetic, GLMMs, and EBLUP Estimators)
Bias	Design unbiased (approximately) by the construction principle	Design biased
		Bias does not necessarily approach zero with increasing domain sample size
Precision (variance)	Variance may be large for small domains	Variance may be large for small domains
	Variance tends to decrease with increasing domain sample size	Variance tends to decrease with increasing domain sample size
Accuracy (MSE)	MSE = variance (or nearly so)	MSE = variance + squared bias
		Accuracy can be poor if the bias is substantial
Confidence intervals	Valid design-based intervals can be constructed	Valid design-based intervals not necessarily obtained

total population. In such sort of mini small areas, especially direct estimators can have large variance. Thus, in many situations, mini small areas/domains are the main reason to prefer indirect model–based estimations to direct design–based model-assisted estimations (Rao 2003; Lehtonen and Veijanen 2012; Rao and Molina 2015). Moreover, by considering the discussion in Sections 2.4.5 and 3.3.4, it is quite evident that in real-world practices, two main approaches are used for design-based SAE: direct estimators that are usually applied for planned domain structures (such as strata whose sample sizes n_i are fixed in the sampling design) and indirect estimators whose natural applications are for unplanned domains (whose domain sample sizes are random). In model-based SAE, indirect estimators that aim at borrowing strength are often used such as synthetic estimators in micro-simulation modeling.

3.4 A Comparison of Three Methods

In view of the earlier discussion, a comparison of the EBLUP, EB, and HB methodologies for estimating explicit models is represented in Table 3.4. The idea is here to summarize the methodological characteristics of each of these statistical approaches for explicit models.

TABLE 3.4

A Comparison of EBLUP, EB, and HB Approaches

Characteristics	EBLUP	EB	HB
Applicable to	The linear mixed models with continuous variables but not to models with binary or count data	The linear mixed models with continuous variables but not to models with binary or count data	The linear mixed models with continuous variables but not to models with binary or count data
Type of approach	Fully frequentist or classical	Hypothetically frequentist, that is, does not depend on a prior density of parameters	Fully Bayesian approach and involving with a (joint) prior density of parameters
Estimate	It is a weighted combination of the direct estimator and the regression synthetic estimator, that is, it is a kind of composite estimator	It is a "Bayes" estimator under quadratic loss function and using the estimates of model parameters	Estimator is the posterior mean of the parameter of interest and fully based on the posterior probability density
Variance assumption	In theory, the variance component of random effects is assumed to be known but may require to estimate from data in many practical situations	The variance component (σ_ε^2) of random effects is unknown and could be estimated from the sample data	When σ_ε^2 is unknown, a prior density is necessary here to take an account of the uncertainty about σ_ε^2
β estimation	The model parameter (β) should be estimated by the WLS approach	β should be estimated from the marginal pdf of data by the ML or REML	β should have a prior distribution with known quantities (e.g., $f(\beta) \infty 1$)
Variance estimation	σ_ε^2 should be estimated by the ML or REML methods/the method of fitting constants or moments	σ_ε^2 should be estimated from the marginal pdf of data by the ML or REML	σ_ε^2 should have a prior density or they should be considered as known
Properties of estimator	Estimator is identical to the James–Stein (J-S) estimator for constant sampling variances	Estimator is identical to the EBLUP (and then J-S) estimator in the case of equal sampling variances for all small areas	Estimator is identical to the BLUP estimator under the assumptions that σ_ε^2 is known and $f(\beta) \infty 1$
	Estimator is model unbiased if ε_i and e_i are normally distributed	Estimator is biased, but more reliable than EBLUP	Estimator is biased, but more reliable than others

(Continued)

TABLE 3.4 (*Continued*)

A Comparison of EBLUP, EB, and HB Approaches

Characteristics	EBLUP	EB	HB
Complexity	Theory is simple and less complicated, but inferences are not very straightforward and exact	Theory is complex and more complicated, but inferences are somewhat straightforward and exact	Theories are more complex and very complicated, but inferences are clear-cut and exact
	This method cannot handle high-dimensional complex problems	This approach may handle high-dimensional complex problems	It can handle high-dimensional complex problems using MCMC (e.g., Gibbs sampler)

3.5 Concluding Remarks

In this chapter, a comprehensive review of literature on different statistical approaches to indirect model–based SAE is provided. The overall methodologies for indirect statistical estimation include two types of statistical models—commonly known as the implicit and explicit models, and a range of statistical techniques for the estimation of small area models.

The review points out that the implicit models–based statistical methods can generate three types of estimators: (1) *synthetic estimators*, (2) *composite estimators*, and (3) *demographic estimators*. All of these indirect estimators can be obtained by borrowing strength from the available auxiliary data, and they have an enhanced precision and statistical reliability compared to the inadequate precision of the direct estimators. Additionally, the comparative summary remarks that each of the implicit model–based indirect estimators has its own features in regard to the theory as well as its advantages, disadvantages, and applications. More exclusively, the implicit model–based estimators are relatively easier to calculate—but they are characteristically biased estimators and limited for applications with the nature and suitability of the applicable data.

Besides this, the explicit models–based approaches to indirect SAE include a range of statistical models that are typically known as *area level, unit level,* and *generalized linear mixed models*. In the first type of models, information on the response variable is available only at the small area level; in the second type of models, data are available at the unit or respondent level; and in the generalized linear mixed models, data are available from the area and/or unit models. Besides, it is apparent from the appraisal that an area level model is characteristically two levels and this type of model can utilize the direct small area estimator and the area level auxiliary data. The area level models under normal error distribution with known variance have been widely used by researchers in various fields of social and economic statistics, such as to find numbers of poor school children and per capita income of households at small area levels. Nevertheless, the unit level model is a nested error and very useful for studying continuous variables. It is remarkable that the nested error models can be applicable to the bivariate and multivariate data. The unit level models have been successfully used in agricultural statistics. Additionally, the generalized linear mixed model is a common form of all types of small area models that are practically useful in many situations where the specific area or unit level models cannot be used. For example, the generalized linear mixed models are applicable to the first-order autoregressive, that is, $AR(1)$ model, time series data, and even when correlation exists between small areas.

Moreover, three statistical approaches, specifically the EBLUP, EB, and HB approaches, have been described in this chapter. It is noted that these approaches have had a significant impact on the area of statistical model–based indirect SAE. During the last few decades, all of these techniques have been extensively used in many countries, particularly in the European countries, the United States, and Canada for studying various explicit models–based indirect estimation. The review suggests that the EBLUP approach is applicable to the linear mixed models that are usually designed for continuous variables. But in practical fields, there are many situations that require dealing with binary or count data where the EBLUP method is not appropriate. Hence, an advantage of the EB and HB approaches over EBLUP is that they are applicable to linear mixed models, as well as models with binary or count data. In addition, under some conditions, these approaches produce identical results. However, they have quite different computational tools and techniques. For instance, in principle, the EB approach is considered as a frequentist approach and does not depend on a prior distribution of the model parameters. In contrast, the HB approach essentially uses a prior distribution of model parameters and it can handle complex problems using MCMC techniques. It is also noticeable that even though these statistical approaches use data from different sources to obtain the small area estimators, they do not involve generating a base microdata file for the small area, which is a significant resource for various further analyses. So, Chapter 4 addresses some other methodologies that are categorized as the *indirect geographic approach* toward SAE.

4

Indirect Estimation:
Geographic Approaches

4.1 Introduction

The geographic approach to small area estimation harks back to the micro-simulation modeling ideas pioneered in the middle of the last century by Orcutt (1957, reprinted in 2007). This approach is fully based on microsimulation modeling technology–based spatial models (Rahman 2009a, 2011; Rahman et al. 2013; Williamson 2013; Rahman and Harding 2014; Rahman and Upadhyay 2015; Rahman n.d.). During the last two decades, microsimulation modeling has become a popular, cost-effective, and accessible method for socioeconomic policy analysis, with the rapid development of increasingly powerful computer hardware, with the wider availability of individual unit record data sets (Harding 1996), and with the growing demand by policy makers (Harding and Gupta 2007) for small area estimates at government and private sector levels.

The starting point for microsimulation models is a microdata file, which provides comprehensive information on different characteristics for every individual on the file. But the survey data may not contain appropriate group indicators or may not include enough sample units in the same groups, which may affect the ability to estimate different domain level effects. Steel et al. (2003) provide a thorough analysis on how aggregate data (such as census data or administrative records) may be used to improve estimates of multi-level populations using survey data, which may or may not contain group indicators. Typically, for microsimulation modeling adjustments, researchers need to create a synthetic population microdata set using information from other reliable sources (such as aggregate data sources); this will ensure that it reflects as fully as possible the population that is being modeled and includes all of the variables that we are interested in.

This chapter explicitly reviews various significant methodologies in the geographic approach toward indirect small area estimation, with a particular emphasis given to the reweighting techniques utilized for microdata generation. The rest of the chapter is organized in the following order. In Section 4.2, the process and types of microsimulation modeling are described with an outline of the advantages of spatial microsimulation models. In Section 4.3, an appraisal of various significant methodologies utilized in spatial micro-simulation modeling is given, along with a detailed description of various synthetic reconstruction methods. In Section 4.4, the combinatorial optimization (CO) reweighting technique is addressed at length. In Section 4.5, the generalized regression weighting (GREGWT) reweighting methodology is described in detail. In Section 4.6, a comparison between the two reweighting techniques is portrayed. Finally, the concluding summary of the chapter is offered in Section 4.7.

4.2 Microsimulation Modeling

Microsimulation modeling was originally developed as a tool for economic policy analysis (Merz 1991; Rahman 2011). Clarke and Holm (1987) provide a thorough presentation on how microsimulation methods can be applied in regional science and planning analysis. In microsimulation models, researchers represent members of a population for the purpose of studying how individual behaviors generate aggregate results from the bottom up (Epstein 1999). This brings about a very natural instrument to anticipate trends in the environment through monitoring and early warning as well as to predict and value the short-term and long-term consequences of implementing certain policy measures (Ferreira et al. 2003; Saarloos 2006; Felsenstein et al. 2007; O'Donoghue et al. 2012a). According to Taylor et al. (2004), microsimulation can be conducted by reweighting a generally national-level sample, so as to estimate the detailed socioeconomic characteristics of populations and households at a small area level. This modeling approach combines individual or household microdata, currently available only for large spatial areas, with spatially disaggregate data, to create synthetic microdata estimates for small areas (Harding et al. 2003; Chin et al. 2006; Rahman 2008a; Christen 2012; Hermes and Poulsen 2012; Rahman and Harding 2014).

4.2.1 Process of Microsimulation

Although microsimulation techniques have become useful tools in the evaluation of socioeconomic policies, they involve some complex subsequent procedures. An overall process involved in spatial microsimulation is described by Chin and Harding (2006) and Rahman (2008a). The authors classified two major steps within this process: (1) create household weights for small areas using a reweighing method and (2) apply these household weights to the selected output variables to generate small-domain estimates of the selected variables. Each of these major steps involves several substeps. In particular, step one involves six substeps: (1) select variables for matching survey and census data, (2) construct benchmarks from census or administrative records, (3) prepare a linkage file for available data via recoding and uprating, (4) prepare the census data by cleansing and balancing, (5) generate the reweighting algorithm using statistical analysis system (SAS) application, and (6) run the reweighting algorithm and check convergence measure and the output of new household weights by small area. Additionally, the second step involves three substeps: (1) run the basic static microsimulation model against the base population for an appropriate base year to produce an output file containing the selected output variables for each household; (2) merge this household-level output file with the new weights file (output from the first major step) to produce, for each small area, a data set containing the output variables and the household weights by household; and (3) the output variables (by household) from the previous substep are aggregated to small area totals to create the synthetic small area estimates for each small area of interest in the following manner:

- For each categorical output variable (e.g., "employment status"), the household weights of each of the categories (e.g., "employed" and "unemployed") are summed to produce the small area total (e.g., summing all household weights in the small area for the "employed").
- For each numerical output variable (e.g., "taxable income"), the value of the output variable is first multiplied by the household weight, and the weighted value is then summed to produce small area totals (or sometimes averaged to produce small area means).

Ballas et al. (2005) outline the following four major steps in the microsimulation process:

1. The construction of a "microdata" set (when this is not available)
2. Monte Carlo sampling from this data set to "create" a microlevel population (or a "synthetic" population by Chin and Harding (2006)) for the interested domain
3. What-if simulations, in which the impacts of alternative policy scenarios on the population are estimated
4. Dynamic modeling to update a basic microdata set

A microdata set contains a large number of records about individual persons, families, or households. In Australia, microdata are generally available in the form of confidentialized unit record files (CURFs) from the Australian Bureau of Statistics (ABS) national-level surveys. Typically, the survey data provide a very large number of variables and adequate sample size to allow statistically reliable estimates for only large domains (such as only at the broad level of the state or territory). In practice, small-area-level estimates from these national sample surveys are statistically unreliable due to sample observations being insufficient, or in many cases nonexistent where the domain of interest may fall out of the sample areas (Tanton 2007; Rahman 2009a). For instance, if a land development agency wants to develop a new housing suburb, then this new small domain should be out of the sample areas. Also, in order to protect the privacy of the survey respondents, national microdata often lack a geographical indicator, which, if present, is often only at the wide level of the state or territory. Thus, spatial microdata are usually unavailable, they need to be synthesized (Chin et al. 2005), and the lack of spatially explicit microdata has in the past constrained spatial microsimulation modeling of social policies and human behavior.

Obtaining better spatial microdata that can be used to test more sophisticated theories of spatial microsimulation necessitates improved empirical methods, such as reweighting techniques and reliability measures, for example, being developed within the field of microsimulation in econometrics or economics. The creation of spatial microdata should be called "simulated estimation" of small area micropopulation, which has the potential to provide a unique source of geographical and individual-level information.

4.2.2 Types of Microsimulation Models

Microsimulation models can be developed in a number of ways based on some basic characteristics related to the problem and data availability. As all microsimulation models start with microdata files, it is essential to have such reliable microdata files. Harding (1993) classified microsimulation models into two major types—static and dynamic, with dynamic microsimulation models being further categorized as *population-based* or *cohort-based* models.

However, the boundaries between these two types of models are becoming increasingly blurred (Gupta and Harding 2007). Recently, a list of different characteristics of microsimulation models is illustrated by van Leeuwen et al. (2007), which is presented in Table 4.1.

Brown and Harding (2002) argue that microsimulation models are by definition quantitative, typically complex, and large and more commonly static, deterministic, nonbehavioral, and nonspatial—although new

TABLE 4.1

Characteristics of Microsimulation Models

Different Characteristics of Microsimulation Models	
Short term (static)	Long term (dynamic)
Deterministic	Probabilistic/stochastic
Spatial	Aspatial
Simple	Complex

Source: van Leeuwen, E.S. et al., *disP*, 170, 19, 2007.

microsimulation models are emerging that are increasingly dynamic and more complicated and probabilistic, encompass behavioral elements, and are designed as spatial models.

Although microsimulation models are traditionally divided into two types—static and dynamic microsimulation—a third type of microsimulation model called "spatial microsimulation" is emerging. Static models examine microunits at one point in time and have been used extensively to examine the differential impact of policy changes (e.g., refer to Lloyd 2007). Dynamic microsimulation models project the population in the base year forward through time by simulating transitions of life events such as fertility and mortality at the individual level. In addition, spatial microsimulation contains geographic information that links microunits with location and therefore allows for a regional or local approach to policy analysis. A brief summary of each type of microsimulation model is given in the following subsections.

4.2.2.1 Static Microsimulation

Static models usually take a cross section of the population at a specified point in time and apply different policy rules to the individual units to measure the effects of policy changes. These models generally do not attempt to incorporate behavioral change of individuals in response to the policy changes. Most static models are used principally to calculate the immediate (or first-round) effects of policy changes, before individuals have had time to adjust their behavior to the changes (Harding 1996). Generally, these models will allow the analyst to vary the rules of eligibility or liability and produce output showing the gains or losses (both to individuals and in aggregate) from the policy change. A new generation of models is also emerging that show the second-round effects of policy changes, typically changes in labor supply (see, e.g., Creedy et al. 2002 and others from the International Microsimulation Association conference [IMA 2007]).

The static microsimulation models have often played a decisive role in determining the final form of policy reforms introduced by governments, and they are now widely used across Europe, the United States, Canada, and Australia (Harding and Gupta 2007). For example, a static microsimulation Static Incomes Model (STINMOD), developed by the National Centre for Social and Economic Modeling (NATSEM) at the University of Canberra, influenced the final shape of the GST tax reform package introduced in Australia in 2000 (Harding 2000; Lloyd 2007). STINMOD is a well-known model of Australia's income tax and transfer system and has been used by a number of commonwealth government agencies including the Department of the Treasury, the Department of Family, Housing, Community Services and Indigenous Affairs, the Department of Education, Employment and Workplace Relations, community organizations, and university researchers. The first version of this was released in 1994, and since then, the model is updated each year in line with the latest changes to the Commonwealth Tax and Transfer System. The model is mostly used to analyze the distributional and individual impacts of income tax and income support policies and to estimate the fiscal impact for the government and the distributional impacts of policy reform for families. A user-friendly, point and click interface version of STINMOD enables researchers to easily analyze the impact of a range of policy changes to most tax and transfer settings. In addition to the GST tax reform in Australia, STINMOD has been used to analyze a wide range of policy input including the changes in effective tax rates and welfare distributions over the last decade and most recently the distributional impact of the recently enacted Welfare to Work and carbon price legislation. It has also been used for research across a wide range of other areas, including housing, legal aid, health benefits, and child support payments. Every year, this is updated to incorporate newly legislated changes in personal income tax and income support policies and to ensure the currency of the models base population and program parameters.

Similarly, the static microsimulation Transfer Income Model (TRIM) has influenced the public policy for many years in the United States (see Harding and Gupta 2007). This model is developed in 1973 and is maintained at the Urban Institute in the United States, with primary funding from the Department of Health and Human Services, Office of the Assistant Secretary for Planning and Evaluation. Since its development, this model has been in continuous use, analyzing the effects of current programs and the potential effects of changes to programs. For example, TRIM was used extensively in the analysis of tax reform options in the 1980s and of both welfare reform and health reform in the 1990s. TRIM-based projects have included analyses related to immigrants' receipt of public benefits, changes in the Temporary Assistance for Needy Families (TANF) caseloads, the government "cost avoidance" attributable to child support payments, and the effect of federal income tax credits. This model simulates the major governmental tax, benefit, and health insurance programs that affect the U.S. population. The programs modeled by TRIM fall into four categories:

1. *Cash transfer programs*: Supplemental Security Income, Temporary Assistance for Needy Families, child support, and the Low Income Home Energy Assistance Program

2. *In-kind transfer programs*: Child care subsidies (primarily through the Child Care and Development Fund), public and subsidized housing, the Supplemental Nutrition Assistance Program, and the Special Supplemental Nutrition Program for Women, Infants, and Children

3. *Health insurance programs*: Medicaid and CHIP (the Children's Health Insurance Program) and employer-sponsored health insurance

4. *Tax programs*: Payroll taxes, federal income taxes (including credits), and state income taxes (including credits)

The latest version of this model is TRIM3, which has four key uses: First, the model estimates the number of households, families, or people eligible for government safety net benefits; these eligibility estimates in turn allow estimation of program participation rates. Second, TRIM3 corrects for under-reporting of benefits in survey data to provide a more complete picture of the current safety net. Third, TRIM3 can estimate the effects of hypothetical or proposed policies. Fourth, the model can analyze families' economic well-being using an aftertax, aftertransfer definition of resources. Recently, TRIM3 has been used by poverty commissions and nonprofit organizations in several states to estimate the antipoverty effectiveness of possible changes to state policies. In particular, TRIM3 is used to compute poverty rates and gaps using expanded poverty measures such as the supplemental poverty measure. Like other microsimulation models, TRIM3 can produce results at the individual, family, state, and national levels. Also, this model is heavily parameterized with detailed program rules to capture interactions across programs, allowing analysts to easily simulate a variety of hypothetical policies.

4.2.2.2 Dynamic Microsimulation

Although, like static models, dynamic microsimulation models often start from the same cross-sectional sample survey data, dynamic ageing procedures are required for updating each characteristic for each individual within the original data set for each time interval (Harding 1996; Brown and Harding 2002). Transition matrices or econometric tools are used to obtain these status shifts through time (Nielsen 2002). The individuals in the base file or original microdata are progressed through time in accordance with probabilities of an event or series of major life events (such as death, marriage, births, and education) occurring. It is a bottom-up strategy for modeling the interacting behavior of decision-makers such as individuals, families, and firms within a larger system. Guy Orcutt's pioneering work from the beginning of the 1960s is the root of one of the most significant contributions within the microsimulation field—DYNASIM, the behavioral Dynamic Microsimulation of Income Model (Orcutt et al. 1961, 1976). DYNASIM is a genuinely dynamic model that simulates the economic and social behavior of American households over time. More specifically, this model can help answer a wide range of policy questions by using the best and most recent data available and by projecting the size and characteristics—such as income and health status—of the U.S. population for the next 75 years. For example, it can help to answer the vital policy question "how will coming generations fare in retirement," given that employer-sponsored pension plans are fading away, social security is changing, costs for health care and long-term services and supports are rising, and the sheer increase in the older population is squeezing younger taxpayers. In addition, how might changes in social security, pension policy, and Medicare, as well as financing options for long-term services and supports, improve retirement security? DYNASIM can also describe "what-if" scenarios, showing how outcomes would likely evolve under changes to public policies, business practices, or individual behaviors. With this model, researchers can show how different groups will fare over time, who is moving ahead and who is being left behind, and which groups would win and lose under various policy options.

These types of model attempt to incorporate individuals' behavioral change. Dynamic microsimulation models are of two types—*dynamic population* models and *dynamic cohort* models. The first type of models involve ageing a sample of an entire population, typically beginning with a comprehensive cross-sectional sample survey for a particular point in time, while the dynamic cohort models involve ageing only one cohort rather than an entire population (Harding 2000; Harding and Gupta 2007; Flood 2007). (Note that an entire population consists of a number of cohort populations.) The Australian Population and Policy Simulator Model (APPSIM) of NATSEM is a dynamic microsimulation model for predicting the future characteristics and retirement income of Australians in the coming years (Harding et al. 2010).

The development of APPSIM was a 5-year project funded by the Australian Research Council and 13 commonwealth agencies. It was delivered in December 2005 with the aims to give the Australian Commonwealth Government the capacity to assess the future distributional and revenue consequences of changes in social security and taxation programs and the future retirement incomes of Australian ageing population. The main development driver of the model is to provide a decision support tool for policy makers allowing them to develop policy. Such policy will help minimize the costs and maximize the benefits in regard to the ageing population into their retirement and aged care. The model must be tailored to represent the particular concerns that arise from the ageing population in Australia. APPSIM provides snapshot output of the characteristics of the population and government programs, with individuals being aged to about 2050. Since its development, this model has been used in evaluating the impact of future social and fiscal policies.

The APPSIM model was first used to simulate the long-term impact of changes in policy levers as the Superannuation Guarantee in research commissioned by the Japanese Cabinet Office in March 2009 (Harding et al. 2009). In addition, Japan is the first country to use APPSIM model in terms of policy levers on aged population, which can be a good testimonial to Australia's policy. APPSIM is a closed, population-based, dynamic microsimulation model that operates in discrete time. The model provides snapshots of the Australian population characteristics and government programs conducted every June 30 of the year from 2001 through to 2051. APPSIM simulates the life cycle of around 200,000 individuals based on a 1% sample of the 2001 census. Individuals are added through births and migration and removed through death and emigration, become disabled, form households, study, work, earn money, buy assets, pay tax, and use aged care and health services. The model projects a 1% representative sample of the Australian population into the future (e.g., from 2001 to 2050). All these details are estimated to up to 50 years, and users can observe how the Australian population grows and develops over time under a number of different scenarios (Harding 2007; Keegan 2011; Lymer and Brown 2012). Furthermore, details of the dynamic microsimulation models are extensively discussed elsewhere (Harding 1993, 2000; Brown and Harding 2002; Harding and Gupta 2007).

4.2.2.3 Spatial Microsimulation

Nowadays, a very fast-growing new field of microsimulation is spatial microsimulation, for estimating the local or small area effects of policy change and future small area estimates of population characteristics and service needs (Landis 1994; Williamson et al. 1998; Laird et al. 1999; Ballas et al. 2003, 2006; Brown et al. 2004; Taylor et al. 2004; Brown and Harding 2005; Chin et al. 2005; Chin and Harding 2006, 2007; Cullinan et al. 2006; Anderson 2007; Anderson et al. 2007; King 2007; Rahman 2009a; Zaidi et al. 2009; Rahman et al. 2013). For instance, spatial microsimulation may be of value in estimating the distributions of different population characteristics such as income, tax and social security benefits, income deprivation, housing unaffordability, housing stress, housing demand, and care needs, at small area level, when contemporaneous census and/or survey data are unavailable (Brown et al. 2004; Taylor et al. 2004; Chin et al. 2005; Anderson 2007; Rahman 2008b; Tanton et al. 2009; Rahman and Harding 2014). This type of model is mainly intended to explore the relationships among regions and subregions and to project the spatial implications of economic development and policy changes at a more disaggregated level.

Fundamental to almost all spatial microsimulation models is the creation of a micropopulation data set containing the spatial distribution of various characteristics on which researchers, decision and policy makers, and stakeholders are interested in. The starting point for this type of models is with unit record data that are available as the CURFs in Australia from the ABS. To generate spatially disaggregated micropopulation data set for analysis, a range of simulation techniques are in use including the reweighting methods.

Spatial microsimulation modelings have been used in many countries across the world. To date, this class of models are successfully applied to a wide range of policy areas including population studies (Van and Post 1998; Ballas et al. 2005; Phillips and Kelly 2006), demography (Holm et al. 2002; Lundevaller et al. 2007), education (Rephann 2002; Wu et al. 2008), economics (Lloyd et al. 2000; Harding et al. 2006), poverty and inequality (Lloyd et al. 2000; Ballas and Clarke 2001; World Bank 2004; Elbers et al. 2008; Harding et al. 2009), crime analysis (Rephann and Ohman 1999; Kongmuang et al. 2006), housing (Hooimeijer and Oskamp 2000; Phillips and Kelly 2006;

Rahman 2010; Rahman and Harding 2014), health (Lymer et al. 2008; Edwards and Clarke 2009; Morrissey et al. 2010), tourism (van Leeuwen et al. 2009; van Leeuwen and Nijkamp 2009), *environment* (Svoray and Benenson 2009), *agriculture* (Hennessy et al. 2007; Hynes et al. 2009; Clancy et al. 2012), land use (Landis and Zhang 1998; Waddell 2002; Miller et al. 2004; Wegener 2004; Moeckel et al. 2007), transportation (Kitamura et al. 1995; Miller and Salvini 2001; Raney et al. 2003; Bradley and Bowman 2006; Birkin et al. 2009; Wu and Birkin 2012), and regional development (Hooimeijer 1996; Veldhuisen et al. 2000; Joshi et al. 2006; Zhao and Chung 2006; Nakaya et al. 2007; Zhou and Kockelman 2010), among many others. A brief overview of a number of applications is given as follows.

From the perspective of policy analysis and planning, knowledge and analysis of geodemographics and spatially disaggregated social and behavioral attributes are fundamentally useful, even without analyzing other policy and/or economic issues. This arena is the subset of a wider field of sociodemographic microsimulation modeling and population projection (Van and Post 1998). Spatial demographic and socioeconomic characteristics are almost invariably part of the model construction process. For example, the System for Visualizing Economic and Regional Influences Governing the Environment (SVERIGE) model dynamically simulates a number of demographic processes (fertility, education, marriage, divorce, leaving home, migration, mortality, immigration, and emigration) in Sweden for use in a range of applications (Rephann 2001; Holm et al. 2002). Lundevaller et al. (2007) using a similar methodology undertook a simulated population for a small area in Sweden. MoSeS (Modeling and Simulation for e-Social Science) is an events-driven model used in the United Kingdom that simulates characteristics of demographic processes and socioeconomic analysis such as education and transport uses (Birkin et al. 2009; Wu and Birkin 2012). This model simulations can form the basis of a wide range of applications in both e-research and public policy analysis, with potentially substantial benefits into threefold such as (1) a big policy impact through the generation of effective predictions, (2) a potential "wind tunnel" or "flight simulator" analogy (i.e., planners can gauge the effects of development scenarios in a laboratory environment), and (3) the use of simulations as a pedagogic tool allows planners to refine understanding of systemic behavior and alternative futures, thus aiding clarity of thinking and improved decision-making. Similarly, Ballas et al. (2005) simulate the Irish population forward from 1991 to 1996 as benchmark comparison against the 1996 census at a county level and projected the population forward into the future. The Australian models in NATSEM meanwhile use static ageing reweighting techniques to update the population (Phillips and Kelly 2006) for use in a large range of analyses.

With a specific focus to education, Wu et al. (2008) in a dynamic demographic model of the Leeds area in the United Kingdom modeled the location and housing choices of university students. Rephann (2002) described an educational participation model of the SVERIGE incorporating the impact of spatial variables on participation. Kavroudakis et al. (2009) studying the hinterlands of 12 UK universities modeled areas of potential students and simulated the impact of changed rules on the eligibility for student loans, living costs, and dropout rates and targeted admission policies.

These methods have been utilized in parallel in many developed countries to consider similar issues. For example, NATSEM in Australia develops a suite of spatial microsimulation models. Their modeling framework was initially built around a regional income and market information focusing on local expenditure and incomes on market clients (Lloyd et al. 2000; King et al. 2002). There have been a number of models built in the United Kingdom for spatial poverty and inequality analyses (Williamson 1999; Ballas and Clarke 2001). For instance, SimLeeds (spatial microsimulation for Leeds) model was developed to examine the labor market in and around the Leeds metropolitan area and to look at changes in poverty and inequality in Leeds and Sheffield between the 1991 and 2001 censuses. This modeling framework was first extended to York and Wales and then later to cover the whole country by developing the SimBritain to look at the income and spatial distributional issues in the United Kingdom (Ballas et al. 2005). Tiglao (2002) extends the SimLeeds method to develop an income model for the developing countries with a case study in Metro Manila, Philippines. Anderson (2007) has also developed a spatial microsimulation model for studying deprivation in England. Similarly, using the Simulation Model of the Irish Local Economy (SMILE), Morrissey and colleagues simulate the spatial distribution of incomes, Labor Force Participation, and factors associated with depression in Ireland (Morrissey et al. 2010; Morrissey and O'Donoghue 2011). SMILE is also linked to a static tax–benefit model to model net incomes and measure spatial redistribution (O'Donoghue et al. 2012c).

Although developing spatial indicators of income inequality is in itself a useful tool to facilitate spatial planning, the addition of tax–benefit microsimulation modeling techniques allows for the spatial impact of policy in reducing poverty and inequality to be assessed. NATSEM's model Synthetic Australian Geo-demographic Information (SYNAGI) has been extended to simulate taxes and benefits (Chin et al. 2005) and has been developed to study issues related to differential spatial poverty and inequality rates (Harding et al. 2006). It has also been used to provide support for a social policy agency (King et al. 2002), developing indicators of child social exclusion and urban poverty and a sensitivity analysis of aspects of spatial poverty (Harding et al. 2009; Tanton et al. 2009). While much of the literature in this area has focused on incidence analyses such as classifications of areas with child social exclusion or spatial income inequality, only a few papers have taken advantage of the capacity of microsimulation models to simulate the impact of policy reform. Harding et al. (2009) simulate the impact of a national family tax–benefit reform. The SimLeeds uses a partial tax–benefit model to simulate a number of tax and pension changes on income, the impact of changes to minimum wage, winter fuel payments, working households' tax credits, and new child and working credits (Ballas et al. 2007).

With regard to crime analysis, Rephann and Ohman (1999) provide an overview of the issues related to modeling crime in a spatial microsimulation model, focusing on socioeconomic correlations with crime and theorizing about the possibility of endogenizing crime within the model. With regard to crime-based applications, the SimCrime model developed in Leeds (Kongmuang et al. 2006) was used to link crime data to spatial socioeconomic data to look at the spatial incidence of crime victimization.

The NATSEM's housing models have been utilized extensively looking at housing issues including housing stress (Rahman 2010; Rahman and Harding 2014) that is related to a combination of low income and a high proportion of rent and housing assistance (Phillips and Kelly 2006), changes in commonwealth rent assistance, and the need for rent assistance (Melhuish et al. 2004). The model incorporating both spatial population and housing weights can project demand for housing assistance to 2015. Analysis examples include the implications for social programs due to demographic and geographic change, the fiscal impact of changes in demand, and the incidence of policy spatially. In the Netherlands, the LocSim model (Oskamp 1997; Hooimeijer and Oskamp 2000) has been developed to model the impact of demographic change on the working of local housing markets and conditional on local housing policy strategies together with a model of housing demand and migration to model the allocation of housing. Related to household, Clarke et al. (1997) use spatial microsimulation methods to model the small area water demand.

Health services provision is a very important policy area that involves both significant expenditure and given the often time-bound requirements for interventions has major spatial issues related to access. A spatial microsimulation modeling that contains microunits in the spatial location with health attributes and the spatial distribution of health services can therefore be a useful planning and analytical tool for health-care services. Morrissey et al. (2010) have utilized the Irish spatial microsimulation model SMILE to examine access issues related to the provision of general practitioner (GP) and psychiatric services. This work involved detailed local modeling of health status and integrating spatial interaction models to assess the access to health services. The onset of disability and ill health may increase the need for caring. Lymer et al. (2008, 2009) using a spatial microsimulation model for New South Wales in Australia modeled the regional incidence of disability and the potential requirement for aged care services at the small area level. In Leeds, the SimHealth model (Smith et al. 2007) links the 2001 census with the 2003 and 2004 Health Survey for England with the objective of simulating type 2 diabetes at a local level. Edwards and Clarke's (2009) SimObesity model, also in Leeds, tried to identify environmental factors that influenced childhood obesity, noting particularly the relationship with poverty and social capital.

Given the fine level of spatial disaggregation provided by spatial micro-simulation models, a significant amount of work has been done looking at the spatial impact of business location. In the Netherlands, Hooimeijer (1996), in a theoretical piece, advocated the combination of event history analysis, microsimulation, and geographic information system (GIS) to investigate how firms and individuals are linked through space. Van Wissen (2000) uses a firm-based demographic model SIMFIRMS to model the life cycle of firms in the Netherlands as an aid to spatial planning. Within the United Kingdom, Ballas and Clarke (2000) describe the early stages of their SimLeeds model focusing on travel to work flows and the impact of a new firm based upon a spatial interaction model, the spatial distribution of different types of work-ers in the Leeds area. Within Ireland, Vega et al. (2014) statistically match a 15% sample of the Irish labor force containing origin and destination data to identify local labor market areas across Ireland and to assess the impact of employment income on the household incomes of the hinterlands of the main gateway and hub towns and cities in the country.

There are a number of examples in this firm–employee nexus of examin-ing the economic impact of a firm closure. Rephann et al. (2005), for exam-ple, used the Swedish SVERIGE model to look at the potential impact of the closure of the SAAB car plant, which given the recent decision of its par-ent company may be more than hypothetical soon. The analysis studies the reemployment rate and the spatial impact on employment. The model was also used to simulate the hypothetical closure of a paper mill by looking at indirect labor market effects (Lindgren 1999) and the local impact of build-ing and operating a nuclear waste repository in a Swedish locality (Lindgren et al. 2007). Ballas et al. (2006) in their SimLeeds spatial model using a jour-ney to work model were able to identify the location of workers of a major engineering plant closure and quantified their loss of income. As an addi-tional step, they also modeled the resulting impact on expenditure and uti-lizing a spatial interaction model simulated the impact on local retailers. They also quantified local interindustry multipliers and long-term employ-ment impacts of the change.

Examining retail store location and consumers demand and preferences, Birkin and Clarke (1985) developed a theoretical framework for using spa-tial microsimulation to look at retail dynamics. Nakaya et al. (2007) and Hanaoka and Clarke (2007) created a spatial microsimulation model for Japan, and they simulated expenditure at the household level and linked it to a spatial interaction model, which is based around lifestyle groups and looked at retail flows from individual households to individual retail units. The analysis was able to look at the spatial distribution of expenditure, aver-age store turnover, and customer profiles. Van Leeuwen et al. (2009) devel-oped a mode SimTown to model spatial retail preferences in a number of Dutch towns. Generating a spatial microsimulation model of the population, they used a multinomial logistic model to estimate the shopping behavior. The microsimulation model was then used to look at the characteristics of shoppers and to simulate future developments such as the creation of a new retail store.

A number of papers have modeled the demand for tourism. Lundgren (2004) using the SVERIGE for domestic tourism in Sweden modeled the number of trips, type of trips, and location of trips. Van Leeuwen and Nijkamp (2009) used spatial microsimulation techniques to generate a population of tourists in Amsterdam and considered their preferences for different types of tourist activity and service. Cullinan et al. (2008) used the Irish SMILE model to model the demand for forest recreation. A parallel area of spatial microsimulation is in the area of transportation planning and analysis. While much of the studies focuses on transport planning and civil engineering issues, their relevance is in modeling social science transportation issues. Fundamental to travel microsimulation is the modeling of activity and an appreciation of the dynamic nature of this activity/travel. Bradley and Bowman (2006) summarize the components of activity-based microsimulation models (AMOS). Ryuichi's AMOS model, prototyped in Washington, DC, is an example of an activity-based model of transportation (Kitamura et al. 1995). This model predicts changes in the travel behavior following changes in the travel environment. The model is a tour-based model rather than a trip-based model in that it models joined up sets of journeys rather than simply single independent trips. It contains five components: (1) a baseline analyzer to ensure baseline diary is complete, (2) a response option generator that models the preferred behavioral response to a change in the travel environment, (3) an activity/travel planner pattern modifier that models the possible changed travel patterns, (4) an evaluation routine that assesses the impact of the change as a function of time-based utility, and (5) an acceptance routine that is the decision routine that determines the optimal travel decision. It is thus methodologically similar to other behavioral analyses used in microsimulation modeling such as the labor supply choice (Van Soest 1995).

The AMOS model calibrates travel diaries based upon a collected travel behavior survey by undertaking rigorous behavioral statistical analysis first, before the microsimulation. In many respects, therefore, the activity-focused travel microsimulation has much in common with social science studies of time use (for a simple example in the spatial microsimulation literature, see Anderson et al. 2007). Another example of a survey-based behavioral simulation is Nielsen's study (2002) that describes a state preference survey combined with a mesoscopic method (groups of agents rather than individual agents) for use in route choices in Copenhagen where alternatives were possible. Doherty and Miller (2000) describe a novel methodology for collecting household scheduling data. Within the field of transport microsimulation modeling, there is a further choice in the level of complexity; the modeling daily activity schedules as per, for example, the TRANSIM model; or the modeling of real-time individual movements such as land changing (Raney et al. 2003). The latter's level of detail allows for an evaluation of the implementation of road management planning, for example.

A clear economic policy analysis using transport microsimulation modeling is the potential impact of instruments such as congestion charging or road pricing systems. The AMOS model was used to simulate the impact of road pricing in Washington, DC. Besides the land use implications and the economic instrument design issues, an obvious potential linkage with the social science–focused spatial microsimulation literature is to quantify the distributional impact of congestion pricing. Santos and Rojey (2004) undertook a partial microsimulation approach to this issue across towns in the United Kingdom; however, linkage to a larger-scale microsimulation model would have allowed for more in-depth distributional analysis. Eliasson and Mattsson (2006) apply a transport model to a specific travel survey in Stockholm; however, it has a relatively limited spatial representation. Another area of interest in regional science relates to environmental issues related to travel, commuting, and transporting goods (Hooimeijer 1996).

The interaction between human activity and the environment is strongly influenced by spatial location. The use of spatial microsimulation models can be useful for modeling socioeconomic–environmental interactions and policies. Svoray and Benenson (2009), in a special issue of the journal *Ecological Complexity* focusing on environmental microsimulation models, highlight the increasing use of microsimulation models within the environmental field, emphasizing the availability of spatial environmental data and models in relation to the interaction between ecological and socioeconomic systems. Kruseman et al. (2008) developed the MAMBO model of livestock and agriculture in the Netherlands to model the impact of tightening environmental policy on phosphate emissions. Petersa et al. (2002) model scenarios relating to wastewater technology in relation to human waste within a Swiss region. Potter et al. (2009) developed an environment-focused model that linked scientific data with farm management data and was used to model carbon sequestration from a cropland in the United States.

In the area of agriculture, Hynes et al. (2009) developed a model of spatial farm incomes as part of the SMILE, which has been used to examine the impact of EU Common Agricultural Policy Changes (Hennessy et al. 2007). Using this income distribution and a less fine spatial scale, O'Donoghue et al. (2012b) extend the farm-focused models to include wider household income sources to be able to assess the wider economic sustainability of farm households. Clancy et al. (2012) utilized SMILE in Ireland to assess the optimal spatial location for the growth of willow and miscanthus for biomass production. Lindgren and Elmquist (2005) linked natural sciences and economics in their Systems Analysis for Sustainable Agricultural production (SALSA) model to evaluate the economic and environmental impact of alternative farm management practices on a site-specific arable farm in Sweden.

A variant of the agri-SMILE (Hynes et al. 2009) focuses on recreational activity in forests within a single city. Also, van Leeuwen et al. (2008) have developed a model exploring the linkages between on- and off-farm employment, which is becoming an increasing part of farmers' income in the EU.

While there have been many examples of aspatial static microsimulation models that have simulated greenhouse gas emissions, the spatial models that have modeled these emissions tend to be those where the spatial context is relevant such as agricultural models (Hynes et al. 2009), land use, and/or transportation issues (Moeckel et al. 2007). In terms of biodiversity-related issues, microsimulation models were used to look at a range of issues including wildlife–recreation interaction (Bennett et al. 2009) and the nonmarket value of wild bird conservation and participation in Rural Environmental Protection Schemes (Hynes et al. 2008).

Microsimulation models have a capacity for use as an experimental platform for examining the impact of rare or extreme events. For example, Brouwers (2005) utilized the SVERIGE model to simulate the transmission of a smallpox virus in Sweden. Brouwers (2005) looked at the incidence of the infections per type of location over a time horizon. The framework could potentially have been used to examine the economic cost of such an outbreak for assistance in crisis management planning and insurance design. Hoshen et al. (2007) developed a gird-based spatial microsimulation model for the United Kingdom and Ireland to look at disease transmission. Each individual was assigned day and night locations (home, school, work), personal schedules (starting and ending work), age, sex, and disease status (susceptible, infected, latent [noninfectious], infectious, and immune). The lengths of the latent and immune stages can also be varied subject to disease type as well as infection and clearance rates.

There is a large overlapping field of land use modeling incorporating a range of techniques including microsimulation modeling and spatial interaction models that treat the environment as endogenous. The field largely focuses on alternative land uses and locational choice. These urban development land use–focused models largely originated out of civil engineering and planning research rather than the social science originating models described earlier. There have been a number of large-scale integrated modeling efforts to link transport microsimulations with land use including the California Urban Futures (CUF) Model (Landis and Zhang 1998), the Integrated Land Use, Transport and Environment (ILUTE) model in Canada (Miller et al. 2004), the Urban Simulation (UrbanSim) model in Seattle (Waddell and Borning 2004), and the models of the Transport and Land Use Model Integration Program (TLUMIP) in Oregon (see Wegener 2004). These studies point to new neighborhood-scale transport policies aimed at promoting public transport, walking, and cycling and the need for environmentally sustainable land use policies with regard to transportation. These policies required detailed information on the precise location of the population and its activities. Wegener (2004) also points to the need for urban models to forecast the environmental impacts and the economic impacts of land use transport policies. It provides a microanalytic theory of urban change and demonstrates how such a model of urban change can function through microsimulation. Related to this work, the ILUMASS modeling system in

Dortmund, Germany, (Strauch et al. 2005) has been constructed with the aim to incorporate "changes of land use, the resulting changes in activity behavior and in transport demand, and the impacts of transport and land use on the environment."

Miller et al. (2004) describe an example of this type of model, the ILUTE modeling system whose aim is to "simulate the evolution of an entire urban region over an extended period of time" using a microsimulation approach of urban land use and transportation. The model links travel choice, car ownership and mode together with locational choice, and a set of businesses and households with trade and monetary flows consistent with an input output matrix. Land development is modeled on both a grid basis and a building basis. It operates in parallel to an activity-based transport model, the Travel/ Activity Scheduler for Household Agents (TASHA).

The UrbanSim modeling framework has been developed for application in different locations with substantial documentation. Felsenstein et al. (2007) have utilized the framework for Tel Aviv, simulating the impact on land use of changes such as the alternative location of shopping malls, while Franklin (2005) developed the framework for use in a congestion charge analysis in Stockholm. Joshi et al. (2006) used the framework to simulate the impact of light rail on urban growth in Phoenix, while Zhao and Chung (2006) evaluated the framework in forecasting land use change in Florida. Nguyen and Moran (2008) used UrbanSim for developing a model SIMAURIF in Paris.

Another application of spatial microsimulation methods in the Netherlands is the regional development and planning model, RAMBLAS, which is based on simulating daily activity patterns in the Eindhoven region (Veldhuisen et al. 2000). The spatial units involved are municipalities (33), zones (400), and postal areas. For each municipality, households are derived from the population according to age, gender, and marital status and distributed over the dwelling stock. From this, a list of individuals in each municipality is created and characterized by age, gender, and marital status and also by the zone/postal area where they live, go to school, or work at. The main aim of RAMBLAS is to estimate the intended and unintended consequences of planning decisions related to land use, building programs, and road constructions for households and firms (Veldhuisen et al. 2000). Finally, while most of the land use and regional development models do not incorporate structural land markets where prices are derived from supply and demand, Zhou and Kockelman (2010) using a spatial microsimulation land model attempt to model property transactions at the plot level.

Furthermore, spatial microsimulation modeling is also a data enhancement technique to allow policy analyses to be undertaken at a spatial level when spatial data are not available for this. A widespread example is the poverty mapping methodology largely developed at the World Bank, which although not labeled spatial microsimulation uses identical methods for similar purposes as analyses within the spatial microsimulation literature (Henninger 1998; Hentschel et al. 1998; Lanjouw et al. 1999; Elbers et al.

2003; World Bank 2004). The method involves statistical matching of micro–household data such as a budget survey to spatial census data to develop poverty maps. This method has been also used to develop models for many countries including the Ecuador (Hentschel et al. 2000), South Africa (Alderman et al. 2002), Malawi (Benson 2005), Azerbaijan (Baschieri et al. 2005), Burkina Faso (Grab and Grimm 2008), and Brazil (Elbers et al. 2008). The Food and Agriculture Organization (FAO) of the United Nations (Davis 2003) and International Food Policy Research Institute (IFPRI) in the United States (Benson 2006) have also used this method to explore food security–related issues.

4.2.3 Advantages of Microsimulation Modeling

One advantage of the microsimulation modeling over the more traditional statistical approaches to indirect small area estimation is that the microsimulation models can be used for estimating the local or small area effects of *policy change* and future small area estimates of socioeconomic and population characteristics. This type of model is mainly intended to explore the relationships among regions and subregions and to project the spatial implications of economic development and policy changes at a more disaggregated level.

Moreover, spatial microsimulation modeling has some advanced features, which can be highlighted as follows (see Rahman 2009a, 2011; Rahman and Harding 2012):

1. Spatial microsimulation models are flexible in terms of the choice of spatial scale.
2. They can combine data from various sources to create a microdata base file at small area level.
3. The models store data efficiently as lists.
4. Spatial microdata have the potential for further aggregation or disaggregation.
5. Models allow for updating and projecting.

Thus, from some points of view, spatial microsimulation exploits the benefits of object-oriented planning, both as a tool and as a concept. Spatial microsimulation frameworks use a list-based approach to microdata representation where a household or an individual has a list of attributes that are stored as lists rather than as occupancy matrices (Williamson et al. 1996). From a computer programming perspective, the list-based approach uses the tools of object-oriented programming because the individuals and households can be seen as objects with their attributes as associated instance variables. Alternatively, rather than using an object-oriented programming approach, a programming language like SAS can also be used to run spatial microsimulation. For a technical discussion of the SAS-based environment used in the development of the STINMOD model and adapted to run other NATSEM regional-level models, see the technical paper by Chin and Harding (2006). Furthermore, by linking spatial microsimulation with static microsimulation, we may be able to measure small area effects of policy changes, such as changes in government programs providing cash assistance to families with children (Harding et al. 2009; Tanton et al. 2008, 2009; Rahman and Harding 2012; Tanton and Edwards 2013). Another advantage of MMTs is the ability to estimate the geographic distribution of socioeconomic variables that were previously unknown (Ballas 2001).

However, spatial microsimulation adds to the simulation a spatial dimension, by creating and using synthetic microdata for small areas, such as statistical local area (SLA) levels in Australia (Chin et al. 2005). Recall that there is often great difficulty in obtaining household microdata for small areas, since spatially disaggregate reliable data are not readily available. Even if these types of data are available in some form, they typically suffer from severe limitations—in either lack of characteristics or lack of geographical details. Thus, spatial microdata are not usually obtainable; they need to be simulated, and that can be achieved by different probabilistic as well as deterministic methods.

4.3 Methodologies in Microsimulation Modeling Technology

As noted earlier, calculating statistically reliable population estimates in a local area using survey microdata is challenging, due to the lack of enough sample observations. To create a synthetic spatial microdata set is one of the possible solutions. Simulation-based methods can deal with such a problem by (re)weighting each respondent in the survey data, to create the synthetic spatial microdata. However, it is not easy to create reliable spatial microdata. Complex methodologies are associated with the process of creating synthetic microdata. This section presents some of the significant methodological issues in spatial microsimulation modeling.

4.3.1 Techniques for Creating Spatial Microdata

In spatial microsimulation processes, to create a simulated spatial micropopulation is difficult. One deterministic method can deal with such problems by (re)weighting each respondent in the survey data to create the synthetic spatial population microdata.

Methods for creating synthetic spatial microdata are mainly classified into two groups: (1) *synthetic reconstruction* and (2) *reweighting* methods (Rahman et al. 2010a). Synthetic reconstruction is a method that attempts to construct synthetic micropopulations at a small area level in such a way that all known constraints at the small area level are reproduced. There are two different ways of undertaking synthetic reconstruction—statistical data matching or fusion (Moriarity and Scheuren 2003; ABS 2004; Tranmer et al. 2005) and iterative proportional fitting (IPF) as part of the Markov Chain Monte Carlo (MCMC) sampling (Birkin and Clarke 1988; Duley 1989; Williamson 1992; Norman 1999). The reweighting method is a relatively new and popular method, which mainly calibrates the sampling design weights to a set of new weights based on a distance measure, by using the available data at spatial scale. The reweighting approach includes a CO process (Williamson et al. 1998; Huang and Williamson 2001; Ballas et al. 2003; Williamson 2007) and the generalized regression (GREG) technique known as GREGWT (Bell 2000; Chin and Harding 2006; Cassells et al. 2010).

Synthetic reconstruction is an older method for generating synthetic microdata. This method attempts to construct synthetic micropopulations at a small area level in such a way that all known constraints at the small area level are reproduced. There are two ways of undertaking synthetic reconstruction and a summary of these techniques is given in the following sections.

4.3.2 Statistical Data Matching or Fusion

Statistical data matching or fusion is a multiple imputation technique often useful to create complementary data sets for microsimulation models. Data collected from two different sources may be matched using variables (e.g., name and address or different IDs) that uniquely identify an individual or household and are recorded in both databases/sources. This type of data matching is commonly known as "exact matching." When matching is without any uncertainty, a single imputation suffices. Multiple imputation is applied when matching is subject to uncertainty, that is, due to data confidentiality constraints, the unique identifier variables may not be available in all cases (e.g., sample units or households in microdata such as CURFs of the ABS used in NATSEM cannot be identified because of the existence of data privacy legislation when gathering data from population). For such a case, records from different data sets can also be "matched" if they share a core set of common characteristics. The notion underlying this sort of data matching (called "statistical matching") is that if two individuals share, for example, the same age, sex, ethnicity, and dwelling, then they are also likely to have other characteristics (such as income, education, and marital status) in common.

In theory, when two data sets A and B are available with information on variables (Y,X) and (X,Z), respectively, we want to generate a composite data set with more inclusive information, say $C = (A \cup B)$, which consists of the variables (Y,X,Z). To do so through the data matching technique, we need a vector of common variables—the matching variables X—shared by both the data files A and B. A very simple representation of data matching is shown in Figure 4.1.

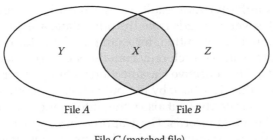

FIGURE 4.1
Data matching. (From ABS, Statistical matching of the HES and NHS: An exploration of issues in the use of unconstrained and constrained approaches in creating a basefile for a microsimulation model of the pharmaceutical benefits scheme, ABS Methodology Advisory Committee Paper, Australian Bureau of Statistics, Canberra, Australian Capital Territory, Australia, 2004.)

In general, the data matching or fusion technique involves a few empirical steps:

- Adjusting available data files and variable transformations
- Choosing the matching variables
- Selecting the matching method and associated distance function
- Validating

A description of these empirical steps and theories behind them are available elsewhere (for instance, Alegre et al. 2000; Rassler 2002). Typically, data that are used for matching can vary in format, structure, and content. Because data matching commonly relies on personal information, such as names, addresses, and dates of birth, it is important to make sure that data sourced from different databases have been appropriately adjusted and standardized with any necessary variable transformations. There are various factors that influence raw data quality, including different types of data entry errors, and the design of databases, such as the format and structure of their attributes. The aim of this first step is to ensure that the attributes used for the matching have the same structure and their content follows the same formats. The raw input data need to be converted into well-defined and consistent formats, and inconsistencies in the way information is represented and encoded need to be resolved. Ideally, the first step is done in such a way that new data files are generated that contain the adjusted and standardized data in such a format and structure that it can be easily used for the next step of the data matching process that is choosing the matching variables.

There is a range of matching methods available to select for the data matching process, such as from the traditional pair classification to collective classification approaches. In pair classification, each record pair is classified independently of all other pairs based only on its comparison vector (or its summed similarity). As a result, a single record from one database can be matched with several records from the other database. In certain applications, this might not be permitted, for example, if it is known that the two databases that are matched each only contain one record per entity (i.e., no duplicate records). The collective classification techniques for data matching aims to overcome this drawback by not only classifying record pairs based on their pair-wise similarities but also using information on how records are related or linked to other records. These approaches apply relational clustering, distance measure, or graph-based techniques to generate a global decision model. Much improved matching results can be achieved with these collective classification techniques.

Once the compared record pairs are classified into matches and non-matches, the quality of the identified matches needs to be assessed. Matching quality refers to how many of the classified matches correspond to true real-world entities, while matching completeness is concerned with how many of the real-world entities that appear in both databases were correctly matched. Both matching accuracy and completeness are affected by all steps of the data matching process, with data adjusting step is helping to make values more similar to each other, choosing the matching variables, filtering out pairs that likely are not matches, and giving the detailed comparison of attribute values providing evidence of the similarity between two records. To evaluate the completeness and accuracy of a data matching project, some form of ground-truth data, also known as gold standard, is required. Such ground-truth data must contain the true match status of all known matches (the true nonmatches can be inferred from them). However, obtaining such ground-truth data is difficult in many application areas. For example, when matching a large taxpayers database with a social security database, it is usually not known which record pair classified as a match refers to a real, existing individual who has a record in both databases.

Further details about data matching techniques are given by Rodgers (1984). Data matching tool is used to create microdata base files by researchers in many countries, such as Moriarity and Scheuren (2001, 2003) in the United States; Liu and Kovacevic (1997) in Canada; Alegre et al. (2000), Tranmer et al. (2001, 2005), and Rassler (2004) in Europe; and ABS (2004) in Australia, among many others.

4.3.3 Iterative Proportional Fitting

The IPF technique initially proposed by Deming and Stephan (1940) is mainly based on the methods of contingency table analysis and probability theory. The authors developed the method for adjusting cell frequencies in a contingency table based on sampled observations subject to known expected marginal totals. This method has been used for several decades as part of the MCMC sampling to create synthetic spatial microdata from a variety of aggregate data sources, such as census or administrative records.

The theoretical and practical considerations behind this method have been discussed in several studies (Fienberg 1970; Evans and Kirby 1974; Norman 1999), and the usefulness of this approach in spatial analysis and modeling has been revealed by Birkin and Clarke (1988), Wong (1992), Ballas et al. (1999), and Simpson and Tranmer (2005). Wong (1992) also considers the reliability issues of using the IPF procedure and demonstrates that the estimates of individual-level data generated by IPF using data of equal-interval categories other than equal-size categories are more reliable and the performance of the estimation can be improved by increasing sample size.

The IPF technique can be defined by the following set of equations (Wong 1992):

$$p_{ij(k+1)} = \frac{p_{ij(k)}}{\sum_j p_{ij(k)}} \times Q_i$$

$$p_{ij(k+2)} = \frac{p_{ij(k+1)}}{\sum_i p_{ij(k+1)}} \times Q_j$$

where
$p_{ij(k)}$ is the matrix element in row i, column j, and iteration k
Q_i and Q_j are the predefined row sums and column sums

Using these equations, the new cell values are estimated iteratively and the iteration process will stop theoretically at iteration m when $\sum_j p_{ij(m)} = Q_i$ and $\sum_i p_{ij(m)} = Q_j$.

The IPF method is well established and appears in a multitude of guises, from balancing factors in spatial interaction modeling through to the RAS method in economic accounting, and its behavior is relatively well known (Simpson and Tranmer 2005). Several recent studies compare and contrast the IPF approach with other methods and also propose improvements (e.g., see Teh and Welling 2003; Rahman 2009a; Scarborough et al. 2009; Anderson 2011; Harland et al. 2012; Hermes and Poulsen 2012; Pritchard and Miller 2012; Lovelace and Ballas 2013; Williamson 2013; Lovelace 2014; Burdett et al. 2015; Lovelace et al. 2015). In a study on small area income deprivation estimation in the United Kingdom, Anderson (2011) reveals that IPF method is a robust and deterministic approach and it performs well in that study context. The property of determinism indicates that variations in input data coding, constraint ordering, or small area table recoding are the only source of variation in the small area estimates. Anderson (2011) also suggests that the ability of the IPF algorithm to produce fractional weights proved crucial to the reconstruction of accurate aggregated estimates for comparison with the original survey data. It also enables the researcher to retain all relevant households in our synthetic small area samples, thus increasing their (weighted) heterogeneity. This creates an opportunity that these data could be used as a basis for the microsimulation modeling and thus small area policy evaluations. Issues to consider when performing IPF include selection of appropriate constraint variables, number of iterations, and "internal" and "external" validation (Scarborough et al. 2009; Pritchard and Miller 2012). It has been noted that IPF is an alternative to "combinatorial optimization" and GREGWT (Hermes and Poulsen 2012; Rahman et al. 2013). One issue with the IPF algorithm is that unless the fractional weights it generates integerized, the results cannot be used as an input for agent-based models (Hassan et al. 2008; Lovelace and Ballas 2013). Lovelace et al. (2015) provide an overview of methods for integeization to overcome this problem. To further improve the utility of IPF, there has been methodological work to convert the output into household units (Pritchard and Miller 2012).

There has been work testing and improving IPF model results (Teh and Welling 2003) but very little work systematically and transparently testing the algorithms underlying population synthesis (Harland et al. 2012). Burdett et al. (2015) demonstrate IPF with a worked example to be calculated using the agriculture data in the United States. A comprehensive introduction to the implementation of a simple IPF example in computer code is provided by Lovelace (2014). The mathematical properties of IPF have been described in several papers (Fienberg 1970; Birkin and Clarke 1988; Rahman 2009a). Similar methodologies have since been employed by Mitchell et al. (2000), Williamson (2013), and Ballas et al. (2005b) to investigate a wide range of phenomena.

Finally, according to Lovelace et al. (2015), it is still desirable if the spatial microsimulation community were able to continue to analyze which of the various reweighting/synthetic reconstruction techniques is most accurate—or to identify whether one approach is superior for some applications while another approach is to be preferred for other applications. Such a research gap is addressed in this book.

4.3.4 Repeated Weighting Method

Repeated weighting is a method that is used to estimate well-defined, relatively simple, table sets from a large database in which all available data (from administrative sources, registrations, and surveys) related to social statistics are linked and stored. The method is developed by the Statistics Netherlands (see Kroese and Renssen 1999, 2000; Statistics Netherlands 2000; Renssen et al. 2001) as part of the development of a large database called "social statistical database" (SSD). This database encompasses all relevant microdata on persons, families, households, jobs, social benefits, and living quarters in the Netherlands. Although outputs related to social statistics at small area are obtained from the SSD and published (for instance, via StatLine on the Internet), an estimate for a certain statistical quantity may be numerically inconsistent with other estimates for the same variable due to the fact that estimates from the SSD are primarily based on data from different sources. These inconsistencies become much more visible due to the large-scale dissemination of statistical information via the Internet and cause confusion among the users of statistical information at local levels. Reliable data are essential for users and any such inconsistencies in estimates must be prevented in the first place. Therefore, Statistics Netherlands designs this indirect estimation method that guarantees—as much as possible—that estimates from a large database are numerically consistent with each other.

Houbiers et al. (2003) provide details of the repeated weighting methodology with an application to search behavior of individuals on the labor market. Simply the method is based on a repeated use of the GREG estimator. It uses the calibration properties of the GREG estimator to enforce numerical consistency between all estimates in the table sets, even though each table may be estimated from different microdata. Basically, repeated weighting consists of three preparation steps (i.e., data linkage, microintegration, and compiling data blocks) and the subsequent application of the method in order to estimate multiple frequency tables consistently (Bakker 2011; Bakker et al. 2014; Statistics Netherlands 2014; Daalmans 2015). Briefly speaking, this method makes a better use of all available auxiliary data and works in following sequences:

- All tables would be estimated from a large database (e.g., SSD) using as many records as possible: depending on the variables of interest, each table may be counted from register data or estimated from survey data from one or more surveys.

- One or more margins of a table may already be known from register counts or may be estimated from different data sources. So for each table it is determined which margins the present table has in common with the tables in the set that are already estimated.

- The table is then estimated by calibrating on these common margins using GREG estimator. This scheme ensures that the table estimate is numerically consistent with all earlier estimated tables in the set.

- At the same time, the estimate can be more accurate, because the margins are estimated from larger data sets or even counted from register data, and as a result, they serve as auxiliary information for the table estimate.

The technique is designed to prevent as much as possible numerical inconsistencies due to survey errors (Boonstra 2004). Although the method may yield estimates with lower variances, it is in the first place applied for cosmetic purposes. It is, however, not applicable in every situation. The method is not suitable when fast estimates or estimates on several subpopulations are required or when numerous edit rules between variables define extra constraints on the estimates. But for estimates for which timeliness is not and numerical consistency is important, the method is a useful tool to produce estimates.

The estimation strategy outline by Statistics Netherlands (2000) is as follows. Suppose a user wants to estimate a set of tables that all relate to the same target group or small area population. From all the data available in the large SSD, the part that is relevant for this table set is then isolated from the SSD: that is, a rectangular data set with all elements of the target population as rows and all variables in the tables as columns is extracted from the SSD. This rectangular data set would typically contain a lot of empty cells. Depending on the particular objectives, a suitable estimation strategy is chosen to obtain estimates from such extracted data set. Researcher can think of imputation (common in, for instance, household statistics), usual reweighting techniques (for fast production of provisional data), or repeated weighting method (when numerical consistency is important).

The basic principle of repeated weighting is that the tables are estimated in sequence (Gouweleeuw and Hartgers 2004). Each table is estimated consistently with all previously estimated tables and with the available registers. In other words, each table is estimated in such a way that the marginal totals it has in common with all already estimated tables would have the same value as in those earlier tables (see Figure 4.2). In this process the starting weights are adjusted. Mathematically, this implies that the calibration properties of the regression estimator (RE) (Särndal et al. 1992) are applied.

A brief description of the basic theory of repeated weighting method is as follows. While a more extensive description of this method can be found in Houbiers (2004). For convenience, all terminology and notations used here are based on Renssen et al. (2001) and Houbiers et al. (2003).

A table T is characterized by one count/quantification variable, specifying what exactly is counted in the table (such as total number of elements in some population or total income of the elements in the population), and one or more dimension/classification variables, specifying the background

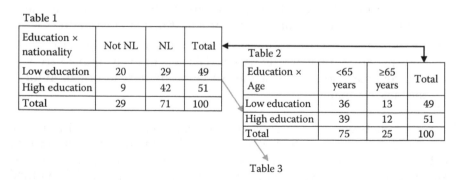

Table 1			
Education × nationality	Not NL	NL	Total
Low education	20	29	49
High education	9	42	51
Total	29	71	100

Table 2			
Education × Age	<65 years	≥65 years	Total
Low education	36	13	49
High education	39	12	51
Total	75	25	100

Table 3

FIGURE 4.2
A simple example of repeated weighting.

variables (such as sex and age class) against which the count variable is counted. Values of a classification variable, *say* X_c, represent categories by which the population is partitioned into an exclusive and exhaustive set of subpopulations. In practice, a classification variable X_c may partition the population in several different ways. For instance, the classification variable "age class" may alternatively consist of 10-year, 5-year, or 1-year classes. With repeated weighting, classification variables are allowed to consist of several hierarchical levels of classification. The degree of classification of a variable can be denoted by a superscript *r*, that is, $X_c^{(r)}$ for $0 \leq r \leq R$ with a larger value of *r*, finer the classification is. The levels of a classification variable X_c are hierarchical if and only if one or more classes of level *r* together make up one class of level $r - 1$ for all $r \in [1, 2,..., R]$.

Suppose that *C* classification variables are distinguished in the population. Each classification variable X_c has R_c hierarchical levels. Then a table *T* can be specified by the expression

$$T = \left[X_1^{(r1)} \times \cdots \times X_C^{(rc)} \right] \times Y,$$

where *Y* represents the count variable of interest in the table. The dimensional part of this table (the part between the square brackets) can alternatively be expressed as a *C* -vector $\mathbf{r} = (r_1,..., r_C)'$, with $0 \leq r_c \leq R_c$. If r_c is zero, the corresponding classification variable X_c is, in fact, not present, in table *T*. For an *M*-way table, only *M* out of *C* components of $\mathbf{r}$ are nonzero.

In repeated weighting, it is necessary to quickly evaluate common margins of tables. With the notation given earlier, the common margin of two tables can easily be derived. Suppose that two tables T_1 and T_2 relate to the same count variable *Y*. The dimensions of T_1 are specified by the vector $\mathbf{r}_1 = (r_{11},..., r_{1C})'$ and the dimensions of T_2 by $\mathbf{r}_2 = (r_{21},..., r_{2C})'$. It is easy to see that the common margin $T_{1,2}$ of these two tables is specified by the *C*-vector $\mathbf{r}_{1,2} = \min(\mathbf{r}_1, \mathbf{r}_2) \equiv [\min(r_{11}, r_{21}), \min(r_{12}, r_{22}),..., \min(r_{1C}, r_{2C})]'$.

Whenever one or more components of $\mathbf{r}_{1,2}$ are nonzero, the two tables have a nontrivial common margin. If, for example, the variable "age class" consists of the three levels, namely, 10-year classes (level 1), 5-year classes (level 2), and 1-year classes (level 3), and T_1 contains, among others, the classification variable "age class" at the level of 5-year classes, whereas T_2 contains this variable at the level of 10-year classes, then these two tables obviously have a common margin related to "age class" at the level of 10-year classes, the minimum of (2, 1). When all components of $\mathbf{r}_{1,2}$ are zero, tables T_1 and T_2 have no common margins, except for the population total of the count variable *Y*.

When deriving approximate variance formulas for repeated weighting estimators, it would be convenient to use a vector notation to indicate tables (Houbiers et al. 2003). Suppose that level *r* of classification variable X_c consists of $P_c \geq 1$ mutually exclusive classes and the score on classification variable

$X_c^{(r)}$ of an element k in the population as a P_c-vector of dummy variables, which is defined as $\mathbf{X}_{ck}^{(r)} = (x_{ck1}, \ldots, x_{ckP_c})'$, where

$$x_{gkp} = \begin{cases} 1 & \text{if element } k \text{ belongs to class } p \text{ of variable } X_g^{(r)} \\ 0 & \text{otherwise.} \end{cases}$$

The score of an element k in the population on the crossing $X_1^{(r_1)} \times \cdots \times X_C^{r_c}$ can be expressed in a similar way. The crossing now consists of $P = P_1 \times P_2 \times \cdots \times P_C$ mutually excluding classes, where each P_c depends on the level r_c of variable X_c in the crossing. A P-vector can be defined as $\mathbf{X}_k^{(r)} = (x_{k1}, \ldots, x_{kP})'$, where

$$x_{kp} = \begin{cases} 1 & \text{if element } k \text{ belongs to class } p \text{ of the crossing } X_1^{(r_1)} \times \cdots \times X_C^{r_c} \\ 0 & \text{otherwise.} \end{cases}$$

Given a population consisting of N_i elements at the ith small area, table counts of a table $T = [X_1^{(r_1)} \times \cdots \times X_C^{r_c}] \times Y$, concerning the count variable Y, can, in vector notation, be expressed as $\mathbf{Y}_T = \sum_{k=1}^{N_i} \mathbf{X}_k^{(r)} y_k \equiv \sum_{k=1}^{N_i} \mathbf{y}_{k,T}$, where y_k denotes the score of element k over the count variable Y. For frequency counts, y_k equals one and could equally well be omitted. For count variables such as "income," y_k may assume any real number. The vector $\mathbf{Y}_T$ consists of P components, corresponding to the $P = P_1 \times P_2 \times \cdots \times P_C$ mutually excluding categories of the crossing $X_1^{(r_1)} \times \cdots \times X_C^{r_c}$, that is, they correspond to the P cells in table T.

Often, the population totals for (the components of) $\mathbf{Y}_T$ are not known: they must be estimated from survey data. The P components in $\mathbf{Y}_T$ can be viewed as a collection of population characteristics that are to be estimated simultaneously. For a sample S of size n drawn from the population, the Horvitz–Thompson estimator for the table counts for table T corresponds to

$$\hat{\mathbf{Y}}_{T[H-T]} = \sum_{k \in S} d_k \mathbf{X}_k^{(r)} y_k \equiv \sum_{k \in S} d_k \mathbf{y}_{T,k},$$

where d_k corresponds to the usual inverse inclusion probability of element k in the sample; see Särndal et al. (1992).

The RE is an important component in repeated weighting. This is designed to increase the accuracy of the estimate of some variable Y by using auxiliary variables $X_1, \ldots, X_J$ that are correlated with Y and whose population totals are known (Rao and Molina 2015). Instead of one specific target variable Y, one can also estimate the P cell totals Y_T of some table T with the RE, using the known population totals Y_{T_M} of one of its margins T_M as the auxiliary variables. Hence, the RE for the table counts can be expressed as

$$\hat{\mathbf{Y}}_{T[RE]} = \hat{\mathbf{Y}}_{T[H-T]} + \mathbf{B}_{T,T_M} \left(\mathbf{Y}_{T_M} - \hat{\mathbf{Y}}_{T_M[H-T]} \right)$$

where $\mathbf{B}_{T,T_M} = \left(\sum_{k \in S} d_k \mathbf{y}_{T_M,k} \mathbf{y}'_{T_M,k} \right)^{-1} \left(\sum_{k \in S} d_k \mathbf{y}_{T_M,k} \mathbf{y}'_{T,k} \right)$ denotes the matrix of regression coefficients; see Särndal et al. (1992). Substituting the **B**-matrix in the expression for the RE, it can be rewritten as

$$\hat{\mathbf{Y}}_{T[RE]} = \sum_{k \in S} d_k \mathbf{y}_{T,k} \left(1 + \mathbf{y}'_{T_M,k} \left(\sum_{k \in S} d_k \mathbf{y}_{T_M,k} \mathbf{y}'_{T_M,k} \right)^{-1} \left(\mathbf{Y}_{T_M} - \hat{\mathbf{Y}}_{T_M[H-T]} \right) \right)$$

$$\equiv \sum_{k \in S} d_k c_k \mathbf{y}_{T,k}$$

$$= \sum_{k \in S} w_k \mathbf{y}_{T,k}.$$

The regression weights w_k are given by the original design weights d_k, times a correction factor c_k. The correction factor c_k is generally close to unity. The important point with this estimator is that when the population totals of the auxiliary variables themselves, that is, the cells in the margin T_M, are estimated with the weights w_k, the known population totals $\mathbf{Y}_{T_M}$ are recovered, that is,

$$\hat{\mathbf{Y}}_{T_M[RE]} = \sum_{k \in S} d_k \mathbf{y}_{T_M,k} \left(1 + \mathbf{y}'_{T_M,k} \left(\sum_{k \in S} d_k \mathbf{y}_{T_M,k} \mathbf{y}'_{T_M,k} \right)^{-1} \left(\mathbf{Y}_{T_M} - \hat{\mathbf{Y}}_{T_M[H-T]} \right) \right) = \mathbf{Y}_{T_M}.$$

In other words, the weights w_k are calibrated on the known population totals $\mathbf{Y}_{T_M}$ (Deville and Särndal 1992). This property of the RE is used in repeated weighting. That is, using the RE, the table estimate is calibrated on one or more margins that are known from register counts or perhaps estimated from different existing data sources. Therefore, by repeatedly weighting the survey design weights, the table is estimated consistently with its margins.

Applications of the repeated weighting method are not yet widespread outside of the Netherlands. Knottnerus and van Duin (2006) applied this method to the Dutch Labour Force Survey, which produces advanced variance estimations. A more general message is that repeated weighting can be applied to very complex estimation problems (Statistics Netherlands 2014). It may also be useful for other applications requiring optimal use of multiple data sources (Daalmans 2015). However, as mentioned in Houbiers et al. (2003), there are some known obstacles of using repeated weighting, especially for large, detailed tables. The first obstacles are computational problems: long computation time and out-of-memory problems. A second problem is the existence of so-called edits rules, that is, consistency rules between the relationships of different variables. A third issue is the zero-cell problem: estimates have to be made from a sample survey that—due to the sample mechanism—may not cover some of the categories that are known to exist in the population, for example, a sample survey that does not include 86-year-old men of an "other than EU" nationality.

Moreover, a fourth obstacle concerns estimation problems, which are caused by the impossibility to satisfy all constraints. After estimating some tables, it may be no longer possible to satisfy all constraints simultaneously. An easy example of this is that one cell value needs to be 100 and 200 at the same time, in order to be consistent with all previously estimated tables. A fifth problem is order dependence. In repeated weighting, tables are estimated one by one. Each table has to be consistently estimated with all previously estimated tables. The estimation order matters: a different order gives a different result. In theory, it is better to estimate all tables simultaneously. Obviously, there is no order problem in that case. In practice however, simultaneous estimation will often be infeasible: an extensive weighting model is needed, which is technically hard to solve (Daalmans 2015). By estimating the tables one by one, the large weighting model is divided into a number of smaller, easier-to-handle, problems.

All five issues are especially relevant for detailed tables, of which some of the cells are covered by only few (or even none) observations. The good news is that the Statistics Netherlands is continuously working on these issues. Some of the recent studies have proposed different solutions to those problems (e.g., see Knottnerus and van Duin 2006; Bakker 2011; Coutinho et al. 2013; Bakker et al. 2014; Statistics Netherlands 2014; Daalmans 2015). Just for an example, Coutinho et al. (2013) advance the applications of repeated weighting by making use of a calibrated hot-deck donor imputation technique that tackles the second and third obstacles. Although repeated weighting can be a powerful method, it may also be worthwhile to consider alternative methods (e.g., imputation or reweighting). Rahman et al. (2010a, 2013) and De Waal (2014) present a comprehensive overview of different methods and their features.

4.3.5 Reweighting

Reweighting is a procedure used throughout the world to transform information contained in a sample survey to estimates for the whole population (Holt and Smith 1979; Smith 1991). For example, the ABS calculates a weight (or "expansion factor") for each of the 6892 households included in the 1998–1999 HES sample file (ABS 2002). Thus, if household number 1 is given a weight of 1000 by the ABS, it means that the ABS considers that there are 1000 households with comparable characteristics to household number 1 in Australia. These weights are used to move from the 6892 households included in the HES sample to estimates for the 7.1 million households in Australia (Chin and Harding 2006). To create a synthetic spatial micropopulation for small area estimation, there are two approaches to reweighting

TABLE 4.2

Synthetic Reconstruction versus the Reweighting Technique

Synthetic Reconstruction	Reweighting Technique
It is based on a sequential step-by-step process—where the attributes of each unit are estimated by random sampling using a conditional probabilistic framework.	It is an iterative process—where a suitable fitting between actual data and the selected sample of microdata should be obtained by minimizing distance errors.
Ordering is essential in this process (each value should be generated in a fixed order).	Ordering is not an issue in this process. Convergence is achievable by repeating the process many times or by some simple adjustment.
Relatively more complex and time consuming.	Although the technique is complex from a theoretical point of view, it is comparatively less time consuming.
The effects of inconsistency between the constraining tables could be significant for this approach due to a mismatch in the table totals or subtotals.	Reweighting techniques can allow the choice of constraining tables to match with researcher and/or user requirements.

survey data: the CO and GREGWT approaches. In the current data-centric world, these two prominent techniques of reweighting are playing central roles for spatial microsimulation modeling toward small area estimation. Therefore, descriptions of these two reweighting methods are provided in Sections 4.4 and 4.5. However, some key issues associated with the synthetic reconstruction and reweighting methods for creating synthetic spatial microdata are presented in Table 4.2.

It is worthwhile to mention here that before the development of "reweighting" techniques, the IPF procedure was a very popular tool to generate small area microdata. A summary of literature using the IPF technique has been provided by Norman (1999, p. 9; Anderson 2011; Lovelace 2014). It appears from this summary that almost all of the researchers in the United Kingdom were devoted to using the IPF procedure in microsimulation modeling. But nowadays most of the researchers are claiming that reweighting procedures have some advantages over the synthetic reconstruction approach (e.g., refer to Williamson et al. 1998; Huang and Williamson 2001; Ballas et al. 2003; Rahman et al. 2013).

4.4 CO Reweighting Approach

The CO reweighting approach was first suggested in Williamson et al. (1998) as a new approach to create synthetic micropopulations for small domains. This reweighting method is to see the fit of an appropriate combination of households from survey data to assess its agreement with the known benchmarks at small area levels using an optimization tool. In the CO algorithms, an iterative process begins with an initial set of households randomly selected from the survey data to see the fit to the known benchmark constraints for each small domain. Then, a random household from the initial set of combinations is replaced by a randomly chosen new household from the remaining survey data to assess whether there is an improvement of fit. The iterative process continues until an appropriate combination of households that best fits known small area benchmarks is achieved (Williamson et al. 1998; Voas and Williamson 2000; Huang and Williamson 2001; Tanton et al. 2007). The overall process involves five steps, which are as follows:

1. Collect a sample survey microdata file (such as CURFs in Australia) and small area benchmark constraints (e.g., from census or administrative records).

2. Select a set of households (e.g., 1%) randomly from the survey sample, which will act as an initial combination of households from a small area.

3. Tabulate selected households and calculate the total absolute difference from the known small area constraints.

4. Choose one of the selected households randomly and replace it with a new household drawn at random from the survey sample, and then follow step 3 for the new set of households' combination.

5. Repeat step 4 until no further reduction in the total absolute difference is possible.

Note that when an array-based survey data set contains a finite number of households, it is possible to calculate all possible combinations of households. In theory, it may also be possible to find the set of households' combination that best fits the known small area benchmarks. But in practice, it is almost unachievable, due to computing constraints for a very very large number of all possible solutions. For example, to select an appropriate combination of households for a small area with 150 households from a survey sample of 215,789 households, the number of possible solutions greatly exceeds a billion (Williamson et al. 1998).

To overcome this difficulty, the CO approach uses several ways of performing "intelligent searching," effectively reducing the number of possible solutions. Williamson et al. (1998) provide a detail discussion about three intelligent searching techniques: hill climbing, simulated annealing, and genetic algorithms. Later on, to improve the accuracy and consistency of outputs, Voas and Williamson (2000) developed a "sequential fitting procedure," which can satisfy a level of minimum acceptable fit for every table used to constrain the selection of households from the survey sample data. The simulated annealing method is addressed here.

4.4.1 Simulated Annealing Method in CO

Simulated annealing is an intelligent searching technique for optimization problems, which has been successfully used in the CO reweighting process to create spatial microdata. The method is based on a physical process of annealing—in which a solid material is first melted in a heat bath by increasing the temperature to a maximum value, at which point all particles of the solid have high energies and the freedom to randomly arrange themselves in the liquid phase. The process is then followed by a cooling phase, in which the temperature of the heat bath is slowly lowered. When the maximum temperature is sufficiently high and the cooling is carried out sufficiently slowly, then all the particles of the material eventually arrange themselves in a state of high density and minimum energy. Simulated annealing has been used in various CO problems (see Kirkpatrick et al. 1983; van Laarhoven and Aarts 1987; Williamson et al. 1998; Ballas 2001).

The simulated annealing algorithm used in the CO reweighting approach was originally based on the Metropolis algorithm, which had been proposed by Metropolis et al. (1953). To simulate the evaluation to "thermal equilibrium" of a solid for a fixed value of the temperature T, the authors introduced an iterative method, which generates sequences of states of the solid in the following way. As mentioned in the book *Simulated Annealing: Theory and Applications* by van Laarhoven and Aarts (1987, p. 8):

> Given the current state of the solid, characterized by the position of its particles, a small, randomly generated, perturbation is applied by a small displacement of a randomly chosen particle. If the difference in energy, ∂E, between the current state and the slightly perturbed one is *negative*, that is, if the perturbation results in a lower energy for the solid, then the process is continued with the new state. If $\partial E \geq 0$, then the probability of acceptance of the perturbed state is given by $\exp(-\partial E/K_B T)$. This acceptance rule for new states is referred to as the *Metropolis Criterion*. Following this criterion, the system eventually evolves into thermal equilibrium, that is, after a large number of perturbations, using the aforementioned acceptance criterion, the probability distribution of the states approaches the Boltzmann distribution, given as
>
> $$p(\partial E) = \frac{1}{c(T)} \exp\left(-\frac{\partial E}{K_B T}\right)$$
>
> where $c(T)$ is a normalizing factor depending on the temperature T and K_B is the Boltzmann constant.

Note that to search an appropriate combination of households from a survey data set that best fits to the benchmark constraints at small area levels is a CO problem, and the solutions in a CO problem are equivalent to the states of a physical annealing process. In the process of CO reweighting by simulated annealing algorithm, a combination of households assumes the role of the states of a solid, while the total absolute distance (TAD) function and the control parameter (e.g., rate of reduction) take the roles of energy and temperature, respectively. According to Williamson et al. (1998), change in energy becomes a potential change in households' combination performance (assessed by TAD) to meet the benchmarks, and temperature becomes a control for the maximum level of performance degradation (% of reduction) acceptable for the change of one element in a combination from sample data. In this case, the control parameter is then lowered in steps, with the system being allowed to approach equilibrium for each step by generating a sequence of combinations by obeying the Metropolis criterion. Additionally, the algorithm is terminated for some small value of the control parameter, for which practically no deteriorations are accepted. Hence, the normalizing constant that is depending on the controlling factor and the Boltzmann constant can be dropped from the probability distribution. In this particular case, we have the equation

$$p(\partial E) = \exp\left(-\frac{\partial E}{T}\right).$$

There are two important features of this probability equation described by Williamson et al. (1998). One is that the smaller the value of difference in energy, ∂E, the greater is the likelihood of a potential replacement being made in a combination. Another feature is that the smaller the value of controlling factor, T, the smaller the change in performance is likely to be accepted.

A typical simulated annealing algorithm is depicted in Figure 4.3. The overall process consists of a series of iterations in which random shifting is occurring from an existing solution to a new solution among all possible solutions. To accept a new solution as the base solution for further iteration, a test of goodness of fit based on TAD is consistently checked.

The rules of the test are if the change of the difference in energy is negative, the newly simulated solution is accepted unconditionally; otherwise, it is accepted as satisfying the aforementioned Metropolis criterion. It is worth mentioning that the simulated annealing algorithm may be able to avoid deceiving outcome at local extremum in the solutions. Moreover, a solution or selected combination of households by simulated annealing algorithm in the CO reweighting approach can generate real individuals living in actual households in a sense that individuals are from modeled outputs and not synthetically reconstructed (Ballas 2001).

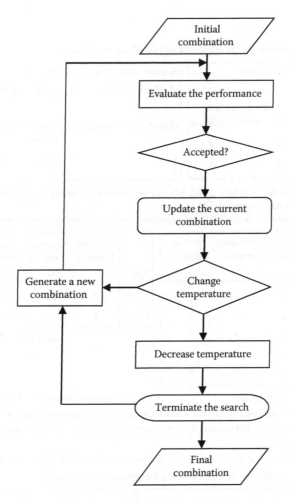

FIGURE 4.3
A flowchart of the simulated annealing algorithm. (From Pham, D.T. and Karaboga, D., *Intelligent Optimisation Techniques: Genetic Algorithms, Tabu Search, Simulated Annealing and Neural Networks*, Springer, London, U.K., 2000.)

4.4.2 Illustration of CO Process for Hypothetical Data

A simplified CO process is depicted in Figure 4.4. Note that in this process when the total absolute difference (aforementioned TAD) is equal to zero, the selection of households' combination indicates the best fit. In other words, in this case, the new weights give the actual household units from the survey sample microdata that are the best representative combination. Thus, it is a selection process of an appropriate combination of sample units, rather than calibrating the sampling design weights to a set of new weights.

Step 1: Obtain sample survey microdata and small area constraints.

Survey Sample Microdata				Known small area constraints			
				1. Household size		2. Age of occupants	
Household	Characteristics						
	Size	Adult	Children	Household size	Frequency	Type of person	Frequency
a	2	2	0				
b	2	1	1	1 2	1 0	Adult	3

Step 2: Randomly select two households from survey sample (for example, *a* and *e*) to act as an initial small area microdata estimate.

Step 3: Tabulate selected households and calculate absolute difference from known constants.

Household size	Estimated frequency (1)	Observed frequency (2)	Absolute difference |(1)–(2)|		Age	Estimated frequency (1)	Observed frequency (2)	Absolute difference |(1)–(2)|
1	0	1	1		Adult	4	3	1
2	1	0	1		Child	1	2	1
3	1	0	1					
4	0	1	1			Sub total:	2	
5+	0	0	0					

Sub total: 4   Total absolute difference = 4 + 2 = 6

Step 4: Randomly select one of selected households (*a* or *e*). Then replace with another household selected at random from the survey sample, provided this leads to a reduced TAD.

Households selected: *d* and *e* (Household *a* replaced by *d*). Tabulate this new combination of households and calculate absolute difference from known constants.

Household size	Estimated frequency (1)	Observed frequency (2)	Absolute difference |(1)–(2)|		Age	Estimated frequency (1)	Observed frequency (2)	Absolute difference |(1)–(2)|
1	1	1	0		Adult	3	3	0
2	0	0	0		Child	1	2	1
3	1	0	1					
4	0	1	1			Sub total:	1	
5+	0	0	0					

Sub total: 2   Total absolute difference = 2 + 1 = 3

FIGURE 4.4
A simplified combinatorial optimization process.   (*Continued*)

Step 5: Repeat step 4 until no further reduction in total absolute difference is possible.

Result: Final selected households are *c* and *d* (since this household combination best fits the small area benchmarks):

Household size	Estimated frequency (1)	Observed frequency (2)	Absolute difference |(1)–(2)|
1	1	1	0
2	0	0	0
3	0	0	0
4	1	1	0
5+	0	0	0

Sub total: 0

Age	Estimated frequency (1)	Observed frequency (2)	Absolute difference |(1)–(2)|
Adult	3	3	0
Child	2	2	0

Sub total: 0

Total absolute difference = 0 + 0 = 0

FIGURE 4.4 (Continued)
A simplified combinatorial optimization process. (From Huang, Z. and Williamson, P., A comparison of synthetic reconstruction and combinatorial optimisation approaches to the creation of small-area microdata, Working Paper 2001/2, Population Microdata Unit, Department of Geography, University of Liverpool, Liverpool, U.K., 2001.)

4.5 Reweighting: The GREGWT Approach

The GREGWT approach has been developed by the ABS for GREG and weighting of sample survey results. It is an iterative GREG algorithm written in SAS macros to calibrate survey estimates to benchmarks. Calibration can be looked at either as a way of improving estimates or as a way of making the estimates add up to benchmarks (see Deville and Särndsl 1992; Bell 2000; Rahman et al. 2010b). That is, the grossing factors or weights on a data set containing the survey returns are modified so that certain estimates agree with externally provided totals known as benchmarks. This use of external or auxiliary information typically improves the resulting survey estimates that are produced using the modified grossing factors.

The GREGWT algorithm uses a constrained distance function known as the truncated chi-squared distance function that is minimized subject to the calibration equations for each small area (Rahman et al. 2013). The method is also known as linear truncated or restricted modified chi-squared (refer to Singh and Mohl 1996) or truncated linear regression method (refer to Bell 2000), and it is somewhat close to model calibration (Estevao and Sarndal 2004; Särndal 2007). GREGWT is an indirect modeling tools in MMT since it uses external sources of data through the indirect modeling approach. The basic feature of this method over the linear regression is that the new weights must lie within a prespecified boundary condition for each small area unit. The upper and lower limits of each boundary interval could be constant across sample units or proportional to the original sampling weights.

4.5.1 Theoretical Setting

Let us assume that a finite population is denoted by $\Omega = \{1, 2, ..., k, ..., N\}$ and a sample $s(s \subseteq \Omega)$ is drawn from Ω with a given probability sampling design $p(\cdot)$. Suppose the inclusion probability $\pi_k = \Pr(k \in s)$ is a strictly positive and known quantity. Now, for the elements $k \in s$, let (y_k, x_k) be a set of sample observations, where y_k is the value of the variables of interest for the kth population unit and $x'_k = (x_{k,1}, ..., x_{k,j}, ..., x_{k,p})$ is a vector of auxiliary information associated with y'_k. Note that data for a range of auxiliary variables should be available for each unit of a sample s. In a particular case, suppose for an auxiliary variable j, the element $x_{k,j} = 1$ in x_k if the kth individual is not in workforce, and $x_{k,j} = 0$ "otherwise." Thus, $\sum_{k \in s} x_{k,j}$ gives the number of individuals in the sample who are not in the workforce. If the given sampling design weights are $d_k = 1/\pi k$ $(k \in s)$, then the sample-based population totals of auxiliary information, $\hat{t}_{x,s} = \sum_{k \in s} d_k x_k$, can be obtained for a p-elements auxiliary vector x_k. But the *true* value of the population total of the auxiliary information T_x should be known from some other sources such as from the census or administrative records. In practice, $\hat{t}_{x,s}$ is far from T_x when the sample s is a bad or poor representative of the population.

To obtain a more reliable estimate of the population total of the variable of interest, we can use this true population total T_x of the auxiliary information. To do so, the main task is to compute new weights w_k for $k \in s$, such that

$$\sum_{k \in s} w_k x_k = T_x \tag{4.1}$$

and the new weights w_k are as close as possible to the sampling design weights d_k.

Equation 4.1 is known as the "calibration equation" or the constraint function, which is used to minimize the distance between two sets of weights. In a survey sampling calibration process, because the prime intention is to minimize the distance or to confirm the closeness of the two sets of weights, it is essential to identify an appropriate distance measure.

In usual notations, let $G_k(w_k, d_k)$ be the distance between w_k and d_k. Then, the total distance over the sample s should be defined as

$$D = \sum_{k \in s} G_k(w_k, d_k). \tag{4.2}$$

Now, the problem is to minimize Equation 4.2, which is subject to the constraints in Equation 4.1. Deville and Sarndal (1992) present a class of well-known distance functions such as the Healing distance, Minimum entropy distance, and Chi-squared distance. They also propose a new distance measure widely known as the Deville–Sarndal distance and defined as $G_k(w_k, d_k) = -d_k \log(w_k/d_k) + w_k - d_k$; $\forall$ $_k$. A discussion about those distance measures minimization is provided elsewhere (Singh and Mohl 1996; Cai et al. 2004). It is notable that the Lagrange multiplier is commonly used as a minimization tool of distance measures to calculate the new weights. The Lagrange equation or Lagrangian for this type of minimization problem would be

$$L = D + \sum_{j=1}^{p} \lambda_j \left(T_{x,j} - \sum_{k \in s} w_k x_{k,j} \right) \tag{4.3}$$

where λ_j; $\forall$ j are the Lagrange multipliers.

Consider a special case: if the distance function defined in (4.2) has a property that the first derivative with respect to w_k can be expressed as a function $f(w_k/d_k)$ and the inverse of this function, i.e., i.e. f^{-1} exists, then after differentiating and applying the first-order minimization condition in (4.3), we have

$$\frac{\partial L}{\partial w_k} = f\left(\frac{w_k}{d_k}\right) - \sum_{j=1}^{p} \lambda_j x_{k,j} = 0 \tag{4.4}$$

It is convenient to write $x_k' \lambda = \sum \lambda_j x_{k,j}$ for a simple representation. Hence, from Equation 4.4 the new weights can be formulated as

$$w_k = d_k f^{-1}\left(x_k' \lambda\right); \quad \text{for } \forall \; k \in s. \tag{4.5}$$

When f^{-1} exists, and for a solution of the Lagrange multipliers vector, λ, the new set of weights can be easily obtained from Equation 4.5. However, to obtain the values of λ, we should use the known relations

$T_x = \sum_{k \in s} w_k x_k = \sum_{k \in s} d_k f^{-1}(x'_k \lambda) x_k$ and $\hat{t}_{x,s} = \sum_{k \in s} d_k x_k$. Hence, these relations can be joined as the following form:

$$T_x - \hat{t}_{x,s} = \sum_{k \in s} d_k \{ f^{-1}(x'_k \lambda) - 1 \} x_k \tag{4.6}$$

where $T_x - \hat{t}_{x,s} = C$ (say) is a known vector, $d_k \{ f^{-1}(x'_k \lambda) - 1 \}$ is a scalar, and the equation is nonlinear in the Lagrange multipliers vector, λ. Hence, (4.6) can be solved by an iterative procedure such as the Newton–Raphson method.

4.5.2 How Does GREGWT Generate New Weights?

The distance measure used in the GREGWT algorithm is known as the truncated chi-squared distance function and it can be defined as

$$G_k^{\chi^2} = \frac{(w_k - d_k)^2}{2 d_k}; \quad \text{for } L_k \le \frac{w_k}{d_k} \le U_k \tag{4.7}$$

where L_k and U_k are prespecified lower and upper bounds, respectively, for each unit $k \in s$.

For a simple special case, the total of this type of distance measure can be defined as

$$D = \frac{1}{2} \sum_{k \in s} \frac{(w_k - d_k)^2}{d_k}.$$

Hence, the Lagrangian for the chi-squared distance function is

$$L = \frac{1}{2} \sum_{k \in s} \frac{(w_k - d_k)^2}{d_k} + \sum_{j=1}^{p} \lambda_j \left(T_{x,j} - \sum_{k \in s} w_k x_{k,j} \right) \tag{4.8}$$

where
$\lambda_j (j = 1, 2, \ldots, p)$ are the Lagrange multipliers
$T_{x,j}$ is the jth element of the vector of true values of known population total for the auxiliary information, T_x.

By differentiating (4.8) with respect to w_k and then applying the first-order condition, we have

$$\frac{\partial L}{\partial w_k} = \left(\frac{w_k - d_k}{d_k} \right) - \sum_{j=1}^{p} \lambda_j x_{k,j} = 0 \tag{4.9}$$

for $k \in s \subseteq \Omega$, along with the pth $(j = 1, 2,..., p)$ constraint conditions in Equation 4.1. Earlier, it is convenient to write $x'_k \lambda = \sum \lambda_j x_{k,j}$ for a simple representation. Hence, the new weights can be formulated as

$$w_k = d_k + d_k x'_k \lambda. \tag{4.10}$$

To obtain values of the Lagrange multipliers, Equation 4.10 can be rearranged in a convenient form. After multiplying the equation by x_k and then summing over k, it can be written as $\sum_{k \in s} w_k x_k = \sum_{k \in s} d_k x_k + \sum_{k \in s} d_k x_k x'_k \lambda$. Now since $\sum_{k \in s} d_k x_k = \hat{t}_{x,s}$ and $\sum_{k \in s} w_k x_k = T_x$ are known, the equation given earlier can be expressed as $\left(\sum_{k \in s} d_k x_k x'_k \right) \lambda = T_x - \hat{t}_{x,s}$, where the summing term in brackets is a $p \times p$ symmetric-square matrix. If the inverse of this matrix exists, using the *Ditto* derivation of the Lagrange multipliers (Ditto et al. 1995), the vector of Lagrange multipliers can be obtained by the following equation:

$$\lambda = \left(\sum_{k \in s} d_k x_k x'_k \right)^{-1} (T_x - \hat{t}_{x,s}); \quad \text{for} \left| \sum_{k \in s} d_k x_k x'_k \right| \neq 0 \tag{4.11}$$

Hence, using the resulting values of Lagrange multipliers, λ, one can easily calculate the new weights w_k from Equation 4.10. Moreover, to minimize the truncated chi-squared distance function in (4.7), an iterative procedure known as the Newton–Raphson method (see Appendix A) is used in the GREGWT program. It adjusts the new weights in such a way that minimizes Equation 4.7 and produces reliable spatial micropopulation data for obtaining small area synthetic estimates of the variable of interest.

4.5.3 Explicit Numerical Solution for Hypothetical Data

Although the GREGWT simulates the new weights through a very complex process, a very simplified and explicit numerical example of the GREGWT approach of simulating new weights is given in the following text using hypothetical data.

Let $x_{k,j}$ be the jth auxiliary variable linked with kth sample unit for which true population values T_x are available from census or other administrative records. Suppose in a hypothetical data set, observations of 25 sample units for a set of 5 auxiliary variables such as *age* (1 = 16–30 years and 0 = "otherwise"), *sex* (1 = female and 0 = male), *employment* (1 = unemployed and 0 = "otherwise"), *income* from unemployment benefits (in real unit values 0, 1, 2, 3, 4, and 5), and *location* (1 = rural and 0 = urban) are available, and its associated auxiliary information matrix, sample design weights, and the known population values vector are accordingly given as follows:

$$X = [\mathbf{x}'_{k,j}] = \begin{bmatrix} 1 & 1 & 0 & 0 & 0 \\ 1 & 0 & 1 & 3 & 1 \\ 0 & 0 & 1 & 2 & 1 \\ 1 & 1 & 1 & 5 & 0 \\ 0 & 1 & 0 & 0 & 1 \\ 0 & 0 & 1 & 1 & 0 \\ 0 & 0 & 0 & 0 & 1 \\ 1 & 0 & 1 & 4 & 0 \\ 0 & 1 & 0 & 0 & 1 \\ 1 & 0 & 0 & 0 & 1 \\ 0 & 1 & 1 & 1 & 0 \\ 1 & 1 & 1 & 3 & 1 \\ 1 & 0 & 1 & 2 & 1 \\ 0 & 0 & 1 & 5 & 1 \\ 0 & 1 & 1 & 4 & 0 \\ 0 & 0 & 0 & 0 & 0 \\ 1 & 0 & 1 & 3 & 1 \\ 0 & 1 & 0 & 0 & 0 \\ 0 & 0 & 1 & 2 & 1 \\ 1 & 0 & 1 & 4 & 0 \\ 0 & 0 & 0 & 0 & 1 \\ 0 & 0 & 1 & 5 & 1 \\ 1 & 1 & 0 & 0 & 0 \\ 0 & 1 & 1 & 1 & 0 \\ 1 & 0 & 0 & 0 & 1 \end{bmatrix}, \quad \mathbf{d} = (\mathbf{d}_k) = \begin{bmatrix} 4 \\ 5 \\ 6 \\ 5 \\ 3 \\ 4 \\ 6 \\ 4 \\ 5 \\ 3 \\ 5 \\ 4 \\ 3 \\ 6 \\ 4 \\ 5 \\ 6 \\ 3 \\ 6 \\ 4 \\ 5 \\ 3 \\ 5 \\ 4 \\ 3 \end{bmatrix} \quad \text{and} \quad T_x = \begin{pmatrix} 50 \\ 45 \\ 70 \\ 200 \\ 65 \end{pmatrix}$$

Note: The first row of matrix X represents a sample unit of *age* between 16 and 30 years, *female*, in *"otherwise"* employment categories that is may be in labor force or employed, with a real unit value of *income* from unemployment is zero dollar, and living in an *urban* area.

Now, we have to estimate $\hat{t}_{x,s}$ and the inverse matrix of $\left(\sum_{k \in s} d_k x_k x_k' \right) = A$ (*say*). By using mathematical formulas, one can easily obtain

$$\hat{t}_{x,s} = \left(\sum_{k=1}^{25} d_k x_{k,1}, \sum_{k=1}^{25} d_k x_{k,2}, \sum_{k=1}^{25} d_k x_{k,3}, \sum_{k=1}^{25} d_k x_{k,4}, \sum_{k=1}^{25} d_k x_{k,5} \right)' = \begin{pmatrix} 46 \\ 42 \\ 69 \\ 206 \\ 64 \end{pmatrix}$$

and

$$A = \begin{bmatrix} A11 & A12 & A13 & A14 & A15 \\ A21 & A22 & A23 & A24 & A25 \\ A31 & A32 & A33 & A34 & A35 \\ A41 & A42 & A43 & A44 & A45 \\ A51 & A52 & A53 & A54 & A55 \end{bmatrix} = \begin{bmatrix} 46 & 18 & 31 & 108 & 24 \\ 18 & 42 & 22 & 62 & 12 \\ 31 & 22 & 69 & 206 & 39 \\ 108 & 62 & 206 & 750 & 120 \\ 24 & 12 & 39 & 120 & 64 \end{bmatrix}$$

where

$$Ajj = \sum_{k=1}^{25} d_k x_{k,j} x_{k,j}' = \sum_{k=1}^{25} d_k x_{k,j}^2$$

$$Aij = \sum_{k=1}^{25} d_k x_{k,i} x_{k,j}'; \text{ for all } i, j \ (=1,2,3,4,5) \text{ and } i \neq j$$

The inverse matrix of the square matrix $A = \left(\sum_{k \in s} d_k x_k x_k' \right)$ can be obtained by using the mathematical operation of the matrix relation $A A^{-1} = I$, where I is an identity matrix of order 5×5 (see Cormen et al. 2001, sec. 28.4, pp. 755–760), and is determined as

$$A^{-1} = \left(\sum_{k \in s} d_k x_k x_k' \right)^{-1}$$

$$= \begin{bmatrix} 0.03661582 & -0.00901288 & 0.00228602 & -0.00429437 & -0.00538212 \\ -0.00901288 & 0.03088625 & -0.01214961 & 0.00183273 & 0.00155596 \\ 0.00228602 & -0.01214961 & 0.09100053 & -0.02239201 & -0.01204764 \\ -0.00429437 & 0.00183273 & -0.02239201 & 0.00794951 & 0.00000656 \\ -0.00538212 & 0.00155596 & -0.01204764 & 0.00000656 & 0.02468079 \end{bmatrix},$$

and then by using the results in the relationship (4.11), the Lagrange multipliers should be calculated for this simple particular example as

$$\lambda' = (0.14209475, 0.03501717, 0.18600019, -0.08176176, -0.00426682).$$

Now, using this result in Equation 4.10, the new weights or calibrated weights for the chi-squared distance measure can be easily obtained. The calculated new weights and its distance measures to the sample design weights are given in Table 4.3.

TABLE 4.3

New Weights and Its Distance Measures to Sampling Design Weights

d_k	w_k	$w_k - d_k$	$\overset{\chi^2}{G}_k$
4	4.70844769	0.70844769	0.06273727
5	5.39271424	0.39271424	0.01542245
6	6.10925911	0.10925911	0.00099480
5	4.77151662	−0.22848338	0.00522047
3	3.09225105	0.09225105	0.00141838
4	4.41695372	0.41695372	0.02173130
6	5.97439907	−0.02560093	0.00005462
4	4.00419164	0.00419164	0.00000220
5	5.15375174	0.15375174	0.00236396
3	3.41348379	0.41348379	0.02849481
5	5.69627800	0.69627800	0.04848031
4	4.45424007	0.45424007	0.02579175
3	3.48091381	0.48091381	0.03854635
6	4.63754748	−1.36245252	0.15468974
4	3.57588131	−0.42411869	0.02248458
5	5.00000000	0	0
6	6.47125708	0.47125708	0.01850694
3	3.10505151	0.10505151	0.00183930
6	6.10925911	0.10925911	0.00099480
4	4.00419164	0.00419164	0.00000219
5	4.97866589	−0.02133411	0.00004551
3	2.31877374	−0.68122626	0.07734487
5	5.88555961	0.88555961	0.07842158
4	4.55702240	0.55702240	0.03878424
3	3.41348379	0.41348379	0.02849481
		TAD = 9.21152591	D = 0.67286721

Sources: After Rahman, A. et al., *Int. J. Microsimul.*, 2010a; Rahman, A. et al., Simulating the characteristics of populations at the small-area level: New validation techniques for a spatial microsimulation model in Australia, *JSM Proceedings, Social Statistics Section*, American Statistical Association, Alexandria, VA, 2010b, pp. 2022–2036; Rahman, A. Small area housing stress estimation in Australia: Microsimulation modelling and statistical reliability, PhD thesis, University of Canberra, Canberra, 2011.

Note: In this table, new weights (w_k) are the main focus in the calculation and is therefore given in bold font.

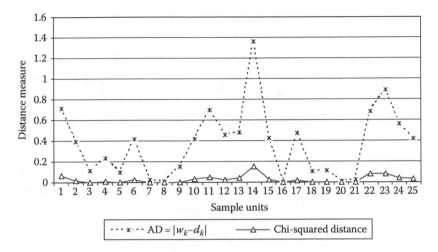

FIGURE 4.5
A comparison of absolute distance (AD) and chi-squared distance measures.

For the 16th unit of our hypothetical data, the new weight remains unchanged to the sampling design weight due to the fact that all entries for this unit are zero. However, this is very rare in GREGWT reweighting. In addition, the TAD indicates higher quantity. While *absolute distance* has a higher value, the corresponding *chi-squared distance* measure also indicates a higher value. Nevertheless, the fluctuations within absolute distances are remarkable compared to chi-squared distance measures (see Figure 4.5).

Furthermore, when the TAD is zero, the total chi-squared distance will also be zero, and in that situation, the calibrated weights will remain the same as the sampling design weights, which indicates the sample data are fully representative of the small area population. Besides, it is interesting to note that the values of a set of *new weights* vary greatly with the changing values of vector for differences between $\hat{t}_{x,s}$ and T_x. These differences can come from the poor representative data and/or the chosen benchmark tables used in the reweighting.

Four random alternative cases of difference vectors $C1 = [4, 3, 1, -6, 1]'$, $C2 = [8, 3, 1, -6, 1]'$, $C3 = [12, 3, 1, -6, 1]'$, and $C4 = [4, 3, 1, 2, 1]'$ (where $Cj = [T_x - \hat{t}_{x,s}]$ for $j = 1, 2, 3, 4$) have been considered for a further analysis, and the resulting sets of new weights are plotted in Figure 4.6. Results show that sets of new weights vary with changing values of the vector of difference between the benchmarks totals and sample-based estimated totals. For instance, in this empirical experiment, the 14th unit of our hypothetical data shows the maximum variation of about 25% for cases $C3$ and $C4$. However, the GREGWT considers the total chi-squared distance measure that could vary from 0 to 1. Modelers using GREGWT reweighting always try to minimize the total chi-squared distance though supplying the most appropriate

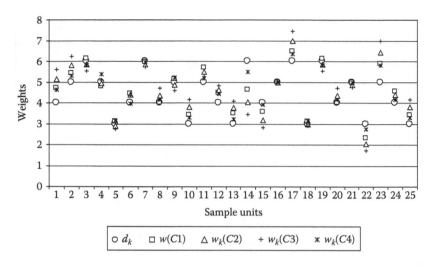

FIGURE 4.6
Plots of sampling design weights and new weights for specific cases. (From Rahman, A. et al., *Int. J. Microsimul.*, 2010a; Rahman, A. et al., Simulating the characteristics of populations at the small-area level: New validation techniques for a spatial microsimulation model in Australia, *JSM Proceedings, Social Statistics Section,* American Statistical Association, Alexandria, VA, 2010b, pp. 2022–2036.)

benchmarks as well as upper and lower limits of the distance function. For these random cases C1, C2, C3, and C4, the *total chi-squared distance* measure $\sum_{k=1}^{25} \overset{\chi^2}{G_k} = \sum_{k=1}^{25} (w_k - d_k)^2/(2d_k)$ of the distance function in Equation 4.7 are determined as 0.23, 0.34, 0.82, and 0.06, respectively. The results show that the case C4 generates a more consistent set of new weights compared to other cases. It is obvious that when the auxiliary information matrix provides quite rich sample data, then the resulting difference vector between $\hat{t}_{x,s}$ and T_x will be fairly close. Hence, the resulting set of calibrated weights will produce more accurate small area estimates.

Nevertheless, the small area estimates obtained from the synthetic spatial microdata that are produced by the GREGWT reweighting may have their own standard errors. The GREGWT algorithm can calculate these standard errors using a "group jackknife" approach, which is a replication-based method (for details, refer to Bell (2000a) and references therein). The key idea of the group jackknife method is to divide the survey sample into a number of subsample replicate groups (practically 30) and calculate the jackknife estimate for each replicate group based on the total sample excluding the replicate group. GREGWT achieves this by computing grossing factors that are also adjusted to meet the benchmarks for each resulting subsample. Then, the difference between these new estimates and the original sample estimates is used to estimate the standard error. The group jackknife approach could use replicate weights in spatial microsimulation modeling, and the problem is basically computational, not statistical. For instance, using SAS v9.1 on a 2.0 GHz PC, it takes about 9 h to run one set of weights, and in that method, researchers would end up with 30 weights for each small area. So, it would take approximately $30 \times 9 = 270$ h to run all replication.

Although the group jackknife approach is generally robust, it has some limitations. One of the major disadvantages is that there has to be a fairly large number of observations in each sample selection stratum. In practice, it is rare in a survey sample to achieve this number of observations, especially at small area levels or in spatial levels microdata. As a result, when there are too few observations in a sample stratum, the resulting standard error estimates should be statistically unreliable. It is of note that about 30 observations per stratum is a good minimum working number and maybe researchers can produce this number of observations by a suitable combination of classes. However, the overall replication process should also deal with the giant file size, which may be a problem for spatial microsimulation. For example, 1,300 columns (SLA) × 30 weights = 39,000 columns; then 39,000 columns × 12,000 households = 468 million cells in the final file.

4.6 Comparison between GREGWT and CO

A comparison of GREGWT and CO methodologies is provided in this section. Although both the approaches are used in the creation of small area synthetic microdata, the methodology behind each approach is quite different. For instance, GREGWT is typically based on generalized linear regression and attempts to minimize a truncated chi-squared distance function subject to the small area benchmarks. CO, on the other hand, is based on "intelligent searching" techniques and attempts to select an appropriate combination of households from a sample that best fits the small area benchmarks.

Tanton et al. (2007) compares these two techniques using a range of performance criteria. The study also covers the advantages and disadvantages of each method. Using the 1998–1999 Household Expenditure Survey data from Australia, the authors conclude that the GREGWT algorithm seems to be capable of producing good results. However, the GREGWT algorithm has some limitations compared to the CO algorithm. One of the drawbacks of the GREGWT approach is that convergence is not achieved for some small areas. That means that the GREGWT algorithm is unable to produce estimates for those small areas, while the CO algorithm is able to do so. In addition, the GREGWT algorithm takes more time to run compared to CO, and it is still unclear whether that extra time is due to the different programming language (GREGWT is written in SAS code and CO uses compiled FORTRAN code) or the relative efficiencies of the underlying algorithms. Moreover, the CO routine has a tendency to include fewer households but give them higher weights—and, conversely, the GREGWT routine has a tendency to select more households but give them smaller weights.

Table 4.4 summarizes features of the GREGWT and CO reweighting approaches. Typically, the focus is here on methodological issues among these two reweighting methods. Although both of the reweighting methods

TABLE 4.4

Comparison of the GREGWT and CO Reweighting Methodologies

GREGWT	CO
An iterative process.	An iterative process.
Use the Newton–Raphson method of iteration.	Use a stochastic approach of iteration.
Based on a distance function.	Based on a households' combination.
Attempt to minimize the distance function subject to the known benchmarks.	Attempt to select an appropriate combination that best fits the known benchmarks.
Use the Lagrange multipliers as minimization tools for minimizing the distance function.	Use CO techniques as intelligent searching tools in optimizing combinations of households.
Weights are in fractions.	Weights are in integers.
Boundary condition is applied to new weights for achieving a solution.	There is no boundary condition to new weights.
The benchmark constraints at small area levels are fixed for the algorithm.	The algorithm is designed to optimize fit to a selected group of tables, which may or may not be the most appropriate ones. There may be a choice of benchmark constraints.
Typically focus on simulating microdata at small area levels and aggregation is possible at larger domains.	Offers a flexibility and collective coherence of microdata, making it possible to perform mutually consistent analysis at any level of aggregation or sophistication.
All estimates have their own standard errors obtained by a group jackknife approach.	There is no information about this issue in literature. May be possible in theory but nothing available in practice yet.
In some cases, convergence does not exist and this requires readjusting the boundary limits or a proxy indicator for this nonconvergence.	There are no convergence issues. However, the finally selected household combination may still fail to fit user-specified benchmark constraints.
Sensitive to disagreements between target benchmarks. For example, if one benchmark totals 1000 households, but other totals 980, the GREGWT may well never converge.	Insensitive to disagreements between target benchmarks, which means optimal solution in this example (in GREGWT section) would be 990, splitting the difference, and thereby minimizing the error.
There is no standard index to check the statistical reliability of the estimates.	There is no standard index to check the statistical reliability of the estimates.
The iteration procedure can be unstable near a horizontal asymptote or at local extremum.	The iteration algorithm may be able to avoid deceiving at local extremum within the solution.

Source: After Rahman, A., Small area estimation through spatial microsimulation models: Some methodological issues, Paper presented at the *Second General Conference of the International Microsimulation Association*, Ottawa, Ontario, Canada, October 6, 2009a.

are based on iterative algorithms, the GREGWT approach uses the Newton–Raphson method of iteration that minimizes a distance function with the Lagrange multipliers minimization tools, and it simulates the new weights in fractions that are constrained by a prespecified boundary condition. In GREGWT algorithm, convergence does not achieve for some cases, and this requires readjusting the boundary limits or a proxy indicator for such non-convergence. On the other hand, the CO approach uses a stochastic method of iteration to select the best fit combination of households to the benchmarks by using intelligent searching tools and simulates the new weights in integers without any boundary condition. It offers a flexibility and collective coherence of microdata.

4.7 Concluding Remarks

This chapter has comprehensively reviewed the overall concept and methodologies for the geographic approach of indirect small area estimation. It is noted that until recently most of the review articles in small area estimation have frequently acknowledged only the methods in the statistical model–based estimation. Nevertheless, another type of method called spatial microsimulation modeling has also been widely used as a geographic approach toward small area estimation. Findings of this review reveal that the spatial microsimulation model–based indirect small area estimation techniques are robust, in the sense that further aggregation or disaggregation is possible on the basis of choice of spatial scales or domains. In addition, since the spatial microsimulation framework uses a list-based approach to microdata representation, it is possible to use the microdata file for further analysis and updating. Also, by linking spatial microsimulation model with static microsimulation models, it is possible to measure small area effects of policy changes. In contrast, the traditional statistical model–based approaches do not have such robustness.

In practice, the spatial microsimulation model–based geographic approaches are associated with very sophisticated methodologies, tools, and techniques. For example, as spatial micropopulation data are not readily available for the world practices, simulating reliable synthetic spatial micropopulation data is challenging and it requires complex methods and computational systems. Different methodologies including the synthetic reconstruction and reweighting approaches demonstrate that reweighting methods are commonly used techniques in spatial microsimulation modeling and they are playing a vital role in producing small area microdata for ultimate small area estimation.

There are two reweighting techniques—CO and GREGWT. The CO technique uses an intelligent searching algorithm *simulated annealing*—which selects an appropriate set of households from survey microdata that best fit to the benchmark constraints by minimizing the "total absolute error" distance function with respect to the *Metropolis criterion*. In this process, change in the total absolute error occurs with potential change in households' combination to meet the benchmark constraints. The new weights give the actual household units that are the best representative combination. Thus, CO is a selection process to an appropriate combination of sample units rather than calibrating the sampling design weights to a set of new weights.

On the other hand, the GREGWT reweighting technique utilizes a truncated chi-squared distance function and generates a set of new weights by minimizing the total distance with respect to some constraint functions. A very simple theoretical view of GREGWT reveals that the minimization tool Lagrange multipliers has been used in this process to minimize the distance function, and it is based on the Newton–Raphson iterative process. Results from an explicit numerical solution show that sets of new weights can vary substantially, perhaps about 25% with changing values of the vector of difference between the benchmarks totals and sample-based estimated totals. Moreover, the chi-squared distance measures used in the GREGWT macroalgorithm show more smooth fluctuations compared to the measures of absolute distance errors used in CO.

The findings of the comparison between CO and GREGWT reveal that they are different reweighting algorithms. The CO routine has a tendency to include fewer households but give them higher weights—and, conversely, the GREGWT routine has a tendency to select more households but give them smaller weights. In GREGWT method, the iteration procedure can be unstable near a horizontal asymptote or at local extremum, and the standard errors can be obtained by a group jackknife approach. Whereas in CO method, the iteration algorithm may be able to avoid deceiving at local extremum within the solution but measures of the standard errors are not obtainable. The overall performances are fairly similar for both the reweighting techniques from the standpoint of use in spatial microsimulation modeling. However, in the context of available data in Australia, the GREGWT reweighting algorithm is capable of producing good results relative to the CO technique. Furthermore, like other methods, the GREGWT technique has some limitations. For instance, the GREGWT reweighting algorithm does not meet "accuracy criterion" for few small areas where the sample size is too small or especially zero, and it also takes more CPU run time compared to CO.

5

Bayesian Prediction–Based Microdata Simulation

5.1 Introduction

Creating a reliable synthetic micropopulation data set at small area level is still challenging in the spatial microsimulation modeling approach to small area estimation. Although a range of methods are used to generate spatial microdata, none of these methods can consider the scenario of whole micropopulation at a small area level. As a result, a newly generated micropopulation data set from those approaches often leads to inaccurate data for many small areas, especially for small areas with a small-sized population. In addition, validating the outputs from a spatial microsimulation model built on such synthetic microdata is also difficult. Presently, there is no straightforward statistical or mathematically robust tool to deal with these sorts of problems.

In this chapter, a new approach to generating synthetic spatial microdata at small area levels is described. Characteristically, the new method takes consideration of the complete scenarios of micropopulation data units at the small area level.

An outline of the chapter is as follows. The basic steps of the new approach are outlined in Section 5.2. The Bayesian prediction theory is described in Section 5.3. The multivariate model toward linking the data is illustrated in Section 5.4. Bayesian prediction distribution for the model is obtained in Section 5.5. Joint density of the observed and unobserved population units is obtained in Section 5.6. Bayesian prediction technology for modeling the unobserved population units is designed in Section 5.7 along with a comparison to other methods. A concluding summary of the chapter is provided in Section 5.8.

5.2 Basic Steps

It is noted that in any area being sampled, a finite population usually has two parts: observed units in the sample called data and unobserved sampling units in the population (Figure 5.1). Suppose that Ω represents a finite population in which Ω_i (say) is the subpopulation of the small area i. Now if s_i denotes the observed sample units in the ith area, then we have $s_i \cup \bar{s}_i = \Omega_i \subseteq \Omega$ for $\forall_i$, where $\bar{s}_i$ denotes the unobserved units in the small area population. Let y_{ij} represent a variable of interest for the jth observation of the population in the ith small area. Thus, we always have the estimate of total population in the ith small area as (see Rahman 2008b)

$$t_{y_i.} = \sum_{j \in s_i} y_{ij} + \sum_{j \in \bar{s}_i} y_{ij}.$$

The main challenge in this approach of microdata simulation is to establish the linkage of observed data to the unobserved sampling units in the small area population. Essentially, it is a sort of prediction problem, where a modeler tries to find a probability distribution of unobserved responses using the observed sample and the auxiliary data. The Bayesian methodology (see Ericson 1969; Lo 1986; Little 2007; Rahman 2007, 2008c; Aitkin 2008; Rahman and Upadhyay 2015; Rahman et al. 2016) can deal with such a prediction problem. The new method is principally based on the Bayesian prediction theory and the Markov chain Monte Carlo (MCMC) method of iteration.

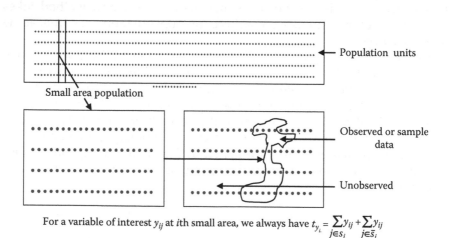

Population units

Small area population

Observed or sample data

Unobserved

For a variable of interest y_{ij} at ith small area, we always have $t_{y_i} = \sum_{j \in s_i} y_{ij} + \sum_{j \in \bar{s}_i} y_{ij}$

FIGURE 5.1
A diagram of new system for generating spatial microdata.

The basic steps involved in this process of spatial microdata simulation are as follows:

1. To obtain a suitable joint prior distribution of the event under research E_i, say, housing stress in the population in the ith small area, that is, $p(E_i)$ for $\forall_i$

2. To find the conditional distribution of unobserved sampling units, given the observed data, that is, $p(y_{ij} : j \in \bar{s}_i | y_{ij} : j \in s_i)$ for $\forall_i$

3. To derive the posterior distribution using Bayes' theorem, that is, $p(\theta|s,X)$; $E_i \subseteq \theta$, where θ is the vector of model parameters and X is an auxiliary information vector

4. To get simulated copies of the entire population from this posterior distribution using the MCMC simulation technique

---

5.3 Bayesian Prediction Theory

The Bayesian prediction theory is very straightforward and mainly based on the Bayes' posterior distribution of unknown parameters (Rahman 2008c). Let y be a set of observed sample units from a model with a joint probability density $p(y|\theta)$, in which θ is a set of model parameters. If a prior density of unknown parameters θ is $g(\theta)$, the posterior density of θ for a given y can be obtained by Bayes' theorem (Bayes 1763) and defined as

$$p(\theta|y) \propto p(y|\theta)g(\theta).$$

Now, if $\bar{y}$ is the set of unobserved units in a finite population, then under the Bayesian methodology the prediction distribution of $\bar{y}$ can be obtained by solving the integral

$$p(\bar{y}|y) \propto \int_{\theta} p(\theta|y)p(\bar{y}|\theta)d\theta,$$

where $p(\bar{y}|\theta)$ is the probability density function of unobserved units in the finite population. Further details of the Bayesian prediction theory for various linear models can be obtained in Rahman (2008c).

5.4 Multivariate Model

The multivariate multiple regression model is a generalization of the multiple regression model in a multivariate or matrix-variate setup. This model represents the relationship between a set of values of several response variables and a single set of values of some explanatory or predictor variables. The multivariate model is used to analyze data from different situations in economics as well as in many other experimental circumstances to deal with a set of regression equations.

There are many experimental situations in real life where we need to study a set of responses from more than one dependent variable corresponding to a set of values from several independent variables. For example, in a country like Australia, if various groups of households in different states or towns are given a single set of social benefits (such as housing assistance, health subsidies, income supports childcare benefits) to observe any responses to the living standard, then for a set of values of the predictor variables, there will be several sets of values of the response variables from various groups of households. More about this model can be found in the literature (e.g., refer to Tiao and Zellner 1964; Guttman and Hougaard 1985; Rahman 2008c).

Suppose $Y_1, Y_2, \ldots, Y_{n_i}$ is a set of vectors of n_i responses in the ith small area and $x_1, x_2, \ldots, x_{k-1}$ is a set of $k-1$ predictor variables. Then a relation between this set of vector responses and the single set of values of predictor variables can be written as a multivariate multiple regression model when each of the n_i responses is assumed to follow the linear model, which is defined as

$$Y_{ij} = \beta_{0[j]} + \beta_{1[j]} x_{j1} + \beta_{2[j]} x_{j2} + \cdots + \beta_{(k-1)[j]} x_{j(k-1)} + e_{ij}; \quad \text{for } \forall i \text{ and } j = 1, 2, \ldots, n_i$$

where

Y_{ij} and e_{ij} are the jth response in the ith small area and its associated error vectors, respectively, each of order $1 \times p$

$\beta_{[j]}$ is a $k \times 1$ dimensional regression parameter vector on the regression line of jth response

$x_{j1}, x_{j2}, \ldots, x_{j(k-1)}$ is a single set of values of the $k-1$ predictor variables for the regression to jth response

Assume e_{ij} follows a p-dimensional Student-t distribution with location parameters being zero, scale parameters Σ, and degrees of freedom v, that is, $e_{ij} \sim t_p(0, \Sigma, v)$. Characteristically, the Student-t distribution modeling is very useful for small sample problems, and it is more robust as the generalization of this distribution becomes normal.

Let $x_j = [x_{j1}, x_{j2}, \ldots, x_{j(k-1)}]$, $Y_j = [y_{j1}, y_{j2}, \ldots, y_{jp}]'$, and $e_j = [e_{j1}, e_{j2}, \ldots, e_{jp}]'$ denote the values of the explanatory, response, and error variables, respectively, for the jth sample or response unit. Then there is an $n_i \times k$ order design matrix for the ith small area

$$X_i = \begin{bmatrix} 1 & x_{11} & \cdots & x_{1(k-1)} \\ 1 & x_{21} & \cdots & x_{2(k-1)} \\ \cdots & \cdots & \cdots & \cdots \\ 1 & x_{n_i 1} & \cdots & x_{n_i(k-1)} \end{bmatrix}.$$

Moreover, the following matrix notations are used to define the respective matrices as

$$\beta = \begin{bmatrix} \beta_{01} & \beta_{02} & \cdots & \beta_{0p} \\ \beta_{11} & \beta_{12} & \cdots & \beta_{1p} \\ \cdots & \cdots & \cdots & \cdots \\ \beta_{(k-1)1} & \beta_{(k-1)2} & \cdots & \beta_{(k-1)p} \end{bmatrix},$$

the unknown regression coefficients matrix of order $k \times p$;

$$Y_i = \begin{bmatrix} y_{11} & y_{12} & \cdots & y_{1p} \\ y_{21} & y_{22} & \cdots & y_{2p} \\ \cdots & \cdots & \cdots & \cdots \\ y_{n_i 1} & y_{n_i 2} & \cdots & y_{n_i p} \end{bmatrix},$$

the values of the responses matrix of order $n_i \times p$; and

$$E_i = \begin{bmatrix} e_{11} & e_{12} & \cdots & e_{1p} \\ e_{21} & e_{22} & \cdots & e_{2p} \\ \cdots & \cdots & \cdots & \cdots \\ e_{n_i 1} & e_{n_i 2} & \cdots & e_{n_i p} \end{bmatrix},$$

the corresponding values of the errors matrix with the same order $n_i \times p$.

Thus, the multivariate linear regression model can be expressed as

$$Y_i = X_i \beta + E_i \tag{5.1}$$

where rank $(X_i) = k$ and $n_i \geq k$ for $\forall i$.

Now, assume that the components of each row in E_i of the model are correlated and jointly follow a multivariate Student-t distribution. Since each row of the errors matrix E_i is uncorrelated with others, the covariance of the errors matrix becomes $(\nu/\nu - 2)[\Sigma \otimes I_{n_i}]$, where ν is the shape parameter or degrees of freedom of the matrix-T distribution for the errors matrix and $\otimes$ denotes the Kronecker product between two matrices Σ and I_{n_i} in which Σ is a $p \times p$ order positive definite symmetric matrix and I_{n_i} is an identity matrix of order $n_i \times n_i$.

For corroboration, it is noted that when a random matrix U of order $n \times p$ follows a matrix-T distribution with the location parameter $\Psi \in \Re^{n \times p}$, both scale factors $\Omega_{n \times n}$ and $\Sigma_{p \times p}$ being positive symmetric matrices, and shape parameter $\nu > 0$ and defined by $U \sim T_{np}(\Psi, \Omega_{n \times n}, \Sigma_{p \times p}, \nu)$, then we have expectation of U, that is, $E(U) = \Psi$, and covariance of U, that is, $\text{Cov}(U) = (\nu/\nu - 2)$ $[\Sigma_{p \times p} \otimes \Omega_{n \times n}]$, where $\otimes$ represents the Kronecker product between matrices (e.g., see Press 1986).

Furthermore, if each row of the matrix U is uncorrelated with others, then the scale factor $\Omega_{n \times n}$ becomes an $n \times n$ order identity matrix that can be defined as I_n, and in such a case, we have $\text{Cov}(U) = (\nu/\nu - 2)[\Sigma_{p \times p} \otimes I_n]$ (see Magnus and Neudecker 1988).

In this case, the errors matrix E_i has a matrix-T distribution with the density

$$f(E_i) \propto \left|I_{n_i}^{-1}\right|^{\frac{p}{2}} \left|\Sigma^{-1}\right|^{\frac{n_i}{2}} \left|I_p + \Sigma^{-1} E_i' E_i\right|^{-\frac{(\nu+p+n_i-1)}{2}}. \tag{5.2}$$

Hence, the responses matrix Y_i from the model (5.1) has also a matrix-T distribution, that is, $Y_i \sim T_{n_i p}(X_i \beta, I_{n_i \times n_i}, \Sigma_{p \times p}, \nu)$, with the following probability density function:

$$f(Y_i \mid X_i \beta, \Sigma, \nu) = C \frac{\left|I_{n_i}^{-1}\right|^{p/2}}{|\Sigma|^{n_i/2}} \left|I_p + \Sigma^{-1}(Y_i - X_i \beta)'(Y_i - X_i \beta)\right|^{-\frac{1}{2}(\nu+p+n_i-1)} \tag{5.3}$$

where the normalizing constant is

$$C = \left[\Gamma\left(\frac{1}{2}\right)\right]^{pn_i} \frac{\Gamma_p[(1/2)(\nu+p-1)]}{\Gamma_p[(1/2)(\nu+p+n_i-1)]},$$

in which $\Gamma_p(\cdot)$ is a generalized gamma function introduced by Siegel (1935) and defined as

$$\Gamma_p(b) = \left[\Gamma\left(\frac{1}{2}\right)\right]^{\frac{1}{2}p(p-1)} \prod_{\alpha=1}^{p} \Gamma\left(b + \frac{\alpha-p}{2}\right); \quad b > \frac{p-1}{2}$$

for the nonzero positive integers p and b with $\alpha = 1, 2, \ldots, p$.

5.5 Prior and Posterior Distributions

Assume that the joint prior distribution of the regression matrix β and the $(1/2)p(p + 1)$ distinct elements of the scale parameters matrix Σ is noninformative or uniform. Moreover, assume that the elements of β as well as Σ are independently distributed, which means if $p(\beta, \Sigma)$ is a joint prior density of β and Σ, then

$$p(\beta, \Sigma) = p(\beta)p(\Sigma).$$

By adopting the invariance theory due to Jeffreys (1961, p. 179), here, we can consider

$$p(\beta) = \text{constant}$$

and

$$p(\Sigma) \propto |\Sigma|^{-\frac{p+1}{2}}.$$

Thus, the joint prior density of unknown parameters' matrices is

$$p(\beta, \Sigma) \propto |\Sigma|^{-\frac{p+1}{2}} \tag{5.4}$$

Note that this uniform prior distribution has been used by many researchers such as Zellner (1971), Bernardo and Rueda (2002), and Rahman (2007, 2009b).

Furthermore, if $[\beta, \Sigma] = \theta$ (*say*) is a matrix of unknown parameters, then by Bayes' theorem the posterior density of θ can be defined as

$$p(\theta|Y_i) = \frac{p(\theta)p(Y_i|\theta)}{p(Y_i)} \tag{5.5}$$

where

$$p(Y_i) = \int_\theta p(\theta)p(Y_i|\theta)d\theta$$

in which $p(Y_i|\theta)$ is the joint probability function or likelihood function for responses Y_i and $p(\theta)$ is a prior density of parameters θ.

Now, by Bayes' theorem a posterior density of parameters β and Σ for the observed responses matrix Y_i can be defined as

$$p(\beta, \Sigma|Y_i) \propto p(Y_i|\beta, \Sigma)p(\beta, \Sigma) \tag{5.6}$$

where
  $p(Y_i|\beta, \Sigma)$ is the joint probability function or likelihood function for Y_i, which is provided in Equation 5.3
  $p(\beta, \Sigma)$ is the prior density of β and Σ, which is defined in Equation 5.4

Hence, we get the following posterior density of unknown parameters' matrices β and Σ given the observed units matrix Y_i as

$$p(\beta, \Sigma|Y_i) \propto |\Sigma|^{-\frac{n_i+p+1}{2}} \left| I_p + \Sigma^{-1}(Y_i - X_i\beta)'(Y_i - X_i\beta) \right|^{-\frac{1}{2}(v+p+n_i-1)} \tag{5.7}$$

in which the normalizing constant of the posterior distribution can be obtained by integrating over the density function with respect to the parameters β and Σ.

To evaluate the normalizing constant, the density in (5.7) can be written as

$$p(\beta, \Sigma \mid Y_i) = \Phi \left| \Sigma \right|^{-\frac{n_i + p + 1}{2}} \left| I_p + \Sigma^{-1}(Y_i - X_i\beta)'(Y_i - X_i\beta) \right|^{-\frac{1}{2}(v + p + n_i - 1)}$$

where Φ represents the normalizing constant. The value of Φ can be obtained by solving the following equation of probability function:

$$1 = \Phi \int_\beta \int_\Sigma \left| \Sigma \right|^{-\frac{n_i + p + 1}{2}} \left| I_p + \Sigma^{-1}(Y_i - X_i\beta)'(Y_i - X_i\beta) \right|^{-\frac{1}{2}(v + p + n_i - 1)} d\Sigma \, d\beta \qquad (5.8)$$

or

$$\Phi^{-1} = \int_\beta \int_\Sigma \left| \Sigma \right|^{-\frac{n_i + p + 1}{2}} \left| I_p + \Sigma^{-1}(Y_i - X_i\beta)'(Y_i - X_i\beta) \right|^{-\frac{1}{2}(v + p + n_i - 1)} d\Sigma \, d\beta \qquad (5.9)$$

Using the matrix transformation $\Sigma^{-1} = \Lambda$ with the Jacobian of the transformation

$$|J| = \frac{d\Sigma}{d\Lambda} = \left| \Lambda^{-1} \right|^{p+1}$$

Equation 5.9 can be expressed as

$$\Phi^{-1} = \int_\beta \int_\Lambda \left| \Lambda \right|^{\frac{n_i}{2} - \frac{p+1}{2}} \left| I_p + \Lambda(Y_i - X_i\beta)'(Y_i - X_i\beta) \right|^{-\left(\frac{n_i}{2} + \frac{v+p-1}{2}\right)} d\Lambda \, d\beta \qquad (5.10)$$

Applying the generalized beta integral for the matrix variables over Λ, this equation becomes

$$\Phi^{-1} = \int_\beta \left| (Y_i - X_i\beta)'(Y_i - X_i\beta) \right|^{-\frac{n_i}{2}} B_p \left(\frac{n_i}{2}, \frac{v+p-1}{2} \right) d\beta$$

or

$$\Phi^{-1} = B_p \left(\frac{n_i}{2}, \frac{v+p-1}{2} \right) \int_\beta \left| S_Y + (\beta - X_i\hat{\beta})'X_i'X_i(\beta - X_i\hat{\beta}) \right|^{-\frac{n_i}{2}} d\beta \qquad (5.11)$$

where
  $\hat{\beta} = (X_i'X_i)^{-1}X_i'Y_i$ is the ordinary least square (OLS) estimate of β
  $S_Y = (Y_i - X_i\hat{\beta})'(Y_i - X_i\hat{\beta})$ of order $p \times p$

Now Equation 5.11 can be written as

$$\Phi^{-1} = B_p\left(\frac{n_i}{2}, \frac{v+p-1}{2}\right)|S_Y|^{-\frac{n_i}{2}} \int_\beta |I_p + S_Y^{-1}(\beta - X_i\hat{\beta})'X_i'X_i(\beta - X_i\hat{\beta})|^{-\frac{\delta+k+p-1}{2}} d\beta$$

where $\delta = n_i - k - p + 1$.

By using the properties of the matrix-T distribution to the integration with respect to β, the result can be expressed as

$$\Phi^{-1} = B_p\left(\frac{n_i}{2}, \frac{v+p-1}{2}\right)|S_Y|^{-\frac{n_i}{2}} \frac{|X_i'X_i|^{-\frac{p}{2}}}{|S_Y|^{-\frac{k}{2}}}\left[\Gamma\left(\frac{1}{2}\right)\right]^{kp} \frac{\Gamma_p((\delta+p-1)/2)}{\Gamma_p((\delta+k+p-1)/2)}$$

or

$$\Phi^{-1} = B_p\left(\frac{n_i}{2}, \frac{v+p-1}{2}\right)|S_Y|^{-\frac{n_i-k}{2}} |X_i'X_i|^{-\frac{p}{2}}\left[\Gamma\left(\frac{1}{2}\right)\right]^{kp} \frac{\Gamma_p((\delta+p-1)/2)}{\Gamma_p((\delta+k+p-1)/2)}$$

or

$$\Phi^{-1} = \left[\Gamma\left(\frac{1}{2}\right)\right]^{kp} \frac{|X_i'X_i|^{-\frac{p}{2}}}{|S_Y|^{-\frac{n_i-k}{2}}} \frac{\Gamma_p((\delta+p-1)/2)}{\Gamma_p((\delta+k+p-1)/2)} B_p\left(\frac{n_i}{2}, \frac{v+p-1}{2}\right) \quad (5.12)$$

Now considering the value of δ and also expressing $B_p((n_i/2),(v+p-1/2))$ as the generalized gamma function, Equation 5.12 can be expressed as

$$\Phi^{-1} = [\Gamma(1/2)]^{kp} \frac{|X_i'X_i|^{-\frac{p}{2}}}{|S_Y|^{-\frac{n_i-k}{2}}} \frac{\Gamma_p((n_i-k)/2)}{\Gamma_p(n_i/2)} \frac{\Gamma_p(n_i/2)\Gamma_p((v+p-1)/2)}{\Gamma_p((v+n_i+p-1)/2)} \quad (5.13)$$

and hence, the normalizing constant in Φ is obtained as

$$\Phi = \frac{|X_i'X_i|^{\frac{p}{2}}|S_Y|^{\frac{n_i-k}{2}}}{[\Gamma(1/2)]^{kp}} \frac{\Gamma_p((v+n_i+p-1)/2)}{\Gamma_p((n_i-k)/2)\Gamma_p((v+p-1)/2)}.$$

5.6 The Linkage Model

The linkage model is set up for the unobserved population units in a small area to connect them with the observed units in the sample. Let $\bar{Y}_i$ be the unobserved responses matrix from the model provided in Equation 5.1 to the

$m_i \times k$ (where $m_i = N_i - n_i$; $\forall$ i) dimensional design matrix $\bar{X}_i$. Then the linkage multivariate model for $\bar{Y}_i$ can be defined as

$$\bar{Y}_i = \bar{X}_i \beta + \bar{E}_i \tag{5.14}$$

where

   β is a $k \times p$ dimensional regression parameters matrix for unobserved responses
   $\bar{Y}_i$ and $\bar{E}_i$ are $m_i \times p$ dimensional unobserved responses matrix and associated errors matrix, respectively

In the linkage model, it is assumed that $\bar{E}_i$ follows m_ip dimensional matrix-T distribution with ν degrees of freedom, which is expressed as

$$\bar{E}_i \sim T_{m_ip}(0, I_{m_i \times m_i}, \Sigma_{p \times p}, \nu).$$

Since the elements in each row of the errors matrix E_i for observed sample units and the errors matrix $\bar{E}_i$ for the unobserved population units are correlated, and n_i rows in E_i as well as m_i rows in $\bar{E}_i$ are uncorrelated, the joint density function of β and Σ for the observed response units matrix Y_i from the model and the unobserved population units matrix $\bar{Y}_i$ from the linkage model can be expressed as

$$p(Y_i, \bar{Y}_i \mid \beta, \Sigma) \propto |\Sigma|^{-\frac{n_i + m_i}{2}} \left| I_p + \Sigma^{-1} Q \right|^{-\frac{1}{2}(\nu + p + n_i + m_i - 1)} \tag{5.15}$$

where

$$Q = (Y_i - X_i\beta)'(Y_i - X_i\beta) + (\bar{Y}_i - \bar{X}_i\beta)'(\bar{Y}_i - \bar{X}_i\beta).$$

5.7 Prediction for Modeling Unobserved Population Units

Using the prior density for β and Σ as given in Equation 5.4 and the joint density function for the matrices Y_i and $\bar{Y}_i$ in (5.15), the joint posterior density of unknown parameters β and Σ given the observed units matrix Y_i and the unobserved units matrix $\bar{Y}_i$ can be determined as

$$p(\beta, \Sigma \mid Y_i, \bar{Y}_i) \propto |\Sigma|^{-\frac{n_i + m_i + p + 1}{2}} \left| I_p + \Sigma^{-1} Q \right|^{-\frac{1}{2}(\nu + p + n_i + m_i - 1)} \tag{5.16}$$

Now the prediction distribution of the set of m_i unobserved population units in the ith small area can be obtained by solving the following integral:

$$f(\bar{Y}_i|Y_i) \propto \int_{\beta} \int_{|\Sigma|>0} p(\beta, \Sigma \,|\, Y_i, \bar{Y}_i) d\Sigma d\beta \tag{5.17}$$

or

$$f(\bar{Y}_i|Y_i) \propto \int_{\beta} \int_{|\Sigma|>0} |\Sigma|^{-\frac{n_i+m_i+p+1}{2}} \left|I_p + \Sigma^{-1}Q\right|^{-\frac{1}{2}(v+p+n_i+m_i-1)} d\Sigma d\beta \tag{5.18}$$

Applying the appropriate matrix transformation $\Sigma^{-1} = \Lambda$ with the Jacobian of the transformation

$$|J| = \frac{d\Sigma}{d\Lambda} = \left|\Lambda^{-1}\right|^{p+1},$$

the probability density in (5.18) can be written as

$$f(\bar{Y}_i|Y_i) \propto \int_{\beta} \int_{|\Lambda|>0} |\Lambda^{-1}|^{-\frac{n_i+m_i-p-1}{2}} \left|I_p + \Lambda Q\right|^{-\frac{1}{2}(v+p+n_i+m_i-1)} d\Lambda d\beta$$

or

$$f(\bar{Y}_i|Y_i) \propto \int_{\beta} \int_{|\Lambda|>0} |\Lambda|^{\frac{n_i+m_i}{2}-\frac{p+1}{2}} \left|I_p + \Lambda Q\right|^{-\left(\frac{n_i+m_i}{2}+\frac{v+p-1}{2}\right)} d\Lambda d\beta \tag{5.19}$$

Now, by using the following properties of the generalized beta integral for the matrix variables case:

$$I = \int_{|W|>0} |W|^{a-\frac{r+1}{2}} \left|I_r + WD\right|^{-(a+b)} dW = |D|^{-1} B_r(a,b)$$

the prediction density can be obtained from Equation 5.19 after integrating with respective to Λ

$$f(\bar{Y}_i|Y_i) \propto \int_{\beta} |Q|^{-\frac{n_i+m_i}{2}} B_p\left(\frac{n_i+m_i}{2}, \frac{v+p-1}{2}\right) d\beta$$

or

$$f(\bar{Y}_i|Y_i) \propto \int_{\beta} |Q|^{-\frac{n_i+m_i}{2}} d\beta \tag{5.20}$$

Now Q can be expressed as a convenient quadratic form in β by the following way:

$$Q = R + (\beta - P)'M(\beta - P) \tag{5.21}$$

where

$$R = Y_i'Y_i + \bar{Y}_i'\bar{Y}_i - P'MP,$$

$$P = M^{-1}(X_i'Y_i + \bar{X}_i'\bar{Y}_i)$$

and

$$M = X_i'X_i + \bar{X}_i'\bar{X}_i.$$

Appling this convenient form of Q in (5.21) to Equation 5.20 and then after integrating with respect to β by matrix-T integral, the prediction distribution can be expressed as

$$f(\bar{Y}_i|Y_i) \propto |R|^{-\frac{n_i + m_i - k}{2}}. \tag{5.22}$$

It is noted that the matrix R is free from the unknown parameters matrix β. Now we have to express the matrix R as an appropriate quadratic form of $\bar{Y}_i$ for obtaining the complete form of prediction distribution.

To simplify the expression of R into a convenient quadratic form, a range of mathematical operations can be performed in the following ways (see Rahman 2009b):

$$
\begin{aligned}
R &= Y_i'Y_i + \bar{Y}_i'\bar{Y}_i - P'MP \\
&= Y_i'Y_i + \bar{Y}_i'\bar{Y}_i - [M^{-1}(X_i'Y_i + \bar{X}_i'\bar{Y}_i)]'M[M^{-1}(X_i'Y_i + \bar{X}_i'\bar{Y}_i)] \\
&= Y_i'Y_i + \bar{Y}_i'\bar{Y}_i - (X_i'Y_i + \bar{X}_i'\bar{Y}_i)'M^{-1}MM^{-1}(X_i'Y_i + \bar{X}_i'\bar{Y}_i) \\
&= Y_i'Y_i + \bar{Y}_i'\bar{Y}_i - (X_i'Y_i + \bar{X}_i'\bar{Y}_i)'M^{-1}(X_i'Y_i + \bar{X}_i'\bar{Y}_i) \\
&= Y_i'Y_i + \bar{Y}_i'\bar{Y}_i - Y_i'X_iM^{-1}X_i'Y - \bar{Y}_i'\bar{X}_iM^{-1}\bar{X}_i'\bar{Y} - Y_i'X_iM^{-1}\bar{X}_i'\bar{Y}_i - \bar{Y}_i'\bar{X}_iM^{-1}X_i'Y_i) \\
&= Y_i'(I - X_iM^{-1}X_i')Y_i + \bar{Y}_i'(I - \bar{X}_iM^{-1}\bar{X}_i')\bar{Y}_i - Y_i'X_iM^{-1}\bar{X}_i'\bar{Y}_i - \bar{Y}_i'\bar{X}_iM^{-1}X_i'Y_i) \\
&= Y_i'(I - X_iM^{-1}X_i')Y_i + \bar{Y}_i'H\bar{Y}_i - Y_i'X_iM^{-1}\bar{X}_i'\bar{Y}_i - \bar{Y}_i'\bar{X}_iM^{-1}X_i'Y_i) \tag{5.23}
\end{aligned}
$$

where $H = I - \bar{X}_iM^{-1}\bar{X}_i'$.

As shown by Zellner (1971, p. 235)

$$H^{-1} = (I - \bar{X}_iM^{-1}\bar{X}_i')^{-1} = I + \bar{X}_i(X_i'X_i)^{-1}\bar{X}_i'$$

which can be verified by the following matrix multiplication:

$$HH^{-1} = (I - \bar{X}_i M^{-1} \bar{X}_i')[I + \bar{X}_i (X_i' X_i)^{-1} \bar{X}_i']$$

$$= I - \bar{X}_i [M^{-1} - (X_i' X_i)^{-1} + M^{-1} \bar{X}_i' \bar{X}_i (X_i' X_i)^{-1}] \bar{X}_i'$$

$$= I - \bar{X}_i M^{-1} [X_i' X_i - M + \bar{X}_i' \bar{X}_i] (X_i' X_i)^{-1} \bar{X}_i'$$

$$= I \qquad (5.24)$$

since $X_i' X_i - M + \bar{X}_i' \bar{X}_i = 0$, by the definition of $M = X_i' X_i + \bar{X}_i' \bar{X}_i$.
Therefore, R can be expressed as

$$R = Y_i'[I - G_1(M,H)]Y_i + [\bar{Y}_i' - G_2(M,H)]'H[\bar{Y}_i' - G_2(M,H)] \qquad (5.25)$$

where

$$G_1(M,H) = X_i M^{-1} X_i' + X_i M^{-1} \bar{X}_i' H^{-1} \bar{X}_i M^{-1} X_i'$$

and

$$G_2(M,H) = H^{-1} \bar{X}_i M^{-1} X_i' Y_i.$$

Now, the following relationships can be established as more suitable forms:

$$G_1(M,H) = X_i M^{-1} X_i' + X_i M^{-1} \bar{X}_i' H^{-1} \bar{X}_i M^{-1} X_i'$$

$$= X_i M^{-1} X_i' + X_i M^{-1} \bar{X}_i'[I + \bar{X}_i (X_i' X_i)^{-1} \bar{X}_i'] \bar{X}_i M^{-1} X_i'$$

$$= X_i M^{-1} [X_i' + \bar{X}_i'\{I + \bar{X}_i (X_i' X_i)^{-1} \bar{X}_i'\} \bar{X}_i M^{-1} X_i']$$

$$= X_i M^{-1} [X_i' + \bar{X}_i' \bar{X}_i \{M^{-1} X_i' + (X_i' X_i)^{-1} \bar{X}_i' \bar{X}_i M^{-1} X_i'\}]$$

$$= X_i M^{-1} [X_i' + \bar{X}_i' \bar{X}_i U] \qquad (5.26)$$

where

$$U = M^{-1} X_i' + (X_i' X_i)^{-1} \bar{X}_i' \bar{X}_i M^{-1} X_i'$$

$$= (X_i' X_i)^{-1} (X_i' X_i) M^{-1} X_i' + (X_i' X_i)^{-1} \bar{X}_i' \bar{X}_i M^{-1} X_i'$$

$$= (X_i' X_i)^{-1} [X_i' X_i M^{-1} X_i' + \bar{X}_i' \bar{X}_i M^{-1} X_i']$$

$$= (X_i' X_i)^{-1} [\{X_i' X_i + \bar{X}_i' \bar{X}_i\} M^{-1} X_i']$$

$$= (X_i' X_i)^{-1} [M M^{-1} X_i'], \quad \text{since } M = X_i' X_i + \bar{X}_i' \bar{X}_i$$

$$= (X_i' X_i)^{-1} X_i'.$$

Applying this value of U to Equation 5.26, we get

$$G_1(M,H) = X_i M^{-1}[X_i' + \bar{X}_i'\bar{X}_i(X_i'X_i)^{-1}X_i']$$

$$= X_i M^{-1}[(X_i'X_i)(X_i'X_i)^{-1}X_i' + \bar{X}_i'\bar{X}_i(X_i'X_i)^{-1}X_i']$$

$$= X_i M^{-1}[X_i'X_i + \bar{X}_i'\bar{X}_i](X_i'X_i)^{-1}X_i'$$

$$= X_i M^{-1}M(X_i'X_i)^{-1}X_i', \quad \text{as } M = X_i'X_i + \bar{X}_i'\bar{X}_i$$

$$= X_i(X_i'X_i)^{-1}X_i'. \tag{5.27}$$

and

$$G_2(M,H) = H^{-1}\bar{X}_i M^{-1}X_i'Y_i$$

$$= [I + \bar{X}_i(X_i'X_i)^{-1}\bar{X}_i']\bar{X}_i M^{-1}X_i'Y_i$$

$$= \bar{X}_i[I + (X_i'X_i)^{-1}\bar{X}_i'\bar{X}_i]M^{-1}X_i'Y_i$$

$$= \bar{X}_i[(X_i'X_i)^{-1}(X_i'X_i) + (X_i'X_i)^{-1}\bar{X}_i'\bar{X}_i]M^{-1}X_i'Y_i$$

$$= \bar{X}_i(X_i'X_i)^{-1}[X_i'X_i + \bar{X}_i'\bar{X}_i]M^{-1}X_i'Y_i$$

$$= \bar{X}_i(X_i'X_i)^{-1}MM^{-1}X_i'Y_i$$

$$= \bar{X}_i(X_i'X_i)^{-1}X_i'Y_i. \tag{5.28}$$

Now applying the relationships in (5.27) and (5.28) to Equation 5.25, the matrix R can be expressed as

$$R = Y_i'[I - X_i(X_i'X_i)^{-1}X_i']Y_i + [\bar{Y}_i' - \bar{X}_i(X_i'X_i)^{-1}X_i'Y_i]'H[\bar{Y}_i' - \bar{X}_i(X_i'X_i)^{-1}X_i'Y_i]$$

$$= Y_i'[I - X_i(X_i'X_i)^{-1}X_i']Y_i + [\bar{Y}_i' - \bar{X}_i\hat{\beta}]'H[\bar{Y}_i' - \bar{X}_i\hat{\beta}] \tag{5.29}$$

where $\hat{\beta} = (X_i'X_i)^{-1}X_i'Y_i$ is the OLS estimates of β.

Again, since we have

$$S_Y = (Y_i - X_i\hat{\beta})'(Y_i - X_i\hat{\beta})$$

$$= (Y_i - X_i(X_i'X_i)^{-1}X_i'Y_i)'(Y_i - X_i(X_i'X_i)^{-1}X_i'Y_i)$$

$$= Y_i'Y_i - 2[Y_i'X_i(X_i'X_i)^{-1}X_i'Y_i] + Y_i'X_i(X_i'X_i)^{-1}X_i'X_i(X_i'X_i)^{-1}X_i'Y_i$$

$$= Y_i'Y_i - 2[Y_i'X_i(X_i'X_i)^{-1}X_i'Y_i] + Y_i'X_i(X_i'X_i)^{-1}X_i'Y_i$$

$$= Y_i'Y_i - Y_i'X_i(X_i'X_i)^{-1}X_i'Y_i$$

$$= Y_i'[I - X_i(X_i'X_i)^{-1}X_i']Y_i.$$

Furthermore, using this result in Equation 5.29, we can express the matrix relation R as the convenient form

$$R = S_Y + [\bar{Y}_i' - \bar{X}_i\hat{\beta}]' H [\bar{Y}_i' - \bar{X}_i\hat{\beta}]. \tag{5.30}$$

Applying the expression of R in (5.30) to Equation 5.22, the prediction distribution of the unobserved population units matrix $\bar{Y}_i'$ for the ith small area, conditional on the observed sample units matrix Y_i, can be expressed as

$$f(\bar{Y}_i | Y_i) = C(Y_i, H) \left| S_Y + [\bar{Y}_i' - \bar{X}_i\hat{\beta}]' H [\bar{Y}_i' - \bar{X}_i\hat{\beta}] \right|^{-\frac{n_i + m_i - k}{2}},$$

where the normalizing constant is given by

$$C(Y_i, H) = \frac{(\pi)^{-\frac{m_i p}{2}} \Gamma_p((n_i - k)/2) |H|^{-\frac{p}{2}}}{\Gamma_p((n_i + m_i - k)/2) |S_Y|^{-\frac{n_i - k}{2}}}.$$

Now, using the value of $m_i = N_i - n_i$, the prediction distribution is obtained as

$$f(\bar{Y}_i | Y_i) = C(Y_i, H) \left| S_Y + [\bar{Y}_i' - \bar{X}_i\hat{\beta}]' H [\bar{Y}_i' - \bar{X}_i\hat{\beta}] \right|^{-\frac{N_i - k}{2}}, \tag{5.31}$$

for

$$C(Y_i, H) = \frac{(\pi)^{-\frac{(N_i - n_i)p}{2}} \Gamma_p((n_i - k)/2) |H|^{-\frac{p}{2}}}{\Gamma_p((N_i - k)/2) |S_Y|^{-\frac{n_i - k}{2}}}.$$

The density in (5.31) is the probability density function of a $(N_i - n_i)p$ dimensional matrix-T with the location matrix $\bar{X}_i\hat{\beta}$, scale factors S_Y, H, and shape parameter $n_i - p - k + 1$. Therefore, the unobserved units matrix $\bar{Y}_i$ for the multivariate model has a $(N_i - n_i)p$ dimensional matrix-T distribution. The location matrix of the prediction distribution is

$$\bar{X}_i (X_i' X_i)^{-1} X_i' Y_i$$

and the covariance matrix is

$$\frac{(n_i - p - k + 1)}{(n_i - p - k - 1)} [S_Y \otimes H],$$

where the symbol $\otimes$ represents the Kronecker product of the matrices S_Y and H.

Moreover, with the result of prediction distribution for unobserved population units $\bar{Y}_i$ in (5.31), the joint posterior density of parameters for the observed sample units Y_i and unobserved population units $\bar{Y}_i$ can be determined as

$$f(\beta, \Sigma \mid Y_i, \bar{Y}_i) \cong |\Sigma|^{-\frac{N_i + p + 1}{2}} \left| I_p + \Sigma^{-1} Q \right|^{-\frac{v + p + N_i - 1}{2}} f\left(\bar{Y}_i \mid Y_i\right) \tag{5.32}$$

where $Q = (Y_i - X_i\beta)'(Y_i - X_i\beta) + (\bar{Y}_i - \bar{X}_i\beta)'(\bar{Y}_i - \bar{X}_i\beta)$.

Now, by applying the MCMC simulation method to Equation 5.32, we can obtain simulated copies of the micropopulation data for the ith small area.

The key feature of this new method is that it can simulate complete scenarios of the whole micropopulation in a small area, which means it can produce more reliable small area estimates and their variance estimations. It is also able to create the statistical reliability measures (e.g., the Bayes credible region or confidence interval) for spatial microsimulation models' estimates (Rahman and Upadhaya 2015). This Bayesian prediction–based microdata simulation technique is a probabilistic approach that is different from the deterministic approach used in GREGWT and the intelligent searching tool *simulated annealing* used in combinatorial optimization. Nevertheless, the new approach can adopt the generalized regression model operated in the GREGWT algorithm to link observed units in the sample and unobserved units in the population. In contrast, from the viewpoint of the combinatorial optimization reweighting method, this new system uses the MCMC simulation with a posterior density–based iterative algorithm. As the Bayesian joint posterior probabilities of the parameters for the observed sample units and the unobserved population units are estimated through the MCMC method, the proposed microdata simulation methodology is somewhat linked with a chain Monte Carlo sampling. However, it is rather different from the multiple imputation technique advanced by Rubin (1987) and other researchers. The basic computation process of the new proposed approach is predominantly associated with a prediction distribution of unobserved population units given the sample units.

Moreover, one of the issues with this new approach is that we need to identify a suitable prior distribution for each interested event, as well as the appropriate model for linking observed sample data and unobserved units for each small area. This can be difficult in practice as they may vary with the real-world problems. Typically, in the Bayesian modeling, there are a range of prior distributions including reference priors, vague priors, informative priors, noninformative priors, conjugate priors, and uniform priors to choose to fit the model. It is also quite common that every modeling approach would have to deal with at least few complex tasks, and perhaps they are related to suitable model selection and/or computations. Nevertheless, the prior distribution based on Jeffreys's (1961) invariance theory and the linkage model chosen in this methodology is working decently by producing appropriate results.

5.8 Concluding Remarks

This chapter has pointed out a new approach in the microsimulation methodology for generating spatial microdata at small area level. The technique is based on the Bayesian prediction theory and can simulate complete scenarios of the whole population in each small area. As a result, the process can yield more accurate and statistically reliable small area estimates compared to the estimates from the other reweighting techniques. For instance, as a probabilistic method, the proposed methodology would be able to produce direct variance estimations of the small area estimates and the Bayes credible region or confidence interval for the estimates to confirm its statistical reliability. Furthermore, since the Bayesian prediction–based microdata simulation method is a probabilistic approach and essentially depends on a prediction distribution of unobserved population units given the observed sample units, it is different from the other reweighting approaches such as the GREGWT and combinatorial optimization. It is also different from the multiple imputation technique but to some extent linked with the chain Monte Carlo sampling method.

5.6 Concluding Remarks

6

Microsimulation Modeling Technology for Small Area Estimation

6.1 Introduction

From the comprehensive appraisals of small area estimation methodologies in the previous chapters, it is clear that microsimulation modeling technology (MMT)-based spatial microsimulation modeling techniques can be very useful and efficient for small area estimation. The review of various methodologies also indicated that spatial microsimulation model–based indirect small area estimation has some significant advantages over the other statistical model–based indirect as well as direct estimation methods. For instance, the spatial microsimulation modeling techniques for small area estimation not only are robust and able to produce a point estimate but also can generate valuable spatial-scale microdatabases for further useful analyses as well as can determine confidence intervals (CIs) of the point estimates. These models can further measure the small area effects of policy changes.

However, developing an effective MMT is dependent upon various factors within the microsimulation model, such as the quality of the initial applicable data, specification of the model, variable selection, simulation of the synthetic microdata, and actual production of small area estimates. One of the key tasks in the spatial microsimulation approach to small area estimation is to *reweight* the survey sampling weights or select individual units from an initial national-level microdata set to "fill in" the micropopulation for small census areas. Moreover, the validation of the model estimates is another significant part of spatial microsimulation modeling for justifying the reliability of the model estimates.

Typically, a spatial microsimulation model that has a reweighting methodology comprises two main phases of computations: (1) generation of small area synthetic weights by reweighting and (2) production of small area estimates by creating synthetic spatial microdata. This chapter gives a detailed description of the entire process involved in building an MMT for small area housing stress estimation in Australia.

The arrangement of this chapter is as follows. Various data sources and significant issues related to the applicable data sets are presented in Section 6.2. Specification of the model including variable selection in the process of generating synthetic weights for spatially disaggregated microdata is described in Section 6.3. The definition of housing stress measures with a comparison of different definitions is summarized in Section 6.4. The procedures for developing synthetic spatial microdata and then the small area housing stress estimates are explained in Section 6.5. The concluding remarks of the chapter are given in Section 6.6.

6.2 Data Sources and Issues

Data are important parts of building a spatial microsimulation model. Reliable and more appropriate initially used data can generate much accurate spatial microdata. To create synthetic microdata at small area level through reweighting is very challenging. This is due to not only the lack of a more suitable reweighting technique but also the availability of useful initial data sets. Characteristically, a range of data sets used for spatial microsimulation modeling come from distinctive sources and with very special formats. For example, the benchmark data tables used in the reweighting process are from the recent census of Australia, whereas the national-level unit record data files essential for the reweighting process and microdata generation come from the survey of income and housing (SIH) conducted by the Australian Bureau of Statistics (ABS) for different years (or other sample surveys). This section gives an account of the various data sources and also highlights some important issues related to the data sets used in the research.

6.2.1 Census Data

A census is a device for counting the entire population in a national territory and recording their characteristics. In Australia, the census is undertaken by the ABS and known as the census of population and housing, which typically collects a range of demographic, social, and economic information on census night from the people and dwellings within Australian states and territories (ABS 2006). Initially, the census was conducted in Australia every 10 years from 1911 with some irregularities due to the world wars, but since 1961, it has been carried out every 5 years on a regular basis. The latest census of population and housing was held on August 8, 2006. It is the most important source of statistical information in the country on various vital topics (see Table B.1).

Although the census utilizes a fairly simple questionnaire, in recent decades, social science researchers are making increasingly sophisticated use of the comprehensive data on national, state, local, and small area populations provided by the census. The topics listed in Table B.1 are covered by the 2006 census of population and housing in Australia, and they allow the generation of results in various formats and for all geographic areas upward from census collection districts that are suitable to apply to different academic research. For instance, social science researchers use census data that are available in a cross-tabulated computer format for a variety of spatial scales, specifically for "small areas" such as statistical local areas (SLAs) or collection districts used in the collection of the census data; for "administrative areas" such as districts, counties, regions, and countries; for "postal areas" such as unit postcodes, postal sectors, postal districts, and postal areas; and for "electoral areas" such as suburbs/wards within local government areas and parliamentary constituencies (Rees et al. 2002). However, the output is modified when some numbers are involved (which may put at risk the privacy of the respondent) and raw microdata itself are not released because of respondent confidentiality. Data are thus released only for a certain level of small areas (e.g., SLA or suburb/ward level) and are not available at the individual or household level except for 1% sample file.

The outputs of the census are counts of persons, families, and dwellings broken down by their demographic, social, and economic characteristics as well as into a range of geographic/spatial scales (ABS 2008a). These are contained in a series of tables or maps on a specific topic or area of interest. The 2006 census of population and housing in Australia provides the following types of aggregate output data sets (ABS 2007a):

- *QuickStats*: The QuickStats data sets provide an overview and summary of the key census data on various important census topics for a chosen area, benchmarked against Australia. These data are available for the full range of geographic scales down to census collection districts used in the 2006 census of population and housing.

- *MapStats*: The MapStats data sets provide a series of thematic maps showing the distribution of census data for a chosen location. The maps are designed to contain information about various key subjects related to particular themes available down to SLA level in Australia. They are not available for all spatial scales such as at census collection districts.

- *Census tables*: The census table data sets provide the standard individual tables of census data, which are available on a range of topics, for a particular geographic location. An individual census table contains key characteristics of persons, families, and dwellings, covering most of the census topics. The available census tables are designed to assist in researching, planning, and analyzing the comprehensive scales of geographic areas down to census collection districts, enabling comparisons to be made between different areas on a number of social, economic, and demographic variables.

- *Community profiles*: The community profiles series include six separate profile data sets, which are the *basic community profile (BCP)*,
  place of enumeration profile (PEP), *indigenous profile (IP)*, *time series profile (TSP)*, *expanded community profile (XCP)*, and *working population
  profile (WPP)*. Each of these data sets is a collection of various tables
  showing census data in detail on various socioeconomic and demographic topics, for a chosen location or small area. In particular, the
  XCP data sets are the most comprehensive community profile data
  in this series, which are providing extended information on the key
  census characteristics of persons, families, and dwellings. All the
  community profile series data are available for the full scales of geographic areas down to census collection districts used in the 2006
  census of population and housing.

All of these data sets are available via *Census Data Online* on the ABS website
(http://www.abs.gov.au/websitedbs/D3310114.nsf/home/Census+data).

Moreover, although the community profile series include six separate profile data sets, this study uses only the *BCP* as well as the *XCP* data sets tables.
Note that the BCP data sets are a collection of 45 tables containing detailed
data on various socioeconomic and demographic topics for a chosen location
or small area, while the XCP data sets consist of 42 tables that are providing
extended information on the key census characteristics of persons, families,
and dwellings. For example, the topics covered in BCP data tables are exhib-
ited in Table B.2. It is noted that the BCP tables B01, *B02*, *B22*, B36, and B38
have been corrected due to errors in the first release. The corrected version
of bold-marked BCP tables has been reissued, and other replacement tables
will be supplied on request.

6.2.2 Survey Data Sets

The SIH produced by the ABS is one of the largest social research surveys,
which collects detailed information about income and personal and household characteristics of persons who are 15 years and older and resident in private dwellings throughout Australia. The SIH was conducted continuously
from 1994–1995 to 1997–1998 and then in 1999–2000, 2000–2001, 2002–2003,
2003–2004, and 2005–2006. The latest available SIH was conducted on a sample of approximately 10,000 dwellings over the period July 2005–June 2006
through a stratified, multistage cluster sampling design. Computer-assisted
interviewing was used to carry out household and personal interviews
for detailed data from the selected dwellings. The interviews were spread
equally over the enumeration period, and all information gathered from
9961 households was included in the final sample.

The content and methodology of the 2005–2006 SIH were largely a repeat
of that used in the 2003–2004 SIH, including the collection of detailed information on assets and liabilities that was first made in the 2003–2004 SIH.

A notable change in the 2005–2006 survey is that there was no methodology for the household expenditure subsample data. However, it was initiated in the 2003–2004 survey for the subsample on household expenditure. More detailed information about SIH including the concepts, definitions, methodology, and estimation procedures used in the survey is given elsewhere (e.g., ABS 2007b).

In the SIH, the respondents are asked about a range of topics, including demographic, socioeconomic, person, and household issues and geography characteristics. The SIH provides an account of different socioeconomic and housing estimates at the national and/or state level. It is also a rich source of information to produce detailed microlevel data. For instance, using SIH, the ABS produces an individual microlevel data set known as "confidential unit record file" (CURF), which covers a selected but still wide range of topics. The 2005–2006 SIH-CURF has covered complete issues of person, income unit, and household characteristics (such as geography or area characteristics, demographics, education, and labor force status [LFS]); detailed housing information; income from various sources; and wealth information (see Table B.3). Various SIH-CURF data are used in our analysis, mainly to build up the spatial microsimulation model for housing stress estimates.

However, there are limitations with the SIH. Mainly, the SIH only surveys people aged 15 years and older in private households. Therefore, it does not include data against people younger than 15 years old and it does not cover the population resident in nonprivate dwellings (NPDs) (such as hospitals, institutions, nursing homes, hotels, and hostels) and dwellings in collection districts defined as very remote. In addition, while the SIH provides a picture of different estimates at the national and/or state level, it cannot provide us information about what is happening in detailed regional levels or at the very small area levels. Therefore, the SIH cannot identify small area hot spots for interesting social issues. Moreover, although CURF data can produce detailed records relating to almost all of the survey respondents at microlevel, it has some more additional restrictions.

First, there is no identification (e.g., names or addresses) of survey respondents on the CURFs. Second, to protect the confidentiality of respondents, the maximum household size is restricted to 6 in the basic CURF and to 8 in the expanded CURF, by removing persons from all larger-size households. This has an impact on age groups, as well as some income unit (mainly single person) records. In addition, the level of detail for many data items has been reduced in the CURFs. For example, the state of usual residence of the Australian Capital Territory (ACT) and the Northern Territory (NT) has been combined as ACT/NT for the basic CURF (but shown individually for the expanded CURFs), and the area of usual residence for the ACT and NT has not been made available on the CURFs. Third, all income items, some relating to housing expenditure and some to loan data, have been perturbed. Moreover, modifications have been made to some records to protect against respondents' classification and to household-level variables and/or person-level variables such as state, area, remoteness, age, educational qualifications, industry, and/or occupation. Therefore, the aggregated data obtained from the CURF are somewhat different from the published data obtained from SIH.

Nonetheless, according to the ABS reports (e.g., ABS 2008b), steps taken to confidentialize the data sets, which are available on the CURF, are undertaken in such a way as to ensure the integrity of the data sets and optimize the content, while maintaining the confidentiality of respondents. Although the ABS can provide their client with specific data on request, purchasers of a CURF should ensure that the data they require at the level of detail they desire are available on the CURF. Data obtained in the survey but not contained on the CURF can be given by the ABS in tabulated form on special request. Note that some data sets used in this study are supplied by the ABS on special requests from the National Centre for Social and Economic Modelling (NATSEM) at the University of Canberra and also through the Australian Research Council research network for "Spatially Integrated Social Science (SISS)" at the University of Queensland, Brisbane, Queensland, Australia.

6.3 Microsimulation Modeling Technology–Based Model Specification

An initial challenging step in MMT is to create synthetic weights for micropopulation data at a spatial scale such as SLAs in Australia, which comprises a list of individuals along with an associated set of exclusive characteristics and which properly fits to the known constraints/benchmarks at that spatial level. The main task here is to *re*weight or allocate appropriate individuals available from unit record data to fill up selected small areas by a unique set of characteristics. In general, the procedure could be done by calibration, which can be looked at as a way of either improving estimates or making the estimates add up to known small area constraints or benchmarks (Bell 2000a). That is, the grossing factors or weights on a data set containing the survey returns are fittingly modified so that certain estimates agree with externally provided known totals. This use of external or auxiliary information typically improves the resulting survey estimates that are produced using the modified grossing factors. Evidence from previous studies (e.g., refer to Taylor et al. 2004; Chin and Harding 2006, 2007; Tanton 2007) has shown that the generalized regression weighting tool (GREGWT) reweighting technique works effectively in terms of calibrating the individual-level sample survey records, which best fits to small area statistical constraints. Therefore, this study utilizes the GREGWT approach of reweighting to generate simulated spatial microdata for populations of SLAs in Australia.

The GREGWT program basically is implemented as a calibration tool over the survey weights to produce more reliable spatially disaggregated population microdata at the small area level. Specially for this study, the implementation of the spatial microsimulation modeling for housing stress estimation in Australia involves appropriately calibrating the survey weights for individuals from 2005 to 2006 SIH-CURF microdata, which best fits the known benchmark constraints in small areas from the 2006 census of Australia. For instance, the ABS calculates a weight (or "expansion factor") for each of the 9961 households included in the 2005–2006 SIH sample file (ABS 2007b). Thus, if the household number 1 is given a weight of 1000 by the ABS, it means that the ABS considers that there are 1000 households with comparable characteristics to the household number 1 in Australia. These weights are used to move from the 9961 households included in the SIH sample to estimates for the 8.4 million households in Australia.

The process can be described in the following way. Let 15,665,300 people be living in 6,262,700 separate house private dwellings as found in the 2006 census of population and housing in Australia, and they are restructured according to the small areas such as SLAs. The GREGWT reweighting procedure involves calibrating the survey weighting records of individuals from 2005 to 2006 SIH-CURF data and redistributing them (repeatedly many times) in small areas by using valuable auxiliary information until the aggregate statistics for each SLA generate a weighted version of SIH-CURF that matches the aggregate SLA figures from the census. By the end of this reweighting process, the product is an individual-level spatially disaggregated database for the small areas constrained by the census benchmark figures. Indeed, the construction of this spatial population database is required if detailed small area population analysis is desired, since the individual-level census data are not available due to the respondents' confidentiality restrictions, as well as there being only a very limited choice of variables. It is also essential for the next stage of the model, for small area estimation by spatial microsimulation modeling.

6.3.1 Model Inputs

To simulate household *synthetic weights* by the GREGWT reweighting algorithm for producing spatially disaggregated synthetic micropopulation data, we have to run the housing stress model. There are two stages of the model: the first stage simulates the small-area-level household synthetic weights and the second stage calculates the ultimate small area housing stress estimates by producing the synthetic small area microdata. The following five groups of files are required to run the first stage of the model:

1. General model file
2. Unit record data files

3. Benchmark files
4. Auxiliary data files
5. GREGWT program file

6.3.1.1 General Model File

The general model file is a statistical analysis system (SAS) file, which links up all other files to run the reweighting process. We term this file *SPATIALMSM*. It contains the program of the path to the unit record data files, benchmark constraints, auxiliary data files, and the GREGWT program file. In addition, the file specifies group variable IDs and filter definitions for the GREGWT program file. The filter definitions utilize specific logic operations and conditions to define the convergence conditions to benchmarks for each small area. Moreover, this file also defines a pathway to outputs folder.

6.3.1.2 Unit Record Data Files

The unit record data available in Australia are produced by the ABS and often known as survey-based national-level *microdata files*. The ABS releases them as CURFs, which are based on national-level sample survey data, and the confidentiality requirements of respondents have been properly maintained. For example, based on the 2005–2006 SIH data, the ABS has released the recent SIH-CURF microdata, which have been used in this analysis. However, to run the GREGWT algorithm, we need to readjust the ABS-provided microdata file to fit it with the benchmark constraints files. To make the study variables from the survey data compatible with the census data, the following selected variables in the 2005–2006 SIH file were checked by its order and classifications with the NATSEM's SIH-linkage file to determine whether an individual/household fits to the benchmark tables from the census data or matches the classifications that exist in the census data. This is necessary for executing the GREGWT algorithm to obtain a reasonable set of new weights.

Age: Age of respondent
   1 = Younger than 15 years
   2 = 15–24 years
   3 = 25–54 years
   4 = 55–64 years
   5 = 65+ years
Sex: Sex of respondent
   1 = Male
   2 = Female

LFS: *Labor force status*

   1 = Employed (full time/part time [FT/PT])

   2 = Not in labor force

   3 = Not applicable (for age group <15 years/child)

Income: Weekly income of respondent

   1 = below 250 AUD

   2 = 250–649 AUD

   3 = 650–999 AUD

   4 = 1000–1699 AUD

   5 = 1700+ AUD

Rent: Weekly rent

   1 = 0–139 AUD

   2 = 140–224 AUD

   3 = 225–349 AUD

   4 = 350–549 AUD

   5 = 550+ AUD

Mortgage: Weekly mortgage repayment

   1 = 1–949 AUD

   2 = 950–1999 AUD

   3 = 2000+ AUD

HHFC: Household family composition

   1 = "Family with kids"

   2 = "Family without kids"

   3 = "Single-parent family"

   4 = "Other family"

   5 = "Alone person"

   6 = "Group"

NPHH: No. of persons in a household

   1 = "One"

   2 = "Two"

   3 = "Three"

   4 = "Four"

   5 = "Five and more"

Dwelling: Dwelling structure

   1 = Separate house

   2 = Semidetached, row, or terrace house

3 = Flat, unit, or apartment

4 = Other dwelling

Type: Tenure type

1 = "O" owner without a mortgage

2 = "P" owner/buyer with a mortgage or repayment

3 = "Rpub" rent at public housing

4 = "Rpriv" rent at private housing

5 = "Other"

6.3.1.3 Benchmark Files

Benchmark files are the constraints tables against which the survey microdata are reweighted. According to Chin and Harding (2006), benchmarks are the census counts that the survey estimates produced by reweighting are constrained to meet. In a reweighting process, using a different set of benchmark files would produce different results. A rule of thumb is that the more constraint variables or benchmarks utilized in reweighting, the better the synthetic microdata produced. However, the selection of benchmarks needs to balance two opposing requirements (Chin and Harding 2006). On one side, we need to choose a maximum number of benchmarks and their classes because they will increase the level of information in our estimates. On the other side, we have to bear in mind that an increase in benchmarks and benchmark classes will increase the complexity of the reweighting process, and this may reduce its ability to meet the census count. On top of these, more time will have to be spent running the model as the more and higher degree of comparisons with the real data will be required for each additional benchmark.

A benchmark can be a 1D variable (e.g., *Sex* of respondent) or a combination of multiple variables (such as *Age by Sex by LFS* of the respondent). Each benchmark contains a number of classes. In particular, for a single-variable benchmark *Sex*, the classes will be "male" and "female," or for the multidimensional/multivariable benchmark *Age by Sex by LFS*, a class would be "employed_25–54 years_male." A variable or a combination of variables chosen in the benchmarks for the reweighting process should be one of the best groups of predictors to the interested outcomes, such as housing stress estimation in our case. It is worthwhile to mention that the regression method of analysis can identify the best set of predictor variables. The benchmarks used in this analysis and its variable classification are provided in Table 6.1.

6.3.1.4 Auxiliary Data Files

As noted earlier, the ABS can also provide user-specific data on special requests. Here, in the NATSEM, we use a supplementary microdata file called

TABLE 6.1

Benchmarks and Benchmark Classes Used in Reweighting

Sex by Age by Labor Force Status	Tenure Type by Household Type	Tenure Type by Weekly Household Income
(1) Male, aged 0–14 years, not applicable	(1) Fully owned, family	(1) Fully owned, weekly income below $250
(2) Male, aged 15–24 years, employed FT/PT	(2) Fully owned, individual	(2) Fully owned, weekly income $250–$649
(3) Male, aged 15–24 years, not in labor force	(3) Fully owned, group	(3) Fully owned, weekly income $650–$999
(4) Male, aged 25–54 years, employed FT/PT	(4) Buyer with a mortgage, family	(4) Fully owned, weekly income $1000–$1699
(5) Male, aged 25–54 years, not in labor force	(5) Buyer with a mortgage, individual	(5) Fully owned, weekly income $1700 and over
(6) Male, aged 55–64, years employed FT/PT	(6) Buyer with a mortgage, group	(6) Buyer with a mortgage, weekly income below $250
(7) Male, aged 55–64 years, not in labor force	(7) Renter public, family	(7) Buyer with a mortgage, weekly income $250–$649
(8) Male, aged 65 years and older, employed FT/PT	(8) Renter public, individual	(8) Buyer with a mortgage, weekly income $650–$999
(9) Male, aged 65 years and older, not in labor force	(9) Renter public, group	(9) Buyer with a mortgage, weekly income $1000–$1699
(10)–(18) Female—repeated as above	(10) Renter private, family	(10) Buyer with a mortgage, weekly income $1700 and over
	(11) Renter private, individual	(11) Renter public, weekly income below $250
	(12) Renter private, group	(12) Renter public, weekly income $250–$649
	(13) Other tenure, family	(13) Renter public, weekly income $650–$999
	(14) Other tenure, individual	(14) Renter public, weekly income $1000–$1699
	(15) Other tenure, group	(15) Renter public, weekly income $1700 and over
		(16) Renter private, weekly income below $250
		(17) Renter private, weekly income $250–$649

(Continued)

TABLE 6.1 (*Continued*)

Benchmarks and Benchmark Classes Used in Reweighting

Sex by Age by Labor Force Status	Tenure Type by Household Type	Tenure Type by Weekly Household Income
		(18) Renter private, weekly income $650–$999
		(19) Renter private, weekly income $1000–$1699
		(20) Renter private, weekly income $1700 and over
		(21) Other tenure, weekly income below $250
		(22) Other tenure, weekly income $250–$649
		(23) Other tenure, weekly income $650–$999
		(24) Other tenure, weekly income $1000–$1699
		(25) Other tenure, weekly income $1700 and over
		Mortgage Repayment by Weekly Household Income
		(1) Mortgage $1–$949, weekly income below $500
		(2) Mortgage $1–$949, weekly income $500–$999
		(3) Mortgage $1–$949, weekly income $1000+
		(4) Mortgage $950–$1, weekly income below $500
		(5) Mortgage $950–$1, weekly income $500–$999
		(6) Mortgage $950–$1, weekly income $1000+
		(7) Mortgage $2000+, weekly income below $500
		(8) Mortgage $2000+, weekly income $500–$999
		(9) Mortgage $2000+, weekly income $1000+
Weekly Rent by Household Income	**Family Composition by Dwelling Structure**	
(1) Rent $0–$139, nil or –ve income	(1) Family with kids in a separate house	
(2) Rent $0–$139, income $1–$649	(2) Family without kids in a separate house	
(3) Rent $0–$139, income $650–$1399	(3) Single-parent family in a separate house	
(4) Rent $0–$139, income $1400+	(4) Other family in a separate house	
(5) Rent $140–$224, nil or –ve income	(5) Lone person in a separate house	
(6) Rent $140–$224, income $1–$649	(6) Group in a separate house	
(7) Rent $140–$224, income $650–$1399	(7) Family with kids in a semidetached, row, or terrace house	
(8) Rent $140–$224, income $1400+	(8) Family without kids in a semidetached, row, or terrace house	
(9) Rent $225–$349, nil or –ve income	(9) Single-parent family in a semidetached, row, or terrace house	
(10) Rent $225–$349, income $1–$649	(10) Other family in a semidetached, row, or terrace house	

(*Continued*)

TABLE 6.1 (*Continued*)

Benchmarks and Benchmark Classes Used in Reweighting

Weekly Rent by Household Income	Family Composition by Dwelling Structure	Mortgage Repayment by Weekly Household Income
(11) Rent $225–$349, income $650–$1399	(11) Lone person in a semidetached, row, or terrace house	
(12) Rent $225–$349, income $1400+	(12) Group in a semidetached, row, or terrace house	
(13) Rent $350–$549, nil or –ve income	(13) Family with kids in flat, unit, or apartment	
(14) Rent $350–$549, income $1–$649	(14) Family without kids in flat, unit, or apartment	
(15) Rent $350–$549, income $650–$1399	(15) Single-parent family in flat, unit, or apartment	
(16) Rent $350–$549, income $1400+	(16) Other family in flat, unit, or apartment	
(17) Rent $550 and over, nil or –ve income	(17) Lone person in flat, unit, or apartment	
(18) Rent $550 and over, income $1–$649	(18) Group in flat, unit, or apartment	
(19) Rent $550 and over, income $650–$1399	(19) Family with kids in other dwelling	
(20) Rent $550 and over, income $1400+	(20) Family without kids in other dwelling	
	(21) Single-parent family in other dwelling	
	(22) Other family in other dwelling	
	(23) Lone person in other dwelling	
	(24) Group in other dwelling	

All Household Types	No. of Persons in Household	Persons in Nonprivate Dwelling	Weekly Rent of Household
(1) Total no. of households	(1) One person	(1) Hotels, motels, boarding houses, and hostels	(1) Weekly rent 0–139 AUD
	(2) Two persons	(2) Homes for the aged and disabled	(2) Weekly rent 140–224 AUD
	(3) Three persons	(3) Hospitals (including psychiatric)	(3) Weekly rent 225–349 AUD
	(4) Four persons	(4) All other NPDs	(4) Weekly rent 350–549 AUD
	(5) Five+ persons		(5) Weekly rent 550 AUD and over

the "SIH-linkage" file that is created by using a range of data files including the SIH-CURFs (basic and/or expanded CURFs), specially requested data files from the ABS, and the useful information from the census data. In brief, the SIH-linkage file is a set of unit record data at household and personal levels that consist of the unit record files of SIH-CURFs (2003–2004 and 2005–2006) compiled with 1% of the recent population and housing census data to provide information about households living in NPDs (see Cassells et al. 2010 for more details). Note that typically the available sample survey data sets do not contain information about NPDs. Moreover, the formation of benchmarks also considerably depends on the variables in the SIH-linkage file.

6.3.1.5 GREGWT File

The GREGWT file is the program file of the GREGWT algorithm written by Bell (2000a) in SAS macro. In practice, this algorithm modifies initial weights used in a survey sampling to create new weights that aggregate to known constraints tables or benchmarks. As described earlier, the GREGWT algorithm is using the generalized regression calibration, and it is based on a truncated chi-squared distance measure. Note that truncation or range restriction on the size of the weights is applied in the GREGWT process to avoid the problem of generating negative weights. Besides, the GREGWT program logically follows the Newton–Raphson method of iteration to achieve convergence against each small area. However, in such a case, if convergence is not achieved after a prespecified number of iteration in the program, then the benchmark figures will not be met, rather than the range restriction being ignored. Indeed using the GREGWT algorithm, the weights should always meet the lower and upper range restrictions. Moreover, the SAS macro file for GREGWT reweighting applies the algorithm in such a way that it can function without any stage storing the intermediate weights produced at each iteration. This certainly allows an efficient computation of all iterations within a single SAS data step, without storing the weights in a computer's internal memory, and this may also minimize the running time of the overall iteration process.

6.3.2 Generating Small Area Synthetic Weights

An overall process of GREGWT reweighting to the creation of small area synthetic weights is depicted in Figure 6.1.

Clearly, the process starts by running the general model file in the SAS language, which contains the path to all input data files and the GREGWT algorithm. It also specifies the link of a folder where all output files could be stored. Main calculations of the iteration process for the GREGWT algorithm

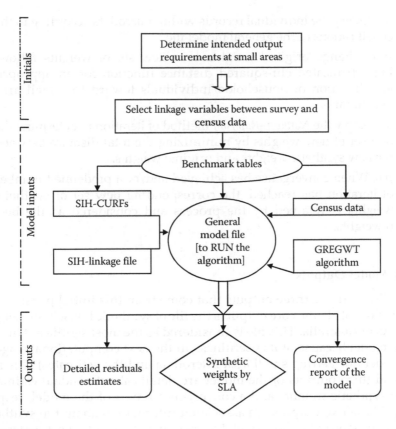

FIGURE 6.1
Overall process for the creation of small area synthetic weights.

go separately with each *id* number of small areas (i.e., SLA codes). This complex process tracks on numerous matrices and/or vectors calculations toward achieving convergence for each SLA by the minimum number of iterations. Besides, it also performs analysis for extreme data units to notice whether the extreme units have effects on the overall calculations. However, the output keeps records on only the top 30 extremes.

Note that although the GREGWT program follows the Newton–Raphson approach of iteration, the entire execution process of the model follows few successive algorithmic steps. In general, these successive steps can be described as follows:

Step 1: First of all, read in general model file.

Step 2: Read in benchmark tables, census data, and microdata records from SIH-CURFs with the SIH-linkage file mentioned in the general model file.

Step 3: Query the individual records within microdata according to the classifications of the general model file.

Step 4: Change original weights to a new set of weights following a truncated chi-squared distance function for an appropriate allocation of households/individuals toward the small area benchmarks.

Step 5: Apply the *Newton–Raphson* method of iteration to determine the best set of new weights by minimizing the total distance between the new synthetic weights and original weights.

Step 6: When convergence has achieved and/or a predefined number of iteration has reached, the corresponding new set of synthetic weights is retained by the process and considered as the best reweights.

6.3.3 Model Outputs

Basically, there are three outputs that come from this initial phase of the model. First of all, the core output is the file of synthetic household weights by SLAs in Australia. This file is considered as the most significant output of the model because of its usefulness in the next computational stage of the model (for getting small area microdata and the estimates). The second and third outputs of the model are details about residual estimates of the synthetic weights and a convergence report of the model, respectively. These two outputs are associated information about the synthetic weights produced by the model. For example, the residual estimates' file shows the accuracy of the new weights according to various benchmark classifications. In the spatial microsimulation process, a modeler's expectation is to minimize the overall residual estimates as much as possible to ensure the consistency and reliability of the synthetic weights. In addition, the convergence report reveals the information about whether the GREGWT reweighting algorithm has converged to the benchmarks or not for a specific SLA. When the convergence rate seems reasonably low, then the modeler may need to revisit the specification of the model for modification.

Moreover, it is noted that the "synthetic weights" file (see Table 6.2) is the central demand in microsimulation modeling approach of small area estimation. The synthetic weights output file is often known as synthetic or simulated spatial microdata new weights, and it is the only output to be used in the next stage of the model for producing ultimate small area estimates. If this stage of the model can generate more accurate synthetic weights at small area levels, then the final small area estimates of interests would likely be statistically more reliable.

TABLE 6.2

Outlook of the Household *Synthetic Weights* Produced by the GREGWT Algorithm for Small Area Microdata

	Turning the National-Level Household Weights in the Data of SIH 03–04 and 05–06 Unit Records Files into...					*Household Synthetic Weights for the SLA-Level Microdata*			
Unit Records	Household ID	Weekly Income	Weekly Rent	Other Variables	Household Weight	NSW SLA1	NSW SLA2	NSW SLA3	Other SLAs
1	1	7	3	.	1029	0	10.2	0	.
2	2	11	4	.	157	0	0	0	.
3	3	11	4	.	157	0	0	0	.
4	4	11	4	.	157	0	0	0	.
5	5	11	0	.	1003	2.45	9.64	16.38	.
6	6	11	0	.	1003	2.45	13.54	16.38	.
7	7	10	4	.	70	0	13.54	12.86	.
8	8	12	4	.	70	0	0	0	.
9	9	12	0	.	703	3.27	0	0	.
10	10	12	0	.	703	3.27	0	9.54	.
11	11	9	2	.	851	4.61	0	9.54	.
.	.	.	.	.	.	.	.	.	.
.	.	.	.	.	.	.	.	.	.
.	.	.	.	.	.	.	.	.	.
53,220	53,220				8.4 million No. of households in Australia	12,465	25,853	27,940	
						No. of households in small areas			

GREGWT reweighting

6.4 Housing Stress

6.4.1 Definition

Typically, *housing stress* describes a financial situation of households where the cost of housing, either as rental or as a mortgage repayment, is considered to be significantly high relative to household income. In other words, households with relatively low income and housing costs greater than a certain proportion of household income (e.g., 30% or more) are said to be in housing stress. It may also be used to describe inadequate housing for a proportion of the population. A range of definitions for describing the situation of housing stress are available in the literature. The following subsections will discuss all methods of measuring housing stress with a comparison of different definitions.

6.4.2 Measures of Housing Stress

Housing stress can be measured by combining two basic quantities: the income and the expenditure of a household. A household can be considered under housing stress when it is spending more than an affordable expected percent of its household income on housing. The affordable expected cutoff point of housing expenditure can vary by the circumstance of households as well as the location of dwelling.

As a general rule of thumb, a household spending 30% or more of its income on housing can be considered under housing stress (see King 1994; Landt and Bray 1997). A different threshold of housing expenditure can be used by researchers, or in many cases, they may restrict the definition to households within different income quintiles. For example, more than 25% of income threshold for housing costs is used by the National Housing Strategy (1991) and Foard et al. (1994). Additionally, a commonly used definition of housing stress is specified in Harding et al. (2004), where more than 30% of threshold of housing costs was used, but only for the households having income in the bottom 40% (lowest two quintiles) of the equivalised income distribution. By this definition, families and singles are considered as being in housing stress if their estimated housing costs are more than 30% of their disposable income and they are in the bottom 40% of the equivalised income distribution. Another definition restricts the designation of "being in housing stress" to those households spending more than 30% of their income on housing and belonging to the bottom 10th to 40th income percentile of the income distribution (ABS 2005). It is noted that any threshold-based definition is an arbitrary slice through a continuum, meaning that small-area-level estimates of a percentage of households in housing stress would be better treated as estimates of small areas with the greatest percentage of households in housing stress. More explicitly, if an area has a very high percentage of households suffering from housing stress with the definition given earlier, the area probably ranks high on the percentage of households suffering from housing stress however defined.

A good presentation of measuring housing stress is given by Nepal et al. (2008). The authors describe the definitions of housing stress by three *rule-based* variants, which are as follows:

1. *30-only rule*: A household is considered to be in housing stress if it spends more than 30% of its disposable or gross income on housing costs.

2. *30/40 rule*: A household is considered to be in housing stress if it spends more than 30% of its disposable or gross income on housing costs and the household also belongs to the bottom 40% of the equivalised disposable income distribution.

3. *30/(10–40) rule*: A household is considered to be in housing stress if it spends more than 30% of its disposable or gross income on housing and falls into the bottom 10th to 40th income percentile of the equivalised disposable income distribution.

Although the cutoff point of housing costs for all these definitions is the same, there are some concerns associated with each of these rules. For example, which one (gross income or disposable income) will be the appropriate base income to calculate housing costs for measuring housing stress? It is worthwhile to mention that gross income is the income of a household from all sources before deducting tax and the Medicare levy, whereas disposable income is the income that remains to a household after deducting the estimated personal income tax and the Medicare levy from gross income. If a researcher uses 30% of gross income as a base, then after possible deductions that figure may turn into around 40%–45% of actual disposable income. Hence, 30% of gross income should equate to a reasonably high proportion of actually received income for housing and other costs. In addition, the *30/40* and *30/(10–40) rules* both restrict the definition to those households that are within the bottom 40% of the equivalised income distribution. The issue here is: "Why is the cutoff point at the lowest 40% of income distribution?" For the latter case, why are households in the bottom 10% of the equivalent income distribution dropping?

There is a common fact that when individuals have a higher income, they have a greater choice of how to spend it. For lower-income households, almost all of their income may be spent on basic necessities, including food, clothing, and housing. This group is at a higher risk of not being able to afford increasing housing costs or they may not have any choice on housing. For the higher-income households, paying more than 30% of income on rent or a mortgage is more likely to be a choice, perhaps to live in a more convenient or desirable area or to pay off extra on mortgage to shorten the term of payment. However, there is a possibility that the households in the third quintile (40th to 60th income percentile) of the income distribution, who usually are known as "middle class earners," may also have financial hardship in meeting high housing costs and may have only limited choices to do with housing. By choosing the bottom 40% of income distribution as the cutoff, the middle class earning households are excluded from the definitions. Although the middle class–income households are at a lower risk of housing stress than the low-income households, they may be at a level of "marginal housing stress" because a substantial rise in interest rates, housing prices, job loss, etc., may cause them to fall into housing stress. Moreover, the 40% cutoff is the same regardless of the area in which the individual or household unit is living. Hence, no account is taken of housing costs that vary with location, for example, the high rents of Canberra and Sydney compared to the low rents of Adelaide are not taken into account in these definitions.

A very severe form of housing stress is the risk of homelessness and may apply to households in the lowest 10% of income distribution. This group is quite vulnerable to rising housing costs. Note that many homeless are homeless due to a situation of financial hardship where individuals are unable to afford housing costs or to keep a place to live. Rapidly increasing housing costs could force more of the lowest earning households into homelessness. Therefore, the exclusion of households within the lowest-income decile from the definition (30/(10–40) rule) may overlook the severe form of housing stress. This definition could also not be used as a means of strategic policy intervention for poverty and housing assistance programs due to its exclusion of mostly disadvantage households. However, some studies do argue that the reported incomes of the households in the bottom 10% of the income distribution do not always accurately reflect their living standards and their inclusion in the definition may overestimate housing stress (see ABS 2005; Rahman and Harding 2014), which is why the ABS argues for the 30/(10–40) rule.

6.4.3 Comparison of Various Measures

A comparison of the three rules of measuring housing stress is provided in Table 6.3. Note that none of these definitions takes into account the fact that housing costs vary according to the area. The specified rules are using the relative income of household and the general rule (30 only) is using the absolute household income. The 30/40 rule is the widely used definition of housing stress in Australia. Although this definition may ignore marginal housing stress, it acknowledges the size of the household income unit by using the equivalised household income distribution.

TABLE 6.3

Comparison of the Different Measures of Housing Stress

30-Only Rule	30/40 Rule	30/10–40 Rule
General definition—"a household is in housing stress if it spends more than 30% of its income on housing costs."	Specified definition—"a household is in housing stress if it spends more than 30% of its income on housing costs and the household also belongs to the bottom 40% of the equivalised income distribution."	More specified definition—"a household is in housing stress if it spends more than 30% of its income on housing and places into the bottom 10th to 40th income percentile of the equivalised income distribution."
It assesses all forms of housing stress in one flag.	It ignores any *marginal* housing stress.	It ignores both the *marginal* and *severe* housing stress.
Only the absolute household income is considered.	The relative income of the household is taken into account.	The relative income of the household is used.
It is free from equivalised household income cutoff.	It is based on equivalised household income cutoff by the bottom 40%.	It is based on equivalised household income between 10 and 40 percentiles.
It has been used in Australia in the past.	It is widely used in Australia.	It is used in a few occasions.
No account is given to the size of the income unit.	Proper treatment is given of the size of the household income unit.	Proper treatment is given of the size of the household income unit.

Although the 30/(10–40) rule is also based on equivalised household income distribution, it is more restricted and occasionally uses definition that ignores both the severe and marginal forms of housing stress. Nevertheless, the availability of suitable data, methodological tools, and specific research interests in each of these definitions is useful.

It is noted that in all the definitions, households with negative and nil incomes have been removed from the analysis. In survey data, few households have reported nil or negative incomes. These are often excluded from any analysis related to income distribution and financial well-being, as research from the ABS has shown that the expenditure of these households is similar to that of households earning much more, so these incomes are considered an unreliable measure of a household's standard of living (ABS 2005).

Moreover, the distributions of housing stress measured by the three different rule-based variants are presented in Figure 6.2. It is obvious from the figure that not only does the percentage of households in housing stress vary under different definitions/variants but also the density of SLAs varies with the percentage of housing stress across Australia.

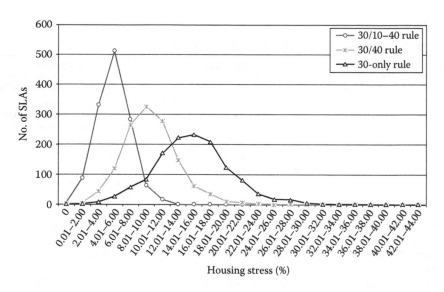

FIGURE 6.2
Distribution of housing stress for three variants in Australia 2006.

The graph of the "30/40 rule"–based variant of housing stress shows that approximately 67% of SLAs have housing stress households of 7%–11% with a mean of 9.52% and the coefficient of variation (CV) measure of 34.95. In addition, the graph of the "30/40–10 rule"–based variant shows that most of SLAs (about 87%) have housing stress households of 3%–7% with a mean of 4.91% and the CV measure of 41.85. The "'30-only rule" variant of housing stress reveals that about 51% of SLAs in Australia have households with a rate of housing stress between 13% and 17% with a mean of 14.68 and the CV measure of 36.71.

According to Karl Pearson, the CV measure is a very powerful tool for comparing the variability of two or more series of variants (Gupta and Kapoor 2008), where a variant having the lowest CV measure is considered to be more consistent than the others. In this regard, since the CV measure for the "30/40 rule"–based variant of housing stress estimation is lowest compared to the CV measures for the other two variants, this variant of housing stress estimation is more consistent than the others. Furthermore, in terms of the distributional pattern of these three curves, the "30/40 rule"–based housing stress variant is also showing a more rational pattern toward the usual normal curve, while the "30/(40–10) rule"– and "30-only rule"–based variants are showing patterns of the leptokurtic and platykurtic curves, respectively. From the statistical point of view, the "30/40 rule"–based housing stress estimation is more consistent and appropriate at small area levels in Australia.

The "30/40 rule"–based definition is also accountable and valid for using socioeconomic policy analyses that link with the housing stress issue. For instance, one of the policy significance of this definition is that this rule is widely used as the basis for determining household eligibility for entry to public rental housing and/or receipt of rent assistance. Moreover, the definition was used by many researchers and public and private organizations, including the National Housing Strategy (1992), ABS (2002b), Harding et al. (2004), and Yates and Gabriel (2006), and recently in estimating figures used by the Australian Prime Minister and subsequently published by the Australian Government Department of Families, Housing, Community Services, and Indigenous Affairs (FaHCSIA 2008). Therefore, this book uses the "30/40 rule"–based variant to define households in housing stress as those with equivalised household gross income in the lowest two quintiles (bottom 40%) of all household incomes in Australia, who are spending more than 30% of their gross household income on either renting costs or mortgage repayments.

6.5 Small Area Estimation of Housing Stress

To produce small area estimates of housing stress, we have to run the second stage of the housing stress model. This section describes various parts in the second stage of the spatial microsimulation model for SLA-level housing stress estimation.

6.5.1 Inputs at the Second-Stage Model

Typically, three input files are essential for the second stage of the housing stage model, which include the following:

1. SIH-CURFs
2. Synthetic weights
3. The consumer price index (CPI) file

These three input files are connected by a SAS program file that is known as the second-stage program file. This SAS file not only contains all the linkage paths toward the input files but also programs the definition of the housing stress measure, various logic operations, and codes of summary statistics for small area estimates. It also indicates a pathway to *outputs folder* where the demanded small area estimates could be stored.

As information about the (1) SIH-CURFs and (2) synthetic weights files is already given in Section 6.3, I will address only the CPI file here.

6.5.1.1 Consumer Price Index File

The CPI file is basically a current social and economic index that aims to measure the change in consumer prices over time. More specifically, the CPI file is constructed to measure changes over time in the general level of prices of consumer goods and services that the households acquire, use, or pay for consumption (see ABS 2009). This may be done by measuring the cost of purchasing a fixed basket of consumer goods and services of constant quality and similar characteristics, with the products in the basket being selected to be representative of households' expenditure during a year or other specified period. The CPI is utilized for uploading the consumer price from previous years to the current year.

As the key focus of this analysis is housing stress associated with households' expenditure on housing, it uses the households' CPI file for housing expenditure indices in different states in Australia. The households' CPI file for housing expenditure has been created from a group of time series data for housing expenditure–related goods and services. It is noted that the overall housing policies vary quite a lot from state to state in Australia, because each individual state has their own social and business policies in terms of housing goods and services. However, in theory, all expenditures by businesses, and expenditures by households for investment purposes, would be out of scope of a CPI. In this regard, expenditure on housing, especially purchasing expenditure, presents particular difficulties, as it can be considered as part investment and part purchase of shelter-related services.

6.5.2 Model Execution Process

The overall execution process in the second stage of the spatial microsimulation model for small area housing stress estimation is depicted in Figure 6.3. It is noted that the program starts by reading in a SAS program file known as the second-stage program file. There are several steps that are followed throughout the execution process once run by this SAS file. First of all, the process reads all the common SIH-CURFs given by the ABS for different years and used in this study. As soon as the SIH-CURFs are ready, the households' CPI file for housing expenditure indices in different states in Australia is applied to upload them into an updated SIH file for the current year. In the next step, the synthetic weights produced at the first stage of the model by the GREGWT reweighting are utilized into the operated SIH file to determine the synthetic spatial microdata for selected variables in the model. Once the reliable synthetic small area microdata are established, a researcher can perform different sorts of valuable analyses using these microdata. However, the analyses have to be consistent with the selected variables used in construction of the model.

As this study is related to housing stress estimation, a definition of a housing stress measure such as the "30/40 rule"–based measure is applied here in the fourth step. Hence, by employing the definition of housing stress, the process adds housing stress flags to the synthetic microdata. In the last step, the process exercises various summary statistics to calculate numbers of

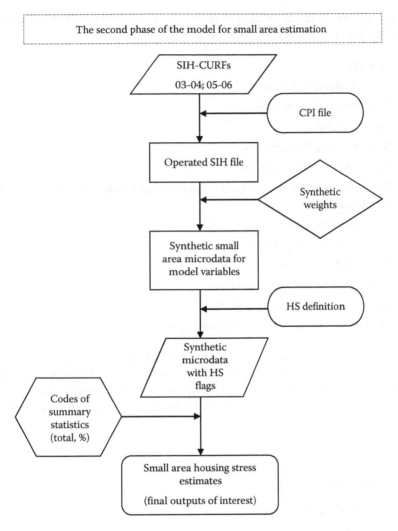

FIGURE 6.3
Computation process of microsimulation modeling technology to produce small area micro-data and synthetic estimates.

households in the numerator and the denominator and then produces small area estimates as a count of total, percentage, and/or ratio of households in housing stress according to selected variables such as tenure types.

6.5.3 Final Model Outputs

The output from the second-stage model is the ultimate file for small area housing stress estimates in Australia. This research is considering the SLA in Australia as a small area. Therefore, the ultimate output file will contain a range of data for the SLA-level housing stress estimation. In particular, the file encompasses data for the following attributes:

- SLA ID
- Total number of households
- Fully owned households
- Buyer households
- Renter public households
- Renter private households
- Other tenure-type households (such as hospitals, hostels, and military tenures)
- Total housing stress
- Owner in housing stress
- Buyer in housing stress
- Renter public in housing stress
- Renter private in housing stress
- Other tenure households in housing stress

Furthermore, the output file provides household-level estimates of total numbers as well as percentages for each of these factors. However, the model can also produce person-level small area estimates for these variables. A full account of the model output results and their discussion will be provided in the next chapter.

6.6 Concluding Remarks

This chapter has presented a detailed description of the sequential procedure in building a spatial microsimulation model for small area housing stress estimation. First, the background information about various data sources and issues related with the applicable data has been discussed. Most of the data sets used in this research were received from the ABS—an internationally recognized institution in Australia for reliable census and various survey data. It is noted that these days although the census data are available in the convenient cross-tabulated format for a variety of spatial scales, they are in modified form to retain the privacy of respondents. In addition, the individual and/or household-level raw microdata are not available at small geographic scales because of the confidentiality agreement with respondents. As a result, some of the data files such as various SIH-CURF files used in this book are supplied by the ABS under special request and specific conditions.

The spatial microsimulation model for small area housing stress estimation has two computational stages: the first stage simulates SLA-level households' synthetic weights and the second stage calculates the ultimate small area housing stress estimates after producing the small area synthetic microdata. The description of the model demonstrates that the five groups of files labeled as the *general model file, unit record data files, benchmark files, auxiliary data files,* and the *GREGWT program file* are required to run the first stage of the model. At this point, the significant output of the computation is the small-area-level "synthetic weights file" for households, having a set of characteristics based on the categories of different variables selected in the model.

Moreover, the definitions of housing stress measured by different variants have been discussed in this chapter. Typically, the concept of housing stress is measured by comprising the following two basic quantities associated with a household: households' income and its expenditure for housing. A household can be considered under housing stress when it is spending more than an expected and affordable percentage of its household income in housing. The expected and affordable cutoff point of housing expenditure can vary by the circumstance of households as well as the location of dwelling. Findings reveal that 30% or more of income threshold for housing costs is the widely used cutoff for the definitions of housing stress. Few studies have also used more than 25% of income threshold as a cutoff point for housing stress. Nevertheless, the 30% of income threshold is popular among researchers. By using this cutoff point, a number of variants for housing stress estimation have been developed. There are three common variants available in the literature for measuring housing stress as follows: (1) "'30-only rule," (2) "30/40 rule," and (3) "30/(40–10) rule." A comparison of these three definitions demonstrates that the "30/40 rule"–based housing stress variant is more consistent and appropriate than others. This definition of housing stress measure is also accountable and valid for use in various socioeconomic policy analyses allied with the housing stress issue.

Finally, the three groups of files called as the "SIH-CURF files," "synthetic weights file," and the "CPI file" are necessary for further computation of the model with the *SAS program file.* The second-stage computational process of the MMT for housing stress estimation creates the synthetic spatial microdata and then determines the small area housing stress estimates in Australia.

7

Applications of the Methodologies

7.1 Introduction

A very simple and stepwise clear description of a microsimulation modeling technology (MMT)–based spatial model has been introduced in the previous chapter. This model is for estimating reliable small area housing stress statistics at the statistical local area (SLA) level in Australia. Essentially, a range of model outputs have been derived by running the spatial microsimulation model. Then using these model outputs, various results are designed to provide valuable reports on housing stress estimates in Australia. This chapter demonstrates the empirical results of the MMT-based small area estimation model along with detailed discussion.

The chapter is divided into five sections. Section 7.2 offers a general view of the results derived from the model estimates. Section 7.3 highlights the housing stress estimates for different states, statistical divisions (SDs), and statistical subdivisions (SSDs) in Australia. Section 7.4 provides a comprehensive report on small area distributions of the estimated numbers of households in housing stress throughout Australia. Section 7.5 presents the spatial distributions of the percentage estimates of housing stress in Australia. A concluding summary of the chapter is given in Section 7.6.

7.2 Results of the Model: A General View

This section highlights some of the general results of the model under a range of relevant subheadings.

7.2.1 Model Accuracy Report

The model is based on an iteration process of reweighting toward the construction of synthetic spatial microdata. An accuracy report of the convergence of the iteration process is important for the model. Table 7.1 presents

TABLE 7.1

Accuracy Report of the Reweighting Algorithm for Creating Synthetic
New Weights

Type of Accuracy	No. of SLA	%	Households in These SLAs (%)
Met accuracy criterion	1270	89.3	99.620
Met benchmarks but an iteration did not	127	8.9	0.377
Not met accuracy criterion[a]	25	1.8	0.003
Total	1422	100	100

[a] Benchmarks population is zero or too small for those SLAs.

the accuracy report of the generalized regression weighting (GREGWT) algorithm used in the MMT for estimating SLA-level housing stress in Australia. There are 1422 SLAs in the Australian territories that include SLAs in the mainland, islands, and offshore land. The *re*weighting algorithm produces reasonable weights for 1270 SLAs. That means for the spatial microsimulation modeling of housing stress estimation, about 89.3% of SLAs were reasonable according to an accuracy index criterion.* In addition, more than 99.6% of households in Australia are located in those SLAs.

The accuracy index is measured by dividing the sum of absolute residuals (which is calculated by the total absolute difference between the new weights produced by the GREGWT algorithm and the sampling design weights) by the corresponding population total benchmark. If the value of the accuracy index is less than one for an SLA, then it is considered as having met accuracy criteria. However, in many cases, the reweighting results show the values of the accuracy index as being more than and/or equal to one—that means for those SLAs the algorithm met benchmark totals, but an iteration of the reweighting procedure did not.

Data in the table also reveal that there are 127 (about 9%) SLAs in such a category where the *re*weighting algorithm met benchmark totals but an iteration did not, and those SLAs have less than 0.4% of households. The synthetic spatial microdata for these SLAs can also produce housing stress estimates, which may or may not be statistically reliable. Further, the GREGWT reweighting algorithm did not produce reasonable weights for only 25 SLAs. It is noticeable that each of those 25 SLAs has a population size of zero or too small.

7.2.2 Scenarios of Housing Stress under Various Measures

Figure 7.1 presents the distribution of housing stress in Australian households for three different rules–based measures of housing stress. The

---

* Accuracy is measured by dividing the total of absolute residuals $\sum_{k \in s} |w_{k,i} - d_{k,i}|$ by the corresponding population total benchmark B_i, for $\forall i$.

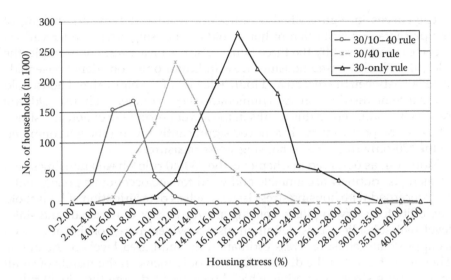

FIGURE 7.1
A scenario of the number of households in housing stress by three different measures.

definitions of different rules have been provided in Chapter 2 (also refer to Nepal et al. 2010 for more details). All three measures appear to be distributed normally with distinct shapes that are associated with the mean values and standard deviations of the corresponding distributions. In contrast with Figure 2.1 (where a number of SLAs were plotted instead of a number of households), interestingly the line graphs in Figure 7.1 reversed to a gradual height on the right side. The patterns of the graphs are quite normal with the "30/40 rule" overestimating the number of households in housing stress relative to the statistics for the "30/(10–40) rule" but underestimating relative to the statistics for the "30-only" rule.

Findings demonstrate that the "30-only" rule–based measure has a relatively wide distribution around the mean value of 18.11% and a higher standard deviation estimate of 4.22% compared to other measures. Additionally, the distribution of housing stress assessed by the "30/40 rule"–based criterion has a mean of 11.73% and a standard deviation of 3.24% of households. However, the "30/(10–40) rule"–based housing stress indicator shows a high density around the mean value of 6.21, with a rather small standard deviation estimate of 1.78%.

The total numbers of households in housing stress vary greatly with the definitions of housing stress. The results indicate that under the "30/40 rule"–based measure, around 773,073 Australian households are in housing stress, whereas under the "30/(10–40) rule"– and "30 only rule"–based measures, about 410,514 households and 1,231,159 households are in housing stress, respectively.

The curves in Figure 7.1 also indicate a relative downward shifting of the percentage distribution of households in housing stress under various income cutoffs used by the three housing stress measures. For example, the "30/(10–40) rule"–based housing stress indicator only considers households who fall between the second and fourth deciles of the equivalized household gross income distribution. As a numerator, these numbers of households are relatively small, and so this results in the lower percentage of housing stress. Thus, the respective curve has moved significantly toward the left compared to the "30-only rule"–based housing stress measure.

Moreover, as observed in Chapter 6, the "30/40 rule"–based housing stress measure is routinely accountable and valid for socioeconomic policy analyses concerned with housing stress issues such as determination of household eligibility for entry to public housing and/or receipt of national- and state-level rent subsidies. This definition also provides more consistent as well as appropriate housing stress estimation at small area levels in Australia. Therefore, the rest of the dissertation will only focus on the results of small area housing stress estimation, which I have derived from the "30/40 rule"–based measure within the detailed discussions.

7.2.3 Distribution of Housing Stress Estimation

A percentage distribution of housing stress at SLAs in Australia is presented in Figure 7.2. The measure of housing stress used in this graph is based on the "30/40 rule."

It appears that the pattern of the percentage distribution of housing stress at SLA levels in Australia is fairly normal. However, the shape of the graph is a bit positively *skewed* that is evidenced by the estimated moment coefficient

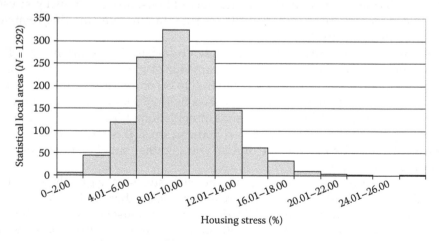

FIGURE 7.2
Distribution of housing stress at small area levels.

of skewness value of 0.43 as well as *leptokurtic* with the moment coefficient of kurtosis estimation of 0.46. This means that the distribution has a slightly longer tail toward the right side and it is somewhat more peaked than the usual normal curve. The rate of average housing stress under this distribution is estimated as 9.52% with a standard deviation estimate of 3.35. It demonstrates that about 1 in 10 households in Australia is experiencing housing expenses–related financial stress whether it comes from very high levels of rent or mortgage payments.

Moreover, the bar graph reveals that about two-thirds of the SLAs (856 SLAs) in Australia have 6.01%–12.00% of households who are in housing stress, and among them 324 SLAs (a quarter of the total small areas) have a housing stress rate from 8.01% to 10.00%. Besides, about one-fifth of the SLAs have a relatively higher proportion of households in housing stress, with the rates between 12.01% and 20.00%. Nonetheless, there are also seven SLAs that boast more than 20% of households in housing stress, which are located within the four major capital cities: Sydney, Brisbane, Canberra, and Melbourne. In particular, the SLAs of *Sydney (C), inner*; *Bankstown (C), north-east*; *Auburn (A)* and *Fairfield (C), east* in Sydney; and *Woodridge* in Brisbane have housing stress rates from 20.01% to 22.00%, and the SLA of *City* in Canberra and the SLA of *Melbourne (C), inner,* in Melbourne have housing stress rates of 23.20% and 27.00%, respectively.

7.2.4 Lorenz Curve for Housing Stress Estimates

Figure 7.3 presents the Lorenz curve diagram of the distribution of low-income households (who fall into the bottom 40% of income distribution) in housing stress against the cumulative percentage of the households (from the entire income distribution) in housing stress. SLAs are first ranked and ordered by their percentages of low-income households in housing stress. Then for each SLA, the cumulative percent of low-income households experiencing housing stress and the cumulative percent of households experiencing housing stress are plotted as the y and x coordinates, respectively, to form the Lorenz curve.

In general, a Lorenz curve lies below the 45° line, the line of equality, representing equal proportions for all spatial units or SLAs (if the vertical and horizontal axes are reversed, the Lorenz curve always lies above this line). The deviation of the Lorenz curve from the line of equality denotes a concentration of low-income households in housing stress within the larger population of households experiencing housing stress.

Typically, a Lorenz curve shows inequality of some sort within the social circumstances. The curve in Figure 7.3 shows that for the SLAs with low prevalence of housing stress, the first 25% of households experiencing housing stress accounts for about 18% of the low-income households. In contrast, in the SLAs with high prevalence of housing stress, the last 25% of households experiencing housing stress accounts for about 30% of the

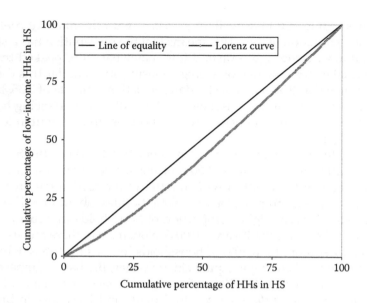

FIGURE 7.3
A Lorenz curve diagram for low-income households in housing stress.

low-income households. That means the SLAs with high proportions of households in housing stress comprise significantly large percentages of lower-income households.

In this particular case, the closeness of the Lorenz curve toward the line of equality demonstrates that most of the small areas in Australia have higher proportions of low-income households in housing stress. The Lorenz curve has become a standard method for analyzing disparities into various fields of social science including income inequalities (e.g., see Lorenz 1905 and Dagum 1980), health inequalities, and disease patterns across geographic units within a large domain (Nishuri et al. 2004; Kerani et al. 2005).

7.2.5 Proportional Cumulative Frequency Graph and Index of Dissimilarity

A proportional cumulative frequency graph can be designed by using a similar approach to the ordinary ogive curve. The main notion here is that rather than a single curve of the cumulative frequency of SLAs in a class interval, two curves are given, showing the cumulative frequency values of the low-income households in housing stress (the interested outcome) and the cumulative frequency values of all possible households experiencing housing stress under the entire income distribution, respectively. The key focus of this graph is on the distribution of the interested outcome estimates versus the distribution of the feasible population within which the estimates fall rather than on the SLAs themselves as in the ogive curve. In this graph, the cumulative frequency curve for low-income households in housing stress will always be above the other curve because the estimates of low-income households are a subset of the estimates of the entire population of households and the order for the graphs is determined by the percentage values that are arranged from low to high along the horizontal (x) axis.

As this proportional graph plots two curves against the percentage estimates for each SLA in Australia, the maximum difference between these curves can be considered as an index of dissimilarity. Typically, an index of dissimilarity measures the evenness with which a variable-like number of households in housing stress are distributed across SLAs. The value of the index ranges from 0, which means perfectly uniform, to 100%, which means perfectly concentrated for an SLA. Such an index of dissimilarity can be added to the proportional cumulative frequency figure by adding a vertical line at the grand rate of households in housing stress. The maximum difference is reached along the x-axis of the graph at the grand average rate for the entire Australia, because the proportion of the total associated with the number of households in housing stress is always greater than that associated with the number of low-income households in housing stress for estimates below the grand rate and vice versa for estimates above the grand rate. However, the geographic setting and size of the SLAs are not taken into account in the calculation of the index. Moreover, two study regions with the same index of dissimilarity may exhibit different patterns of spatial autocorrelation in the distribution of SLAs, which have high concentrations or clusters of low-income households in housing stress.

Figure 7.4 represents a proportional cumulative frequency graph and an index of dissimilarity. The housing stress estimates show approximately

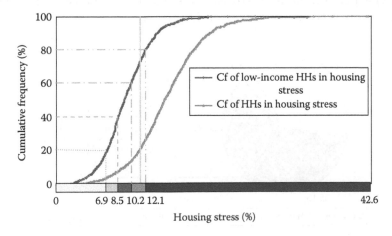

FIGURE 7.4
A proportional cumulative frequency graph with the index of dissimilarity.

normal (s-shaped) distributions for both groups of households. However, the distribution is much skewed for the low-income households, compared to the household populations from the entire income distribution. The curves also follow each other relatively with a significant distance except the edges. The grand average percentage of housing stress estimate is about 11.7%, and the index of dissimilarity is about 55%. These results suggest that the estimates of housing stress for low-income households are significantly unevenly spread across households categorized by the SLAs in Australia. Moreover, the graph demonstrates that nearly 80% of low-income households are experiencing housing stress and have resided in the SLAs with relatively small rates of housing stress (12.1% or less).

7.2.6 Scenarios of Households and Housing Stress by Tenures

The percentage distribution of Australian households by tenure types is given in Figure 7.5. The results reveal that most of the households (about 70%) in Australia are living in the tenure in which they fully own it or have purchased it, and among them half of the households (almost 35%) fall in each category.

About 27% of Australian households are living in the renting tenures. Among the renter households, 22.47% are in private housing and only 4.46% are in the public housing. These figures indicate that a greater proportion of renters are renting their tenures from private landlords. Moreover, only about 3% of Australian households are living in the other type of tenure such as hospital beds, military housing, and hotels/hostels.

To get a glimpse about those who are in housing stress, Figure 7.6 presents the percentage distribution of households in housing stress by their tenure type. The results reveal that 3 in 5 households who are private renters (59.55%) experience housing stress, while just 1 in 15 households who are public renters (6.88%) experience housing stress.

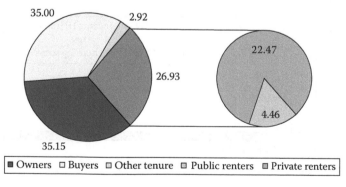

FIGURE 7.5
Percentage distribution of households by tenure types.

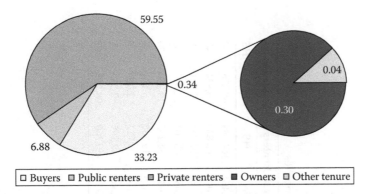

59.55

0.04

0.34

0.30

6.88

33.23

□ Buyers □ Public renters □ Private renters ■ Owners □ Other tenure

FIGURE 7.6
Percentage distribution of households in housing stress by tenure types.

In general, the public renters, which mean households renting from a state or territory government housing authority, enjoy lower housing costs and greater security of tenure than those renting from a private landlord. Moreover, one-third of Australian households (33.23%) who are purchasing their dwelling are in housing stress. The combined proportion of housing stress for owners and other tenure households is very small (i.e., 0.34% including 0.04% for other tenure) and may be regarded as negligible for further detailed discussion.

7.3 Estimation of Households in Housing Stress by Spatial Scales

The model estimates a total number of 7,128,035 households in Australia of which 773,073 households are experiencing housing stress. In this section, detailed discussions of various results for the households in housing stress are presented with appropriate subheadings.

7.3.1 Results for Different States

A relevant distribution of an aggregate of the estimated figures for different states and territories of Australia is presented in Table 7.2. The results show that nearly one-third of the Australian households (2,328,200 households) live in New South Wales, and among them, 269,329 households (about 11.6%) are experiencing housing stress that include 816 owners, 83,894 buyers, 20,417 public renters, 164,089 private renters, and 92 other tenure-type households. The second populated state Victoria is the residence of about a quarter of Australian households (1,781,601 households), and among them

TABLE 7.2

Estimates of the Numbers of Households and the Housing Stress by Tenures for States and Territories

States and Territories	Overall Total		Owners		Buyers		Public Renters		Private Renters		Other Tenure	
	HH[a]	HS[b]	HH	HS	HH	HS	HH	HS	HH	HS	HH	HS
NSW	2,328,200	269,329	836,696	816	760,241	83,894	114,423	20,417	548,464	164,089	68,376	92
VIC	1,781,601	185,646	665,595	492	649,015	71,397	57,158	9,846	364,009	103,835	45,824	47
QLD	1,387,069	156,607	452,587	576	480,441	47,078	49,455	7,746	362,374	101,085	42,211	60
WA	701,116	69,458	226,922	198	270,603	24,071	29,681	4,426	151,063	40,701	22,847	35
SA	583,284	61,455	208,924	134	208,090	20,693	42,311	6,428	104,603	34,167	19,356	20
TAS	181,666	18,373	70,923	42	62,269	6,418	10,912	1,532	32,428	10,363	5,134	7
ACT	116,911	7,700	35,567	3	45,761	2,070	9,453	1,895	24,101	3,722	2,027	0
NT	48,188	4,505	8,432	36	18,174	1,287	4,533	876	14,668	2,298	2,380	1
Australia	7,128,035	773,073	2,505,646	2297	2,494,594	256,908	317,926	53,166	1,601,710	460,260	208,155	262

[a] Number of households.
[b] Number of households in housing stress.

185,646 households (about 10.4%) are in housing stress in which most of them are buyers and renters. In Queensland, the estimated number of total households is 1,387,069; among them, about 11.3% of households (156,607 households including 576 owners, 47,078 buyers, 7,746 public renters, 101,085 private renters, and 60 other tenure-type households) are in housing stress.

Moreover, although the state Western Australia is occupying 701,116 households of which 69,458 households (about 9.9%) are in housing stress, both the estimates of households and their housing stress for public renters are higher in South Australia compared to their corresponding estimates in Western Australia. It is noted that 583,284 households are living in South Australia; among them, about 10.5% (61,455 households including 6,428 public renters) are in housing stress. An estimated number of 181,666 households are living in Tasmania, of which 18,373 households are experiencing housing stress with the highest number in private renters. In the Australian Capital Territory (ACT), only about 6.6% of households (7,700 households out of the estimated total of 116,911) are in housing stress. Although there are 2,027 households within the ACT living in other tenure types, none of them is experiencing housing stress. Nevertheless, in the Northern Territory, the number of total households is estimated as 48,188, and among them 4,505 households are experiencing housing stress including 1,287 buyer and 2,298 private renter households.

The estimated number of households experiencing housing stress at the SLA level in Australia ranges from 0 (in *Yalgoo (S)*, Central WA) to 7852 households (in *Canterbury (C)*, Sydney, NSW). There are other 15 SLAs having zero estimated number of household in housing stress, and most of them are the *offshore areas* and *migratory* SLAs. Typically, these SLAs are nonconvergent, and for this reason, they have been excluded from the analysis. There is an average of 598 households in housing stress per SLA, with a median of 242 households. Three hundred and 55 SLAs have less than 100 estimated households in housing stress, while 187 SLAs have less than 50 households and 80 SLAs have less than 20 households in housing stress. On the other side, 243 SLAs have more than a thousand estimated households in housing stress: 105 SLAs have more than 2000 households and just 5 SLAs have more than 5000 estimated households in housing stress including two SLAs that have more than 7000 households in housing stress. Besides, over 0.77 million households in Australia are experiencing housing stress under the "30/40 rule," and this estimated total figure may be higher than the actual figure as the model may overestimate the number.

7.3.2 Results for Various Statistical Divisions

Table 7.3 presents the results by aggregating the SLA-level housing stress estimates to the corresponding SDs in Australia. An estimated number of 163,655 (about 21.2%) households experiencing housing stress are in SD Sydney and 135,702 (17.6%) are in Melbourne. Nonetheless, a relatively

TABLE 7.3

Number of Households in Housing Stress by Statistical Divisions

ID	SD[a] Name	HS[b]	%	ID	SD Name	HS	%
105	Sydney	163,655	21.17	340	Mackay	4,368	0.57
205	Melbourne	135,702	17.55	155	Murray	4,292	0.56
305	Brisbane	66,718	8.63	135	Northwestern	4,204	0.54
505	Perth	53,766	6.95	620	Mersey–Lyell	3,912	0.51
405	Adelaide	46,749	6.05	230	Mallee	3,404	0.44
307	Gold Coast	25,787	3.34	245	Ovens–Murray	3,339	0.43
110	Hunter	24,764	3.20	215	Western district	3,203	0.41
115	Illawarra	17,058	2.21	705	Darwin	3,171	0.41
125	Mid-north coast	15,777	2.04	250	East Gippsland	3,016	0.39
309	Sunshine Coast	14,261	1.84	312	West Moreton	2,825	0.37
120	Richmond–Tweed	12,680	1.64	420	Murray Lands	2,657	0.34
315	Wide Bay–Burnett	11,991	1.55	435	Northern	2,637	0.34
210	Barwon	9,783	1.27	425	Southeast	2,153	0.28
350	Far north	9,055	1.17	535	Central	1,870	0.24
320	Darling Downs	8,011	1.04	515	Lower Great Southern	1,848	0.24
605	Greater Hobart	7,856	1.02	415	Yorke & lower north	1,612	0.21
510	Southwest	7,742	1.00	225	Wimmera	1,486	0.19
145	Southeastern	7,716	1.00	525	Midlands	1,423	0.18
805	Canberra	7,700	1.00	710	Northern Territory Bal	1,334	0.17
240	Goulburn	7,339	0.95	610	Southern	1,266	0.16
235	Loddon	6,794	0.88	530	Southeastern	1,245	0.16
130	Northern	6,654	0.86	430	Eyre	1,147	0.15
345	Northern	6,654	0.86	160	Far west	727	0.09
140	Central west	6,568	0.85	545	Kimberley	685	0.09
255	Gippsland	5,959	0.77	325	Southwest	575	0.07
220	Central Highlands	5,621	0.73	355	Northwest	529	0.07
330	Fitzroy	5,609	0.73	540	Pilbara	449	0.06
615	Northern	5,339	0.69	520	Upper great southern	430	0.06
150	Murrumbidgee	5,234	0.68	335	Central west	224	0.03
410	Outer Adelaide	4,500	0.58	000	Australia	773,073	100

[a] Statistical division.
[b] Total number of households in housing stress.

smaller but significant number of households experiencing housing stress are in some other major capital city SD such as Brisbane, 66,718 (about 8.6%); Perth, 53,766 households (nearly 7.0%); and Adelaide, 46,749 households (about 6.1%). Thus, Sydney, Melbourne, Brisbane, Perth, and Adelaide collectively account for about 60.5% of the estimated total number of households in housing stress in Australia.

In addition, compared to these five SDs, a very small number of households are in housing stress in other major capital city SDs such as in Greater Hobart, 7856 households; in the nation's capital Canberra, 7700 households; and in Darwin, 3171 households, which add up to only 2.4% of households. Therefore, about 37.1% of households in housing stress are living in SLAs outside of the eight major capital city SDs.

The results also reveal that seven SDs (which are the noncapital major coastal cities) with a relatively higher number of households estimated in housing stress range from 11,991 to 25,787 households (see Table 7.4). These SDs are Hunter, Illawarra, mid-north coast, and Richmond–Tweed in the NSW and Gold Coast, Sunshine Coast, and Wide Bay–Burnett in Queensland.

It is noted that these seven southeast coastal SDs encompass almost 15.8% of households in housing stress. Moreover, further 21.3% of households in housing stress are distributed across the SLAs in other SDs, including Barwon (9783 households) in Victoria, outer Adelaide (4500 households) in South Australia, and Pilbara (449 households) in Western Australia.

7.3.3 Results for Various Statistical Subdivisions

To get a much better sense at regional levels where most households experiencing housing stress are residing, the small area estimates are aggregated to the SSD level. The results of the 35 SSDs with the greatest estimated number and the highest percentage of households in housing stress are presented in Table 7.4. Essentially, the table is separated into two parts. The left part of the table focuses on the results for estimated numbers and the right part highlights the results for percentage estimates. Besides, it is worth mentioning that in the case of the estimated numbers of households experiencing housing stress, the listed 35 SSDs with the highest numbers are the regions for more than half (54.7%) of the total counts. Some of these SSDs (e.g., Fairfield–Liverpool, Canterbury–Bankstown, central western Sydney, Blacktown, Gold Coast east and west, Sunshine Coast, inner Melbourne) also have the highest percentage estimates. Nevertheless, the top listed 35 SSDs with the highest percentage estimates are the residences of 30.1% of the total households in housing stress.

The SSD of Newcastle contains the most number of 20,990 households in housing stress. This finding highlights that significantly large numbers of households experiencing housing stress are located in the noncapital port city Newcastle. Whereas, at the SLA level, estimates for this region are not remarkable compared to the estimates of many SLAs located within capital cities such as Sydney, Melbourne, Perth, and Adelaide.

The other SSDs with the highest counts of households in housing stress point out that there are several main geographic regional parts where housing stress is concentrated in Sydney, Melbourne, Perth, Adelaide, and coastal regions in New South Wales and Queensland. In particular, prominently the 12 SSDs making up the western, southwestern, and

TABLE 7.4

Thirty-Five SSDs with the Highest Housing Stress Estimates

ID	SSD Name	HS[a]	%	ID	SSD Name	HS	%[b]
11005	Newcastle	20,990	11.4	10525	Fairfield–Liverpool	17,464	16.9
10525	Fairfield–Liverpool	17,464	16.9	12501	Coffs Harbour	3,055	16.7
20510	Western Melbourne	17,098	11.5	10520	Canterbury–Bankstown	15,935	16.1
50515	North metropolitan	16,090	10.1	30710	Gold Coast east	10,889	15.5
10520	Canterbury–Bankstown	15,935	16.1	12007	Lismore	1,758	15.4
40505	Northern Adelaide	15,626	11.9	10540	Central western Sydney	15,352	15.2
10540	Central western Sydney	15,352	15.2	12005	Tweed Heads and Tweed Coast	3,611	15.1
10515	St George–Sutherland	14,748	9.8	20575	Greater Dandenong City	6,384	14.9
10505	Inner Sydney	14,589	12.1	12010	Richmond–Tweed SD Bal	7,311	14.9
10570	Gosford–Wyong	14,365	13.0	12503	Port Macquarie	2,338	14.6
20505	Inner Melbourne	14,264	12.3	20535	Hume City	6,453	14.1
50525	Southeast metropolitan	13,417	11.0	30715	Gold Coast west	11,732	14.1
20565	Southern Melbourne	13,338	9.1	12505	Clarence (excl. Coffs Harbour)	5,146	14.0
40520	Southern Adelaide	12,689	10.0	31507	Hervey Bay City part A	2,589	14.0
20550	Eastern middle Melbourne	12,316	8.3	30905	Sunshine Coast	11,195	14.0
30715	Gold Coast west	11,732	14.1	30705	Gold Coast north	2,533	13.9
10545	Outer western Sydney	11,640	11.2	30520	Caboolture Shire	6,324	13.8
10553	Blacktown	11,322	13.2	30545	Redcliffe City	2,806	13.6
30905	Sunshine Coast	11,195	14.0	12510	Hastings (excl. Port Macquarie)	5,238	13.5
11505	Wollongong	11,142	11.6	30530	Logan City	7,670	13.4
50520	Southwest metropolitan	11,003	9.9	10553	Blacktown	11,322	13.2
30710	Gold Coast east	10,889	15.5	31505	Bundaberg	2,954	13.2
20580	Southeastern outer Melbourne	10,446	11.9	14515	Lower south coast	3,362	13.0
40510	Western Adelaide	9,800	11.6	30910	Sunshine Coast SD Bal	3,066	13.0
30507	Northwest outer Brisbane	9,339	8.4	10570	Gosford–Wyong	14,365	13.0

(Continued)

TABLE 7.4 (*Continued*)

Thirty-Five SSDs with the Highest Housing Stress Estimates

ID	SSD Name	HS[a]	%	ID	SSD Name	HS	%[b]
20530	Northern middle Melbourne	9,199	10.1	14003	Bathurst	1,381	12.7
10555	Lower northern Sydney	9,140	8.2	20585	Frankston City	5,484	12.6
50510	East metropolitan	8,934	10.1	11507	Nowra–Bomaderry	1,433	12.6
10530	Outer southwestern Sydney	8,837	11.9	35005	Cairns City part A	5,485	12.5
10560	Central northern Sydney	8,815	6.6	23005	Mildura Rural City part A	2,110	12.4
40515	Eastern Adelaide	8,634	9.8	30720	Gold Coast SD Bal	633	12.4
10510	Eastern suburbs	8,568	9.8	30501	Inner Brisbane	4,227	12.4
30511	Southeast outer Brisbane	8,345	10.5	20505	Inner Melbourne	14,264	12.3
60505	Greater Hobart	7,856	10.3	24005	Greater Shepparton City part A	1,948	12.1
20555	Eastern outer Melbourne	7,826	9.1	10505	Inner Sydney	14,589	12.1

[a] Arranged by the number of households experiencing housing stress.
[b] Arranged by the percentage of households experiencing housing stress.

northern Sydney parts (Fairfield–Liverpool, Canterbury–Bankstown, central western Sydney, St George–Sutherland, inner Sydney, Gosford–Wyong, outer western Sydney, Blacktown, lower northern Sydney, outer southwestern Sydney, central northern Sydney, and eastern suburbs) collectively contain an estimated number of 150,775 households in housing stress, or 19.5% of the total estimated numbers in Australia.

Moreover, in Melbourne, the SSDs around western, inner, eastern middle, southern, and northern outer regions have considerably larger numbers of households in housing stress. Western Melbourne has the third highest estimated numbers of 17,098 households. The top listed seven SSDs from Melbourne jointly have a number of 84,487 households in housing stress. Besides, several SSDs located in the north, east, and southeast metropolitan of Perth and the northern, southern, western, and eastern parts of Adelaide also have a noticeably large number of households experiencing housing stress. For example, the north metropolitan SSD in Perth has the fourth highest count of 16,090 households and northern SSD in Adelaide has the sixth highest count of 15,626 households with housing stress.

Besides, not only Newcastle but also some SSDs located around the other major coastal centers in New South Wales (Wollongong, Richmond–Tweed,

Hastings, and Clarence) contain considerable numbers of households in housing stress. Wollongong has the estimated count of 11,142 households. Additionally, in Queensland coastal regions, the SSDs across northwest and southeast Brisbane, Logan City, and Gold Coast and Sunshine Coast, as well as Wide Bay–Burnett and Cairns City have much higher numbers of housing stress estimates. In particular, the Gold Coast west and east SSDs together have a number of 22,621 households experiencing housing stress. Furthermore, the Greater Hobart SSD in Tasmania also shows an estimated number of 7856 households with housing stress.

The results of the percentage estimates (in the right part of Table 7.4) demonstrate that a large number of SSDs spread across various coastal cities in Queensland, New South Wales, and Victoria have the highest values. There are six SSDs from Sydney with more than 12.1% of households in housing stress. Fairfield–Liverpool has the highest rate compared to all SSDs in Australia. Some non-Sydney SSDs such as Coffs Harbour, Lismore, Tweed Heads, Richmond–Tweed, Port Macquarie, Clarence, Hastings, lower south coast, Bathurst, and Nowra–Bomaderry have also a high proportion of households in housing stress. However, all of these SSDs provide relatively lower estimated numbers. Although most of the Queensland coastal SSDs have the highest estimated rate of housing stress, only some of them are with higher counts. It is noticeable that low-income households residing in tourist attraction Gold Coast regions are more prevalent (an average rate of 14.0%) in housing stress. This may be because of a very high level of house price or rent in Gold Coast areas. Moreover, the Greater Dandenong, Hume, and Frankston cities and inner Melbourne have housing stress rates of 14.9%, 14.1%, 12.6%, and 12.3%, respectively. Also, there are few outer Melbourne SSDs, namely, Mildura Rural and Greater Shepparton cities, including Greater Bendigo that also show significantly higher percentage estimates of housing stress.

7.4 Small Area Estimates: Number of Households in Housing Stress

The model produces two types of estimates: the estimated number of households experiencing housing stress at the SLA levels in Australia and their corresponding proportional percentage rate of housing stress for the total number of households living in the SLA. Typically, the distribution of housing stress by an estimated absolute number can be heavily influenced by the varying population sizes of SLAs. Thus, both the variables are considered for the geographical mapping to provide better and clearer information of the housing stress estimates.

The housing stress estimates have mapped across the whole SLAs of Australia and individual major capital city SDs in different states and territories. The maps depict overall housing stress estimates for households by SLAs and also the estimates by tenure types. In general, the geographic maps are illustrative of distinctive spatial patterns of housing stress variables. For both types of estimates, the synthetic spatial data are mapped according to the quantile classification (except the map in Figure 7.7) in which an equal

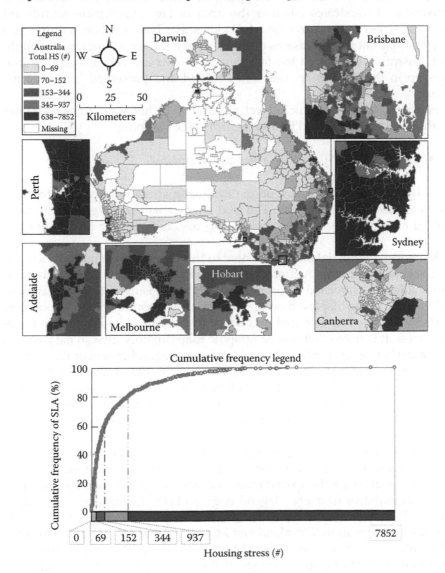

FIGURE 7.7
A spatial distribution of the small area estimates of housing stress with an alternative cumulative frequency legend.

number of SLAs in each data class interval is used to portray the spatial distribution of the housing stress variables using the following option:

- *Distribution of all SLAs in Australia excluding the SLAs, which did not meet the accuracy criterion at the reweighting stage*

It is noted that this choice of mapping describes relativity to all SLAs in Australia. It illustrates whether the area or city experiences significant housing stress as measured on a broader scale. The variation within cities is less apparent in this case as the scale is general for all regions throughout Australia. In view of the fact that city areas are very condensed and unseen in the main map, they are zoomed and presented into separate boxes.

However, to see the micro-level variability in housing stress estimates within a city, there should be another option of mapping:

- *Distribution of SLAs in the major capital city itself*

This option shows the distribution of housing stress relative only to other SLAs located in the same city. The next chapter will consider such an individual city-scale mapping to discuss the results for major capital cities in Australia.

Moreover, the spatial maps presented in this chapter are commonly known as the choropleth map. A choropleth map is a thematic map in which areas are shaded or patterned in proportion to the measurement of the statistical variable being displayed on the map, such as population density, income distribution, poverty rate, and number or percentage of households in housing stress. It is noted that this geographic map can only give an easy way to visualize how a measurement varies across a geographic area or it can show the level of variability within a region. However, it cannot thoroughly clarify the statistical distribution patterns of small areas within a very large area. In this case, an alternative cumulative frequency legend can be designed and attached to each spatial map for obtaining vital information of the map more clearly.

Typically, a cumulative frequency legend can be useful to depict the cumulative percentage of SLAs associated with each data interval of the estimates of housing stress on the spatial map. The data range for each class interval in the cumulative frequency legend is given both the numbers and length of distribution, which is typically a high-level perceptual task for visually decoding information (Cleveland and McGill 1987). While the range for each class interval in the standard legend within the geographic map is only given by numbers. Further, an advantage of the cumulative frequency legend is that it can reveal the patterns of the distribution of housing stress estimates by small areas.

7.4.1 Estimated Numbers of Overall Households in Housing Stress

Figure 7.7 shows a spatial distribution of the numbers of low-income households in housing stress by SLAs in Australia. In this mapping, the quantile classification (in which an equal number of SLAs are in each data class) is used to portray the spatial distribution. The graph reveals that a large number of SLAs in the southeast coastal regions (mostly in the New South Wales coastal cities) have high numbers of estimated households in housing stress (ranging from 938 to 7852 households). In some inland regions in the New South Wales and Victoria, the fast-growing mining regions in Queensland (e.g., Fitzroy, Mackay, far northern Mareeba, and northwestern Mount Isa), Western Australia (Kimberley, Pilbara, Central Gascoyne, and southeastern Johnston), and South Australia (northern Flinders Ranges) have SLAs with a relatively higher number of estimated households. Moreover, almost all SLAs in Sydney and Perth SDs have shown estimated numbers in the highest quantile. Significantly large numbers of small areas in Melbourne, Adelaide, and Hobart also have higher estimates. Since the sizes of SLAs in Brisbane, Canberra, and Darwin are much smaller, they have lower estimated numbers. However, many SLAs in Brisbane have estimates between 345 and 937 households with few southwest SLAs having estimates at the highest quantile.

Although from the quantile classification a map reader can understand that in each interval there are the same numbers of SLAs, the reader cannot get the idea about the distributive patterns of the SLAs by the estimated numbers of households in housing stress. The cumulative frequency legend for the map of estimated numbers shows a very different and highly positively skewed distribution. Most of the low-income households with housing stress are in SLAs that have a large number of households residing in them and located within urban centers. Furthermore, it is clear from the legend that more than a half of the SLAs in Australia have less than about 300 households in housing stress.

Furthermore, for a more explicit account, the results of 50 SLAs with the greatest numbers of households in housing stress are listed in Table C.1. Data demonstrate that out of these 50 SLAs, 43 are located in the four major capital city SDs, with 23 in Sydney, 14 in Melbourne, 5 in Perth, and 1 in Brisbane. Besides this, among the seven noncapital city SLAs, six are located in other (outside of Sydney) coastal centers of New South Wales such as in Wollongong, Coffs Harbour, Tweed Heads, Newcastle, and Shoalhaven, and one SLA is located in Launceston, Tasmania.

The SLA with the largest estimated number of 7852 households in housing stress is *Canterbury (C)*—part of the Canterbury–Bankstown SSD in the midwestern Sydney. Another SLA, *Fairfield (C), east* in western Sydney, also has a very high number of 7219 households in housing stress. It is noted that although *Canterbury (C)* has the highest estimated number of housing stress that is more than all of the estimated number of 7700 households experiencing housing stress in the ACT, the percentage rate of housing stress for Canterbury is smaller than that for Fairfield. This is due to the fact that the number of total households (i.e., the denominator of percentage) living in *Canterbury (C)* is considerably bigger than the total number of households living in *Fairfield (C), east*.

For the next eight SLAs with the highest housing stress count, numbers range from 4223 to 5294 households. Six SLAs are specifically located in western and southwestern Sydney, one in inner Wollongong (a southern coastal center in New South Wales), and one (*Stirling (C)—Central*) within central Perth in Western Australia. In addition, the subsequent 30 highest housing stress count SLAs are located in Sydney (13 SLAs with the highest count of 3962 households in *Blacktown (C), southeast*), Melbourne (11 SLAs with the highest count of 3869 households in *Brimbank (C), Sunshine*), Perth (3 SLAs with the highest count of 3475 households in *Swan (C)*), and three outer capital SLAs in the NSW coast, namely, *Wollongong (C) Bal* (with 3259 households); *Coffs Harbour (C), Pt A* (with 3055 households); and *Tweed (A), Tweed Heads* (with 3030 households). It is worth mentioning here that typically the size of size of SLAs in New South Wales and Victoria is relatively larger than the size of SLAs in other states and territories. More specifically, the sizes of the SLAs within Brisbane in Queensland and within Canberra in the ACT are significantly small, compared to the sizes of the SLAs in Sydney and Melbourne.

Moreover, among the last 10 SLAs out of the greatest count 50, the *Canning (C)* in Perth has the estimate of 2930 households, just following the *Ipswich (C), Central* in Brisbane City, with 2933 households in housing stress. There are four SLAs located into the New South Wales, two (*Penrith (C), west*, 2929 households and *Bankstown (C), northwest*, 2917 households) are in Sydney and the other two (*Newcastle (C), inner city*, 2826 households and *Shoalhaven (C), Pt B*, 2811 households) are in other coastal cities outside of Sydney. Besides this, three SLAs, namely, *Casey (C), Berwick; Gr. Dandenong (C), Dandenong;* and *Wyndham (C), north*, are all in Melbourne City with more than 2900 households in housing stress. The final SLA called *Launceston (C), Pt B*, located within Greater Launceston in Tasmania has an estimated total of 2798 households in housing stress.

7.4.2 Estimated Numbers of Buyer Households in Housing Stress

Housing stress estimates for only buyer households demonstrate somewhat different scenarios. Findings from the estimates of buyer households reveal that more than three quarters of the SLAs in Australia have less than 200 buyer households in housing stress and half of the SLAs have an estimate of less than 100 households experiencing housing stress. In addition, about 52% (133,850 households) of the buyer households in housing stress are residing in upper 150 SLAs with the largest estimated numbers, and about 26% (66,654 households) are living in the top 50 SLAs.

Table C.2 presents the results of the 50 SLAs with the greatest numbers of estimates of housing stress for the buyer households. Out of these 50 SLAs, 48 SLAs are located within the three major capital cities—Sydney (23 SLAs), Melbourne (19 SLAs), and Perth (6 SLAs). The top two SLAs with more than 2000 buyer households in housing stress are *Fairfield (C), east* (with 2455 households), and *Canterbury (C)* (with 2175 households) in Sydney. The next four SLAs having the largest estimated numbers are all in Melbourne, where an average of 1761 households purchasing their house is experiencing housing stress. In addition, the estimated number of housing stress for the rest of the SLAs in Sydney ranges from 1012 (in *Wyong (A), south* and *west*) to 1683 (in *Blacktown (C)*) households and in Melbourne ranges from 975 (*Gr. Dandenong (C), Dandenong*) to 1582 (*Hume (C), Craigieburn*) households, whereas the six SLAs of Perth have a number of households in housing stress that range from 1149 (in both of *Joondalup (C), south*, and *Cockburn (C)*) to 1710 (in *Swan (C)*) households.

Moreover, in the top 50 SLAs' list, there are only two SLAs located in non-capital city, namely, *Wollongong (C) Bal* with 1089 (about 10%) and *Shoalhaven (C), Pt B*, with 1008 (about 17%) buyer households in housing stress that are both located within "Illawarra" in the NSW south coast. This result indicates that there is no SLA in the highest 50 class lists from the states of Queensland, South Australia, Tasmania, ACT, and Northern Territory. However, there are many SLAs in those states that may have significant numbers of buyer households in housing stress. Finally, the findings also reveal that among these top 50 SLAs, the proportional rate of buyer households experiencing housing stress is the lowest at 6.7% for *Sutherland Shire (A), west*, and the highest at 25.3% for *Bankstown (C), northeast*, which are both geographically located in Sydney.

7.4.3 Estimated Numbers of Public Renter Households in Housing Stress

The housing stress estimates for public renters show that a total of 53,166 households under the public renting are in housing stress; among them, 18,444 households (about 34.7%) are living in the top 50 SLAs with the highest estimated numbers. There are more than a thousand SLAs with less than 100 estimated numbers of public renter households in housing stress, of which more than three quarters of these SLAs have less than 25 estimated numbers.

In theory, although households living in public housings are paying up to a level of less than 30% of their assessable income in housing rent (AIHW 2009), in the equivalized household gross income amount, they are perhaps paying more than 30% of their income in housing costs. The Commonwealth Rent Assistance eligibility is dependent on recipients being on some form of government transfer payment, which is also the primary source of income for public housing households. However, as very low–income households, these tenure groups are likely to be in housing stress. For instance, in 2005–2006, the proportions of public housing households in Australia with an older resident were 28% and with a member with disability were 29%, while substantial percentages (about 29% and 33% of households with older tenant and member with disability, respectively) of them were still in housing stress, after the Commonwealth Rent Assistance has been received (e.g., see SCRGSP 2007; AIHW 2008).

The results of the 50 SLAs with the greatest numbers of housing stress estimates for the public renter households show that the top six SLAs with the highest estimated numbers are located in Sydney (see Table C.3). The largest estimated number of 909 households is observed in the SLA of *Blacktown (C), southwest*, a part of the inner western Sydney SSD, in the western Sydney. Further, about 3350 public renter households are experiencing housing stress in the subsequent five SLAs. It is remarkable that the estimated numbers of housing stress in *Canterbury (C)* (with 560 households) and *Fairfield (C), east* (with 503 households), are comparatively lower for public renter households, even as for the private and total renters, these two are in the top listed SLAs in terms of the largest number of households in housing stress.

There are 28 SLAs with the highest estimated housing stress numbers located in New South Wales; among them, only six are located outside of Sydney. These non-Sydney city SLAs with significant numbers of public renter households in housing stress are *Wollongong (C), inner* (with 580 households); *Wollongong (C) Bal* (with 420 households); *Newcastle (C), inner city* (with 301 households); *Lake Macquarie (C), east* (with 291 households); *Newcastle (C), Throsby* (with 274 households); and *Shellharbour (C)* (with 252 households).

Besides, among the rest of these 50 SLAs, there are 11 SLAs in Victoria with 10 in the Melbourne SD (and 1 SLA: *Corio, inner*, within "Greater Geelong City Part A" SSD of Barwon in northwestern Victoria), 6 in South Australia with 5 in Adelaide SD (and 1 within Whyalla SSD in northern South Australia), 2 within Perth SD in Western Australia, 2 in Tasmania with 1 in Hobart SD (and 1 in Launceston), and 1 (i.e., the SLA of *Inala* with 233 households) within Brisbane SD in Queensland. More specifically, the 10 SLAs located in Melbourne are located in inner Melbourne (5 SLAs: *Yarra (C), north*, with 393 households; *Melbourne (C), Remainder*, with 301 households; *Yarra (C), Richmond*, with 262 households; *Stonnington (C), Prahran*, with 236 households, and *Port Phillip (C), west*, with 235 households), western Melbourne (2 SLAs: *Moonee Valley (C), Essendon*, with 551 households and *Maribyrnong (C)* with 288 households), northern middle Melbourne (2 SLAs: *Darebin (C), Preston*, with 328 households and *Banyule (C), Heidelberg*, with 234 households), and Hume City (1 SLA: *Hume (C), Broadmeadows*, with 234 households) SSDs. In addition, after the Sydney and Melbourne SDs, Adelaide appears to host a higher number of SLAs with relatively higher estimates of housing stress for public renters. The SLAs that are located in Adelaide are *Playford (C), Elizabeth* (with 379 households), within northern Adelaide; *Marion (C), central* and *north* (with an average of about 270 households), within southern Adelaide; *Port Adel Enfield (C), Park* (with 243 households); and *Charles Sturt (C), northeast* (with 231 households), within western Adelaide SSDs. The two SLAs, *Stirling (C), Central* (with 423 households), and *Swan (C)* (with 221 households), in Perth are located, respectively, within the east and north metropolitan SSDs.

7.4.4 Estimated Numbers of Private Renter Households in Housing Stress

The housing stress estimates for private renter households significantly dominate the estimates of renter households and then further add up to the overall results of housing stress estimation. The results demonstrate that the estimated numbers of housing stress (460,260 households) for private renter households are much higher (almost 30%) than the sum of the estimated numbers for buyers (256,908 households) and public renters (53,166 households) in Australia. More than a half, that is, about 52% (240,704 households), of the private renter households in housing stress are living in the top 150 SLAs with the greatest estimated numbers, and about 25% (114,369 households) are living in the selected top 50 SLAs.

Among the selected 50 SLAs with the largest estimated numbers of private renter households in housing stress, 23 of these SLAs are located in Sydney, 12 in Melbourne, 4 in Perth, and 1 in Brisbane. The other 10 SLAs are located in the noncapital coastal centers in the states of New South Wales, Tasmania, and Queensland. More specifically, these SLAs are distributed as two within Wollongong, two within Newcastle, one within Coffs Harbour, one within Port Macquarie, one within Tweed Heads, one within Illawarra SSD in New South Wales, one within Launceston in Tasmania, and one within Hervey Bay in Queensland (see Table C.4).

The results show that the percentage rates of housing stress for private renter households are significantly higher in these selected SLAs. For example, there are 20 SLAs in the top 50 lists that have more than 40% of households living at private rental properties experiencing housing stress and five of these SLAs (e.g., *Fairfield (C), east; Tweed (A), Tweed Heads; Blacktown (C), southwest; Bankstown (C), northeast;* and *Wollongong (C) Bal*) have more than 50% of their households in housing stress.

Furthermore, as the housing stress estimates for private renter households dominate the overall estimates, the general sequential pattern of the selected 50 SLAs for private renter households is fairly similar to the top 50 lists for overall (public + private) renter households. Findings reveal that the first six SLAs are in the same order, with *Canterbury (C); Fairfield (C), east;* and *Randwick (C)* in Sydney being the top three. To point out a specific change in the order of SLAs, the two SLAs, *Marrickville (A)* with 2791 households and *Rockdale (A)* with 2773 households, in Sydney jump toward the more estimated numbers and come up above the *Liverpool (C), east,* whereas the SLA of *Newcastle (C), inner city,* has the estimated numbers of 2064 households come down under the SLA of *Penrith (C), east.* Moreover, compared to the lists of 50 SLAs for renter households, there are four new SLAs that joined in the lists for private renter households, which are *Waverley (A)* (with 1788 households) in eastern suburbs, Sydney; *Hervey Bay (C), Pt A* (with 1769 households), in Hervey Bay City; *Shoalhaven (C), Pt B* (with 1739 households) in Illawarra; and *Hastings (A), Pt A* (with 1669 households), in Port Macquarie.

7.4.5 Estimated Numbers of Total Renter Households in Housing Stress

The estimates of housing stress for renter households significantly dominate the overall results of housing stress estimation. In contrast with the estimates for buyer households, a significantly large number of households living in renting dwellings are experiencing housing stress; among them, approximately 90% are in the private rental dwellings. The results of the 50 SLAs with the largest estimated numbers of renter households in housing stress aggregate a number of 129,109 households—that is almost 2.4 times the total estimated numbers (53,166 households) for buyers in Australia. Additionally, more than a half (nearly 53%) of the total renting households in housing stress are located within the top 150 SLAs with the greatest estimated numbers, and about 25% are living in the top 50 SLAs.

Of the 50 SLAs with the largest estimated numbers of renter households in housing stress, a half of these SLAs are located in Sydney and almost a quarter (13 SLAs) in Melbourne, four in Perth, and one in Brisbane, and the other seven SLAs are within the noncapital coastal cities: Wollongong, Newcastle, Coffs Harbour, and Tweed Heads in New South Wales and Launceston in Tasmania (see Table C.5). The results demonstrate that the percentage rates of housing stress for renter households are significantly higher in some of the noncapital coastal city SLAs. For example, about 43% of renter households (the highest figure among the estimates of 50 SLAs) in the SLA of *Tweed (A), Tweed Heads*, and 39% of renter households in the SLA of *Coffs Harbour (C), Pt A*, are experiencing housing stress. Also, *Launceston (C), Pt B*, the only Tasmanian SLA that comes into the top 50 lists, has about 28% or 2053 renter households in housing stress.

The SLAs of *Canterbury (C); Fairfield (C), east*; and *Randwick (C)* in Sydney are the top three, in which the first one stands alone at the peak of the hill of 5665 renter households in housing stress with a significant difference between the estimated numbers. Estimated numbers of the additional SLAs in Sydney are ranging from 1878 households (in *Penrith (C), west*) to 3452 households (in *Gosford (C), west*), and they are distributed across the city with most of them in inner Sydney, eastern suburbs, St George–Sutherland, Canterbury–Bankstown, Fairfield–Liverpool, outer western Sydney, Blacktown, lower northern Sydney, and Gosford–Wyong SSDs. Besides, among the 13 SLAs in the Melbourne SD, most are located in inner Melbourne, southern Melbourne, western Melbourne, and Greater Dandenong City SSDs. The average estimated number of housing stress for these SLAs is about 2286 renter households with a range from 1896 households within the SLA of

Moreland (C), Brunswick, located in Greater Dandenong City SSD to 2785 households within the SLA of *Melbourne (C), Remainder,* located in inner Melbourne.

Moreover, the SLAs with higher estimated numbers in Perth are *Starling (C), Central* (with 4011 households), located within north metropolitan, *Rockingham (C)* (with 2255 households) located within southwest metropolitan, *Canning (C)* (with 2017 households) located within southeast metropolitan, and *Bayswater (C)* (with 1918 households) located within east Metropolitan SSDs. The only SLA that comes from Brisbane in the top 50 lists is *Ipswich (C), Central* (with 2123 renter households in housing stress), and is located within the Ipswich City SSD in far southwest Brisbane.

7.5 Small Area Estimates: Percentage of Households in Housing Stress

Estimated numbers or counts are not typically useful in mapping to draw inferences about the likelihood of housing stress because of size variations of populations with the SLAs. It is expected that SLAs with a larger number of households would have a higher count of housing stress. So a standardization of size variations can be performed by percentage estimates, and the maps of percentage estimates may mask the spatial pattern in the likelihood of incidences if the rates are based on population households of very different sizes. The small numbers problem arises whenever a small number at risk is involved as the denominator of the percentage estimates. Tension also exists between the relative and absolute views of housing stress in which the possibility of a household in housing stress must be weighted against the frequency of occurrence of the numerator of the percentage estimates. Besides this, more prevalent small areas are a warning to the local- and national-level policy makers (as well as to households) and a prospect to understand the context of the socioeconomic issue. Therefore, this section is designed to offer the results of percentage estimates of housing stress in Australia by tenure types.

7.5.1 Percentage Estimates of Housing Stress for Overall Households

Figure 7.8 displays a spatial distribution of the percentages of low-income households in housing stress by SLAs throughout Australia. An equal interval classification is considered in this mapping to portray the percentage estimates data for showing a further prospect of the cumulative frequency legend. The map reveals that most of the SLAs in the east coast and some SLAs in the west coast regions have a relatively higher percentage (above 11.2%) of households in housing stress.

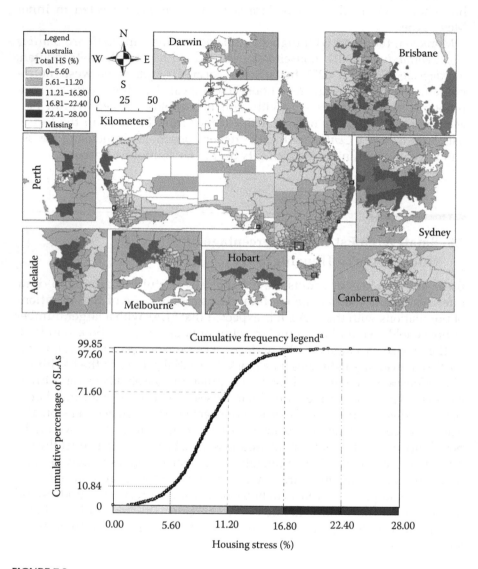

FIGURE 7.8
A spatial distribution of the percentage estimates of housing stress with an alternative cumulative frequency legend. *Note:* An equal interval classification is used to portray the spatial distribution of the percentage of low-income households in housing stress by SLA in Australia. Note that this classification is considered here only as a special case to see the further features of cumulative frequency legend. All other figures will consider the quantile classification.
[a]A cumulative frequency legend to depict the cumulative percentage of SLAs associated with each data interval on the spatial distribution map.

Large numbers of small areas in inland remote regions throughout Australia have the lowest rates of housing stress. Additionally, small areas across the mining regions in Queensland and Western Australia illustrate relatively higher percentages of low-income households in housing stress. Further reasons could be a significant lack of supply of housing in the market in these quickly growing mining areas, which create a high demand of housing and then increase of housing costs for households. The figure also reveals that a number of SLAs located within some major capital cities of Australia have significantly high rates of housing stress (ranging from 16.81% to 28.00%). Normally, in these major city SLAs, supply of housing is practically limited to low- and middle-income households with growing populations and a typical increase in housing costs.

Basically, a geographic map can only give an easy way to visualize how a measurement varies across small areas or can show the level of variability within a big region. However, it cannot thoroughly clarify the statistical distribution patterns of small areas within a very large area. For instance, from the spatial map in Figure 7.8, as well as its ordinary equal interval classification of legend, a reader cannot find exact information about the numbers or percentages of SLAs in Australia in each range of housing stress classification. The alternative cumulative frequency legend demonstrates that information of the map more clearly. The data range for each class interval in the cumulative frequency legend is given in both the numbers and length of distribution.

Moreover, the cumulative frequency legend can reveal the patterns of the distribution of estimates by small areas. For example, in the earlier map, the percentage estimates of housing stress demonstrate a pattern of normal (s-shaped cumulative frequency legend) distribution for low-income households. Additionally, since the cumulative frequency legend is based on the cumulative percentage of observational units such as the percentage of SLAs, the map reader can easily discern the information that 26% of the small area geographic units or SLAs in Australia are placed in the third class interval. This can be calculated by examining the projection lines associated with the upper and lower limits of the class interval for the percentage of housing stress, that is, 97.60 − 71.60 = 26.

The usual choropleth map legend gives no indication of the percentage or number of SLAs in any class interval. In which case, without an alternative legend or any other specific indication, a reader may have to count the number of small areas on the map shaded into the same interval color to calculate the percentage of small areas falling within the specific interval. In practice, such a count of geographic small units would be time-consuming and ineffective for a big geographic domain like Australia, which encompasses a large number of SLAs. Although the quantile classification may give an answer to this counting problem, still the patterns of the distribution would be unveiled by the cumulative frequency legend for better understanding of the geographic map.

For a further perception, Table C.6 presents the results of 50 SLAs with the highest percentage estimates of housing stress. These results are quite different from the results highlighted in the estimated numbers in Table 7.4 in the previous section. Data demonstrate that few SLAs in the inner locations of Melbourne, Canberra, and Adelaide have the highest percentage estimates of housing stress. For example, small areas of inner city in Melbourne and Canberra are, respectively, with 27.0% and 23.2% of households experiencing housing stress. In addition, a large number of SLAs from Brisbane and Sydney and some others from coastal cities in Queensland and New South Wales also have the highest rates of housing stress.

It is evident that several SLAs in Sydney that contain a significantly high proportion of households in housing stress also have a significantly large estimated number in housing stress. In fact, this is because of not only insufficient supply of housing in the high-demand markets but also a large number of households live in those Sydney SLAs with a sizable representation of them from the low-income households. Perhaps the small numbers problem exists within many SLAs in Brisbane, where the estimated number of households experiencing housing stress is very low, but the percentage estimate is significantly high due to a small value of denominator.

Most of the SLAs with high percentage estimates of housing stress in Queensland are located around inner Brisbane, far south Logan City, coastal Redland Shire, and Redcliffe City in Brisbane, in Gold Coast and Richmond–Tweed, and in the entire coastal regions from the Caboolture Shire to Hervey Bay including Sunshine Coast and in far north Fitzroy, Mackay, Cairns City, and the mining regions of Herberton and Croydon. However, the whole middle and west parts of Queensland have SLAs with the lowest estimated values for housing stress (see also Figure 7.8).

As a general view, the SLAs across New South Wales show relatively higher percentage estimates than the SLAs of other states in Australia. The entire coastal areas of New South Wales contain housing stress rates from 11.2% to 22.4%. Besides, there are a few SLAs to be found in the central New South Wales—Goulburn Mulwaree, Bathurst, central Macquarie, Armidale, and Inverell—that also have percentage estimates like high values in coastal regions.

Essentially, these growing populated inland cities are with somewhat inadequate socioeconomic conditions, and perhaps households residing there have income at lower levels. Also, rapidly growing mining regions within Western Australia (e.g., Gascoyne, Northampton, Nungarin, and Vasse) may also have similar situations as most other mining areas have in Australia. Nonetheless, a number of SLAs in Perth have shown the highest estimated numbers of households in housing stress, but none has rated more than the 16% required to be included at the top 50 SLAs' lists.

7.5.2 Percentage Estimates of Housing Stress for Buyer Households

The percentage estimates of housing stress for buyer households reveal that about 80% of SLAs in Australia have up to 14% buyer households in housing stress and half of the SLAs have less than 10%. Most of the SLAs with a remarkably small number of buyer households have very high rates of housing stress due to the small number effect into denominator. Nearly all SLAs around the central New South Wales, Victoria, and Tasmania, in southeast and far north of Queensland, and in the areas around the capital cities of Adelaide and Perth show relatively higher percentage estimates. In contrast, small areas within middle to central regions of Queensland and south and Western Australia have low rates of housing stress estimates for buyer households.

Moreover, a large number of SLAs across Brisbane, Canberra, northern Sydney, and greater inner parts of Melbourne, Adelaide, Perth, Darwin, and Hobart have percentage estimates at lower levels, compared to the estimates for SLAs located in other parts of these cities. In particular, SLAs with significantly high rates of buyers in housing stress are mostly clustered in Redland Shire and the southern region of Brisbane, in north and south Canberra, in greater western Sydney, in outer Melbourne, and in northern Adelaide.

Table C.7 offers the results of the 50 SLAs with the highest percentage estimates of buyer households experiencing housing stress. Findings reveal that most of these 50 SLAs are located outside the major capital cities in Australia, and these SLAs have fairly small numbers of estimated households purchasing their house. In addition, results illustrate that the SLAs from Sydney (e.g., the northeast and west Bankstown, south Parramatta, east Fairfield, and Auburn) and Melbourne (e.g., Hume–Broadmeadows and Brimbank–Sunshine) show significantly high percentage values with very large numbers of buyer households in housing stress. It may seem predictable that the proportion of low-income households buying their house in these SLAs is significant with a high house price, and the socioeconomic situation of these areas is somewhat lower compared to that of other areas.

7.5.3 Percentage Estimates of Housing Stress for Public Renter Households

The percentage estimates of housing stress for public renter households demonstrate that about 96% of SLAs in Australia have an estimate less than or equal to 33% and two-thirds of these SLAs have estimates between 10% and 20%. As with the estimates of buyer households, most of the SLAs with too small a number of public renter households have remarkably high rates of housing stress. In addition, the SLAs in capital cities (especially in Sydney, Melbourne, Brisbane, Canberra, Perth, and Darwin) and major coastal centers in the eastern and western parts of Australia have significantly higher rates of housing stress estimation for public renter households. For example, lots of SLAs in Brisbane, Canberra, and Darwin and almost all SLAs in Sydney have a higher percentage of public renter households with housing stress. This is due to significant proportions of low-income households receiving welfare payments such as CRA, disability allowance, and aged pension are resident of these areas (see discussion in Section 7.4.3). In addition, nearly all small areas across the greater central regions (excluding the far mid-north mining areas of Queensland) of each individual state in Australia reveal relatively much lower percentage estimates.

Furthermore, the results of the 50 SLAs with the highest percentage estimates of public renter households in housing stress are represented in Table C.8. Data show that these 50 SLAs are located in various parts of Australia (mostly in Queensland) and the majority of which have very small numbers of households at public renting tenures. However, for many of these SLAs, the estimated results of the percentage of households in housing stress could be ignorable due to the persistence of the small numbers problem.

7.5.4 Percentage Estimates of Housing Stress for Private Renter Households

The percentage estimates of housing stress for private renter households reveal that most of the SLAs (more than 80%), which are geographically located across the east and west coasts in Australia, have a higher range of estimates from 18.1% to 58.8%. Among these SLAs with higher percentage estimates, half of the areas have estimated rates of 24.3%–33.2%. In general, SLAs in New South Wales and Victoria show much higher estimates than the SLAs in other states. Variations in the estimates are also obvious for small areas within different capital cities. For example, a majority of SLAs in Brisbane, Canberra, and Darwin have somewhat lower rates, compared to SLAs in other capital cities. Particularly, the greater western part of Sydney (e.g., Fairfield–Liverpool, Canterbury–Bankstown, and Parramatta regions), most part of Melbourne and Perth (except the central suburbs), and about the entire Adelaide and Hobart region contain SLAs with very high percentages of private renter households in housing stress.

Moreover, the results of the 50 SLAs with the highest percentage estimates of housing stress for private renters demonstrate that these 50 SLAs are geographically located in different capitals and noncapital cities in Australia (see Table C.9). Some of the SLAs with very high rates within capital regions are *Hall* in Canberra; *Pinjarra Hills, Redland (S) Bal, Bribie Island, Caboolture, Hinterland,* and *Woodridge* in Brisbane; *Playford (Elizabeth), west* and *west central,* and *Port Adelaide Enfield, park* and *inner,* in Adelaide; *Fairfield (east), Bankstown (C) (northeast* and *northwest),* and *Parramatta (south)* in Sydney; *Hume, Broadmeadows,* and *Yarra Ranges, central,* in Melbourne; *Sorel (M), Pt A,* in Hobart. On the other hand, most of the noncapital city SLAs with very high levels of percentage estimates are from the coasts of New South Wales: Tweed Heads, Byron Bay, Richmond Valley, Great Lakes, Clarence Valley, Coffs Harbour, Lismore, Kempsey, Hastings, Shoalhaven, and Wollongong with some SLAs from the coastal cities in Queensland. Additionally, *Nungarin* in Western Australia, *West Tamar (M), Pt B,* in Tasmania, *Hepburn, east* in Victoria, and *Alexandrina (DC), Coastal* in South Australia, also have considerably higher estimated rates, which have entered into the list of top 50 SLAs.

7.5.5 Percentage Estimates of Housing Stress for Total Renter Households

The percentage estimates for total renter households show fairly similar results to the estimates for private renter households. Most of the SLAs across the east and west coasts have a high level of estimates up to the maximum value of 58.8%. Three-fifths of the total SLAs (about 776 SLAs) in Australia have estimated rates between 18.1% and 30.6% with a relatively heavy density into the fourth quantile SLAs having estimates from 27.0% to 30.6%. Besides, a majority of SLAs in New South Wales and Victoria have higher-level estimates than the SLAs in other states, and small area estimates for different capital cities show much variation. Like the estimates for private renter households, SLAs in the greater western regions of Sydney, outer parts of Melbourne, and Adelaide demonstrate very high percentage estimates for total renter households experiencing housing stress.

Moreover, the results for 50 SLAs with the highest percentage estimates of total renter households in housing stress are presented in Table C.10. A number of these SLAs are found within major coastal centers in New South Wales and Queensland. For example, various SLAs within the regions of Richmond–Tweed, Clarence–Coffs Harbour, Great Lakes, Hastings, Shoalhaven, Kempsey, Wyong, Fairfield, and Bankstown in New South Wales have very high proportions of renter households with housing stress. Additionally, several SLAs within the southeast outer and coastal Brisbane regions including various SLAs across Gold Coast, Wide Bay–Burnett, Sunshine Coast, Fitzroy, Townsville City, and far north regions of Queensland show very high rates. It is noted that these are some of the famous tourist destinations in Australia where the demand of renting houses is very high but supply of housing is quite low. The central Yarra Ranges, east Hepburn, Mount Alexander, west Moorabool, and Greater Bendigo in Victoria and Hall in the ACT also have percentage estimates at the highest levels. Overall, the findings demonstrate that although many SLAs in the east coast of Australia have significantly high proportions of low-income households (particularly those who are renting) in housing stress, a number of these SLAs in Queensland and the ACT have very small estimated counts. Therefore, at various SLAs, the highest percentage estimates have been observed as a result of the small numbers problem.

7.6 Concluding Remarks

In this chapter, a thorough discussion about various empirical results of small area estimation derived from the model outputs has been provided. The results have shown that housing stress in Australia varies with geographic units and by tenure types of households. About 1 in 10 households in Australia is experiencing housing stress and large numbers of these households are living in the east coast states of New South Wales, Victoria, and Queensland. Additionally, at the level of SD, most of the households in housing stress are residing in Sydney, Melbourne, Brisbane, Perth, Adelaide, and some other SDs mainly located across the coastal centers of New South Wales and Queensland. Canberra and other capital city SDs have shown relatively low estimates of housing stress.

There are several major geographic regional areas throughout Australia where housing stress is concentrated in capital cities and noncapital major coastal centers. In particular, Newcastle has shown the peak estimated numbers of households with housing stress at SSD level in Australia. Some other noncapital coastal cities such as Wollongong, Richmond–Tweed, Hastings, and Clarence in New South Wales and Gold Coast, Sunshine Coast, Wide Bay–Burnett, and Cairns City in Queensland have spatial subdivisions with much higher values in housing stress estimates. Moreover, many SSDs within capital cities have also demonstrated large estimated figures. Basically, these regional subdivisions are located in the greater western and northern regions of Sydney; the western, inner, eastern middle, southern, and northern outer regions of Melbourne; the northwest, southeast, and Logan City regions of Brisbane; the north, east, and southeast metropolitan regions of Perth; and the northern, southern, western, and eastern regions of Adelaide.

Findings have revealed that SLAs with the highest numbers of households in housing stress are mostly located in the New South Wales coastal cities. The rapidly growing mining areas around inland locations in different states have many SLAs with relatively higher estimates of housing stress. Besides, small areas in Sydney and Perth have shown very high estimated numbers. High levels of estimates are also observed in most of the SLAs in Melbourne, Adelaide, and Hobart. In contrast, significantly large numbers of SLAs in Brisbane, Canberra, and Darwin have much lower estimates. These SLAs are not only small in sizes but also have relatively much smaller population households. Additionally, the results of the percentage estimates have demonstrated that many SLAs with very low estimated numbers across Australia have very high rates of housing stress. Nonetheless, various SLAs in capital cities have confirmed significantly large values in housing stress estimation for both count and rate.

Moreover, the private renter households in Australia are more prevalent in housing stress compared to all other tenures. Typically, in most cities, the small area housing stress estimates for private renters have significant influence on the estimates for total renters and then overall households. Only several SLAs in Canberra and Darwin have shown that public renter households are more in housing stress. Besides, findings for the estimates of buyer households have indicated that a significant number of buyer households in Australia are also experiencing housing stress, and most of these households are residing in SLAs within various major cities. As observed, about a quarter of the total buyer households in Australia are concentrated into several SLAs located within some specific regions in Sydney, Melbourne, and Perth. Finally, a comprehensive account of the estimates for individual capital cities will be presented in the following chapter.

8

Analysis of Small Area Estimates in Capital Cities

8.1 Introduction

The various empirical results from the model have been discussed in the earlier chapter. Essentially, those discussions were fully based on the statistical local area (SLA)-level housing stress estimates throughout Australia. However, it is natural that most of the SLAs located in the major city areas are relatively smaller in geographic size but typically compacted with vast numbers of households and/or populations. Also, major capital city areas are usually considered as the places where issues of housing stress are more acute. The characteristics of different cities (as well as various regions within a city) also vary with many factors including the geographic location, population size, socioeconomic conditions, governances, and social welfares. Taking an Australia-wide view, it is unlikely to clarify the actual scenarios of housing stress for all small areas situated within the major cities. Therefore, zooming in to view different capital cities, a specific close look on the model estimates of housing stress is needed.

This chapter is designed to offer a thorough representation of the results of housing stress estimates for the eight major capital cities, that is, Sydney, Melbourne, Brisbane, Perth, Adelaide, Hobart, Canberra, and Darwin in Australia. The outline of the chapter is as follows. The rest of "Introduction" gives a glimpse of the aggregated results for the eight capital cities, trends in housing stress for some cities, and a specification of mapping the small area estimates. Besides, Sections from 8.2 through 8.7 present the reports and spatial distributions of the SLA-level housing stress estimates by households' tenure types for the six cities—Sydney, Melbourne, Brisbane, Perth, Adelaide, and Canberra. The results and discussions for the other two cities, Hobart and Darwin, are included in Appendix E in order to manage the length of the chapter. Finally, Section 8.8 provides a concluding summary of the chapter.

8.1.1 Scenarios of the Results for Major Capital Cities

A synopsis of the aggregated numbers of households and their housing stress estimates is presented in Table 8.1 for the eight cities: Sydney, Melbourne, Brisbane, Perth, Adelaide, Hobart, Canberra, and Darwin in Australia. The first part of the table shows the estimated numbers of households living in each city by their tenure types. Altogether there are 4,521,339 households (about two-thirds of the total estimated households in Australia) living in these eight cities—among them, about 60% of households are living in Sydney and Melbourne. Sydney alone has more than 30% of households. Moreover, the estimated number of public renter households for Sydney is almost twice the number for Melbourne. Considering the total number of households for each city, Adelaide has also a high number of 32,338 households living in public renting.

TABLE 8.1

Scenarios of Households and Housing Stress for Eight Capital Cities

Major Cities	Buyers	Renters			Owners	Other (*Tenure*)	Total	
		Public	Private	Total				
Sydney	482,594	72,337	362,453	434,790	468,979	37,177	1,423,540	No. of
Melbourne	477,808	37,738	277,603	315,341	459,968	30,211	1,283,328	HH
Brisbane	232,546	25,309	164,870	190,179	191,146	14,319	628,190	
Perth	214,453	19,505	111,647	131,152	168,750	13,744	528,099	
Adelaide	156,662	32,338	79,690	112,028	149,057	13,026	430,773	
Hobart	27,649	5,136	14,172	19,308	27,674	1,730	76,361	
Canberra	45,761	9,453	24,101	33,554	35,567	2,027	116,911	
Darwin	13,813	3,156	9,827	12,983	6,268	1,073	34,137	
Sydney	51,154	13,717	98,444	112,161	312	23	163,655	No. of
Melbourne	51,592	6,903	76,896	83,799	284	25	135,702	HH
Brisbane	19,091	4,140	43,281	47,421	156	9	66,718	in *HS*
Perth	18,525	3,249	31,873	35,122	99	23	53,766	
Adelaide	14,768	5,059	26,834	31,893	73	10	46,749	
Hobart	2,498	744	4,599	5,343	12	1	7,856	
Canberra	2,070	1,895	3,722	5,617	3	0	7,700	
Darwin	952	607	1,592	2,199	19	1	3,171	
Sydney	10.60	18.96	27.16	25.80	0.07	0.06	11.50	% of
Melbourne	10.80	18.29	27.70	26.57	0.06	0.08	10.57	HH
Brisbane	8.21	16.36	26.25	24.93	0.08	0.06	10.62	in *HS*
Perth	8.64	16.66	28.55	26.78	0.06	0.17	10.18	
Adelaide	9.43	15.64	33.67	28.47	0.05	0.08	10.85	
Hobart	9.03	14.49	32.45	27.67	0.04	0.06	10.29	
Canberra	4.52	20.05	15.44	16.74	0.01	0.00	6.59	
Darwin	6.89	19.23	16.20	16.94	0.30	0.09	9.29	

Abbreviations: HH, households; HS, housing stress.

The number of households residing in Canberra is more than the total number for Hobart and Darwin. But Hobart alone has an estimated number of 7856 households in housing stress, which is more than the estimate for Canberra (see in the second part of the table). In addition, the numbers of households experiencing housing stress for owners and other tenures are very low, and thus, they are negligible in further analysis.

A comparatively small estimate of 6.59% of households is experiencing housing stress in Canberra (see in the last part of the table), whereas other cities have percentage estimates from 9.29% for Darwin to 11.50% for Sydney. Housing stress depends upon both housing *demand* with regard to household income and housing *supply* in the local markets. If an area has limited supply and high-income households, housing costs will be driven up due to high demand and even high-income households could end up in housing stress. On the other hand, if an area has enough housing supply and low-income households, housing costs will be quite steady or slowly increasing due to low demand, but the low-income households could end up in housing stress. Typically, supply of housing is adequate in Canberra, and most of the households living in Canberra have much better levels of household income. Perhaps, these are good reasons for the low rate of households in housing stress in Canberra. Nonetheless, other cities may have a much higher representation of low-income households along with limited supply of housing (at least in some parts of the city) like in Sydney, and those may be reasons for the higher percentage of households in housing stress.

Besides, the results indicate that housing stress is more prevalent within public renter households residing in Canberra, Darwin, Sydney, and Melbourne (see also Figure 8.1 for clear visibility). However, the percentages of housing stress households for private renters as well as total renter households are much lower in Canberra and Darwin but greatly higher in Adelaide and Hobart. Also, the estimates are rather higher in Perth, Melbourne, Sydney, and Brisbane, compared to Canberra and Darwin.

It is also clear from the graph that in Darwin and Canberra the housing stress estimates for public renter households dominate the estimates for total renter households, while, in all other cities, the estimates for private renter households dominate the estimates for the total renters. Further, the estimates for buyer households demonstrate that a significantly low proportion of households purchasing their house in Canberra are experiencing housing unaffordability, compared to buyers in all other cities.

8.1.2 Trends in Housing Stress for Some Major Cities

Trends in the number and percentage estimates of housing stress for households residing in four major capital cities, namely, Sydney, Melbourne, Brisbane, and Canberra, are presented in Figure 8.2a and b. Statistics for the year 2000–2001 are considered from the earlier study by Taylor et al. (2004), where the authors only considered four eastern states. The graph

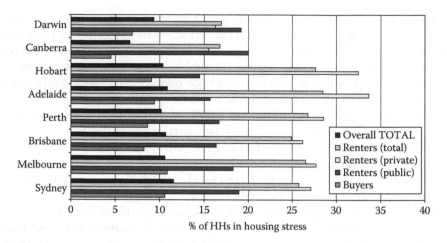

FIGURE 8.1
Distribution of housing stress by tenure in the eight capital cities.

for estimated numbers reveals that the counts of households experiencing housing stress have increased in all of these four cities over the 5-year period. In Sydney and Melbourne, the estimated numbers have dramatically increased by 57,955 and 38,875 households, respectively, whereas Canberra and Brisbane show a rather steady rise of housing stress incidence between 2000–2001 and 2005–2006 estimates.

Moreover, during this time period, the prevalence rate of housing stress households has increased by about 4.5% in Sydney, 3.4% in Melbourne, 1.8% in Brisbane, and 0.9% in Canberra (see Figure 8.2b). It is noticeable that the percentage estimates for Sydney have increased radically from 7.0% in 2000–2001 to 11.5% in 2005–2006 and jumped over the rates for Brisbane and Melbourne. Some of the reasons for this dramatic increase in housing stress estimates in Sydney and Melbourne are perhaps increase in populations, housing costs, interest rates, income risk, etc., and relative decrease in social housing stock, supply of housing in the market, and per capita income of the households. Considering the time phase and population growth of Brisbane, the estimate shows some steadiness in the increasing trend. Nonetheless, the percentage estimate of housing stress in Canberra is always much lower than that in the other cities. This indicates a big representation of relatively higher-income households living in the national capital region and/or that housing-related policies including new land development strategies for adequate supply of housing and the Community Housing Canberra (CHC) affordable housing program for the Australian Capital Territory (ACT) are better functioning for its community.

8.1.3 Mapping the Estimates at SLA Levels within Major Cities

As indicated in Chapter 6, the geographic maps are more descriptive for observing the spatial distribution patterns of housing stress estimates in

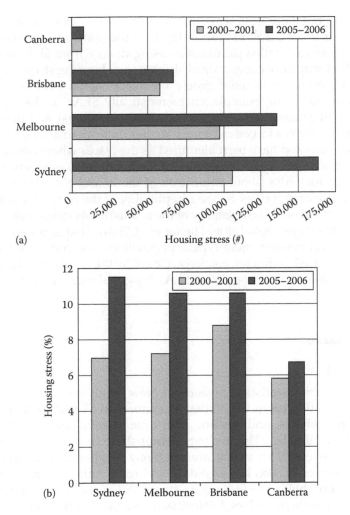

FIGURE 8.2
Trends in housing stress estimates for some major cities of Australia (a) for estimated numbers and (b) for percentage estimates. (From Taylor, E. et al., *Aust. J. Reg. Stud.*, 10(3), 279, 2004 for 2000–2001 figures.)

various cities. SLA-level data for both the percentage and number estimates of housing stress have been mapped in different major capital cities with the quantile classification and under the following option:

- *Distribution of SLAs located within the individual city* (considering the SLAs that did not meet an accuracy criterion as being in the missing category).

Typically, the scope shows the distribution of housing stress relative only to all SLAs located within the same city. These maps reveal the relationships within cities, but there is no sense of relativity to other cities or regions. The notion here is to see the actual variability in housing stress estimates *within* a city. This means a detailed picture of housing stress within SLAs in a particular city that may have comparatively high or low housing stress as a whole.

Moreover, zooming in more closely on major capital cities is important for many reasons. For example, characteristically SLAs in the capital city are smaller in geographic sizes (but with big populations), and they are not clearly observable in a large-scale map like the whole of Australia. These areas are spontaneous and have been identified as the places where housing stress issues are more acute. In addition, most of the households experiencing housing stress in Australia (about two-thirds of the households) are residing in the eight major capital cities. So the acceptability of the convergence measure (and, hence, the expected consistency of the results) tends to dissipate with distance from the major capital cities (Taylor et al. 2004). That is, urban city areas have larger and possibly more typical populations, meaning that GREGWT reweighting-based estimates converge closer to the target benchmarks for urban areas than for low-population SLAs typical of rural city areas.

8.2 Sydney

Sydney, the capital statistical division of New South Wales, is divided into 14 statistical subdivisions (SSDs) that encompass 64 SLAs. The distributions of SLAs, households, and housing stress for various SSDs in Sydney are presented in Table D.1. The findings demonstrate that the "inner Sydney" consists of seven SLAs, which are relatively smaller in size. The aggregated number of 121,056 households is living in "inner Sydney" of which 12.05% of households are in housing stress. All of the northern SSDs such as "Central Northern Sydney," "Northern Beaches," and "Lower Northern Sydney" have low percentages of households in housing stress. This should be because most of the households living in these areas may have higher income. Although the highest numbers of 150,204 households have resided in "St George–Sutherland," the rate of housing stress for this SSD is comparatively lower. Besides, data show that most of the Western Sydney's SSDs have around 11.50% of households in housing stress except the "Central Western Sydney" that has an estimate of 15.19%. Further, a very high proportion of households (an average of 16.50%) experiencing housing stress are residing in the SSDs' Canterbury–Bankstown and Fairfield–Liverpool.

8.2.1 Housing Stress Estimates for Overall Households

Figure E.1 presents the spatial distributions of the estimates of housing stress for overall households in Sydney. The map for the percentage estimates shows that housing stress rates are highest within the southwest Fairfield–Liverpool region of mideastern Sydney; in inner and Western Sydney; in Parramatta, Canterbury, and Bankstown; and in Gosford–Wyong located far northeast of Sydney. More specifically, the SLAs, that is, *Fairfield (C), east*; *Auburn (A)*; *Sydney (C), inner*; *Parramatta (C), south* and *north*; *Bankstown (C), northeast*; and *Canterbury (C)*, have high prevalence rates of households in housing stress. The SLA of *Wyong (A)—south* and *west*—located in the Gosford–Wyong region is also in the highest quantile. In fact, this SLA has a much lower percentage rate compared to the rates of other SLAs in the fifth quantile. The rest of the SLAs in Sydney may perhaps be the residence of a significant number of high-income households, and thus, most of these SLAs fall in the lower quintiles (see Figure E.1a). However, Sydney as a whole, compared to all other cities in Australia, does contain a vast number of housing stress households.

Moreover, the usual legend of quantile classifications shows that each class interval has approximately equal numbers of SLAs. But it does not give any idea about the distribution pattern of the SLAs within the quantile interval. Nonetheless, the cumulative frequency legend adds more specific and clear information. For instance, the alternative cumulative frequency legend illustrates clearly that at the bottom quantile there is only one SLA that has exactly 4.30% of households in housing stress and all other SLAs are nearest to the upper range of 7.10%. Also, at the fifth quantile, the distribution pattern of the SLAs is not only very wide but also the SLAs are clustered around the rates of 16.50% and 20.50%.

The spatial distribution of the estimated numbers in Figure E.1b shows somewhat different results than that of the estimated percentages. The findings demonstrate that SLAs in South and Central Western Sydney, Fairfield–Liverpool, Canterbury, Eastern Suburbs, and Northern Beaches have very higher numbers of households experiencing housing stress. As observed earlier (see Section 7.4.1), the cumulative frequency legend clearly further indicates that the SLAs of *Canterbury (C)* and *Fairfield (C), east*, are closest to the right end of the fifth quantile and far above to the top from the other SLAs in Sydney. Additionally, the Gosford–Wyong region in the far northeast of Sydney also has very high estimated numbers. For example, *Gosford (C), west*, and *Wyong (A), northeast*, both fall in the fifth quantile. However, for the percentage distribution, these two SLAs were in the fourth quantile. The inner Northern and the far west Southern Suburbs, on the other hand, have relatively low estimated numbers of households in housing stress.

In terms of the housing stress measures (both numbers and percentage), the rupture of east/west and north/south in Sydney is fairly apparent. The influence of income distribution in determining these estimations is perhaps strong. Like the distribution of gross estimates of housing stress, almost half of Sydney seems to be a place that is highly unaffordable for low-income households relative to the other SLAs in Australia. Nevertheless, most of the SLAs located across the Northern Sydney's SSDs (including lower

Northern Sydney, Central Northern Sydney, and Northern Beaches) and some SLAs located in Central and outer Western Sydney's SSDs have less than 8.7% of families in housing stress; they are thus in the lower 40% of all SLAs in Australia.

8.2.2 Small Area Estimation by Households' Tenure Types

8.2.2.1 Estimates for Buyers

Figure E.2 provides the spatial distributions of the estimates of housing stress for buyer households. Most of the SLAs located in the mid-southwest regions of Sydney have a very high proportion of buyer households in housing stress. It is noted that all SLAs in Canterbury–Bankstown SSD are in the highest quantile. Additionally, the SLAs of *Fairfield (C), east* and *west* in Fairfield–Liverpool; *Parramatta (C), south; Auburn (A);* and *Strathfield (A)* in inner and Central Western Sydney and a number of SLAs in Blacktown and Rockdale regions also have higher rates for the fourth quantile (as does the far northeastern Wyong area). However, most of the SLAs in inner Sydney and north-coastal, west, and south-coastal regions have low rates of households in housing stress. Findings reveal that about 60% of Sydney SLAs show a relatively lower rate (up to 10.9%) of households experiencing housing stress, whereas three SLAs (such as *Fairfield (C), east; Parramatta (C), south;* and *Bankstown (C), northeast*) show higher rates (close to 25.3%). Moreover, the spatial distribution for the estimated numbers displays a fairly similar pattern (see Figure E.2b). The whole middle Sydney region (from the eastern St George–Sutherland to the western Blue Mountains) and Gosford–Wyong are the places with most counts of households estimated as experiencing housing stress.

8.2.2.2 Estimates for Public Renters

The spatial distributions of the housing stress estimates for public renter households in Sydney are presented in Figure E.3. At least geographically most of the regions in Sydney are at the bottom two quantiles. The SLAs with higher percentages of housing stress are scattered across the Sydney statistical division. Specifically, these hot spots' small areas are mostly located in lower Northern Sydney, inner Sydney, Parramatta, outer Southwestern Sydney, Fairfield, Blue Mountains, and Northern Beaches, and also in the east and northeast of the Gosford–Wyong region. Besides, more than 95% of Sydney's SLAs show the housing stress rates approximately from 16% to 22% (see the alternative legend). It is remarkable that the range of the fifth quantile is so large. This is because there is one SLA called *Pittwater (A)* in Northern Beaches, which has only three public renter households and all of them (100%) are estimated to be in housing stress. Furthermore, estimated numbers of households in housing stress are higher in the SLAs of

Fairfield (C), east; Canterbury (C); Blacktown (C), southeast; Campbelltown (C), north; Randwick (C); Liverpool (C), east; Sydney (C), south; and *Blacktown (C), southwest,* located in the middle east and southwestern Sydney (see in Figure E.3b). In contrast, the SLAs in mid-north and far southwest regions of Sydney have very low estimated numbers.

8.2.2.3 Estimates for Private Renters

Figure E.4 displays the spatial distributions of the housing stress estimates for private renter households. It is notable that the prevalence of housing stress is more common in private renter households compared to the buyer and public renter households. Also, the division of east/west and north/south in Sydney is fairly apparent in the Figure E.4(a) for percentage estimates. The entire southwest part (except *Camden (A)* and *Penrith (C)—West*) and the northeast part of Sydney have SLAs in the higher quantiles. In particular, almost all SLAs in Canterbury–Bankstown, Fairfield–Liverpool, and Gosford–Wyong and some SLAs in Central Western Sydney and Blacktown have the highest percentages of private renter households in housing stress, whereas the middle (east to west) part and southeast part of Sydney have SLAs in the lower quantiles, and the SLAs located in these areas are presumably those with the greater representation of high-income households living in private tenure.

Moreover, the map (b) in Figure E.3 demonstrates that SLAs in inner eastern and mid-southwestern Sydney very obviously have higher numbers of households in housing stress for private renters (as does Gosford–Wyong in the far north). As estimated earlier, *Canterbury (C); Fairfield (C), east; Randwick (C);* and *Gosford (C), west,* are some of the higher-ranked SLAs for estimated numbers of private renter households experiencing housing stress. The mid-northwestern and far Southern Suburbs, on the other hand, have relatively low estimated numbers.

8.2.2.4 Estimates for the Total Renters

Figure E.5 portrays the spatial distributions of the housing stress estimates for total renter households in Sydney. In terms of the percentage estimates, Sydney is typically divided into two parts—the east to west middle regions are in lower quantiles and the north and south regions are in higher quantiles. In particular, almost all SLAs in the entire middle part of Sydney from the east *Randwick (C)* located in Eastern Suburbs to the west *Hawkesbury (C)* located in outer Western Sydney (just without *Sydney (C)—inner*) and two SLAs in the St George–Sutherland such as *Sutherland Shire (A), east* and *west,* have lower proportions of renter households in housing stress. On the other hand, all SLAs in Gosford–Wyong in the north, most of the SLAs in the southwestern regions including Blue Mountains, and small areas in Fairfield to Canterbury suburbs have higher percentage estimates.

Although the map presents the SLAs in different quantiles, typically SLAs in Sydney have high percentage estimates for total renter households ranging from 11.10% to 40.10%. It is notable that the prevalence of housing stress is more common in renter households, compared to the buyer households. In renter households, some of the most visible hot spots for housing unaffordability in Sydney are *Bankstown (C), northwest* and *northeast; Strathfield (A); Gosford (C), west; Canterbury (C); Parramatta (C), south; Wyong (A), southwest* and *northeast; Auburn (A);* and *Fairfield (C), east.*

Moreover, the map of estimated numbers for total renters shows virtual consistency with the map of the estimated numbers for the private renter households as well as the total households. This finding supports the results that private renters in different parts of Sydney (such as Canterbury–Bankstown, Fairfield–Liverpool, Blacktown, central Holroyd, Ryde and Randwick, and west Gosford) are more prevalent in housing stress, and the estimated numbers for this tenure group greatly influence the estimated numbers for total renter households, and finally, the overall housing stress counts in Sydney.

In general, the spatial distributions of housing stress estimates again point to the middle inner west and southwest parts of Sydney (and to Gosford–Wyong) where housing supply is very limited and the demand as well as costs of housing is high for a significant proportion of low- and middle-earner households. Although the micro-level scenarios within various spatial maps somewhat vary with these patterns in distributions (indicated by cumulative frequency legend), the SLAs in these areas have many of the highest percentages and numbers of households in housing stress in relation to both Sydney itself and the entire Australia.

8.3 Melbourne

Melbourne, the capital statistical division of Victoria, is divided into 16 SSDs, which includes 79 SLAs. The distributions of SLAs, households, and housing stress for various SSDs in Melbourne are presented in Table D.2. The results reveal that the top five SSDs, namely, western Melbourne, eastern middle Melbourne, southern Melbourne, inner Melbourne, and northern middle Melbourne, host more than half of the households (about 50.77% or 651,488 households). In particular, the inner Melbourne contains eight smaller-sized SLAs, and the aggregated number of households living in this SSD is 116,387, of which 14,264 households (12.26%) are in housing stress.

Moreover, the SSDs such as eastern middle and outer Melbourne, southern Melbourne, northern outer Melbourne, Boroondara City, and the Yarra Ranges Shire Part *A* have low percentages of households in housing stress. Among these SSDs, the Boroondara City has the lowest prevalence. This may be because a very high number of households living in this SSD have relatively higher income. Furthermore, comparatively smaller numbers of households have been residing in the Hume City (with 45,678 households) and Greater Dandenong City (with 42,725 households), but the rates of housing stress for these SSDs are very high. All other Melbourne's SSDs have around 11.50% of households in housing stress except Frankston City, which has an estimate of 12.59% of households experiencing housing stress.

8.3.1 Housing Stress Estimates for Overall Households

Figure E.6 provides the spatial distributions of the estimates of housing stress for overall households in Melbourne. The map for percentage estimates shows that the SLAs in the highest quantile are scattered across different SSDs in Melbourne. For example, SLAs with higher prevalence of housing stress are located in Melton, Hume City, western and inner Melbourne, eastern middle Melbourne, Dandenong City, Frankston City, southeastern outer Melbourne, and central Yarra Ranges Shire. In contrast, most of the SLAs in eastern middle and outer Melbourne, northern outer Melbourne, southeastern outer Melbourne, Boroondara City, southern Melbourne, and southwest Yarra Ranges Shire have estimates in lower quantiles.

Additionally, the SLA of *Port Phillip (C), west,* in inner Melbourne has a much lower percentage rate compared to the rates of other SLAs in this SSD. However, the alternative cumulative frequency legend indicates that the SLA of *Melbourne (C), inner,* stands alone at the right top end with 27.00% of households in housing stress. Also, this estimate makes the length of the fifth quantile so long, as more than 80% of SLAs in Melbourne have prevalence rates of up to about 13%. Thus, compared to the SLAs within some other cities in Australia, Melbourne as a whole has a large number of SLAs with an estimated lower proportion of households in housing stress, even though it contains an overall higher number of housing stress households.

The spatial distribution of the estimated numbers of households in housing stress for Melbourne is presented in Figure E.6b. The findings reveal that almost all SLAs located around the outer Melbourne (except *Hume (C), Craigieburn,* and *Mornington Peninsula Shire*) and some SLAs in Boroondara City, in southern and western Melbourne, and even in inner Melbourne have very lower numbers of households in housing stress. On the other hand, the middle northwest region, southern and eastern regions (including some SLAs in the Greater Dandenong City), and inner Melbourne have very high estimated numbers. For example, *Gr. Dandenong (C), Bal,* and *Gr. Dandenong (C), Dandenong,* are both in the fifth quantile classification with a density of the number of estimates between around 1500 and 2500 households.

Moreover, in terms of both the number and percentage estimates, the split of inner to middle and outer regions in Melbourne statistical division is somewhat clear. Except for a very few SLAs, most of the outer Melbourne regions and some parts of inner and middle Melbourne are seemingly affordable places for relatively low-income households.

8.3.2 Small Area Estimation by Households' Tenure Types

8.3.2.1 Estimates for Buyers

Figure E.7 displays the spatial distributions of the estimates of housing stress for buyer households in Melbourne. The map for percentage estimates clearly indicates that almost all of the SLAs (except two SLAs: *Melbourne (C), inner*, and *Melbourne (C), S'bank-D'lands*, in inner Melbourne) located in the inner southwest regions, eastern (including Boroondara City), and western middle to outer Melbourne have very low percentages of buyer households in housing stress. Also, the *Wyndham (C), south* in Melton–Wyndham, has a lower rate. In contrast, most of the outer regions of Melbourne show higher proportions of households experiencing housing stress. More specifically, SLAs in the highest quantile located around the outer regions are *Mornington P'sula (S), south* (in Mornington Peninsula Shire); *Cardinia (S), south; Casey (C), Hallam,* and *Casey (C), Cranbourne* (in southeastern outer Melbourne); *Yarra Ranges (S), central* (Yarra Ranges Shire Part A); Hume (C), Craigieburn, and Hume (C), Broadmeadows (in Hume City); *Whittlesea (C), southwest* (in northern outer Melbourne); and *Melton (S) Bal* (in Melton–Wyndham). Furthermore, few SLAs in western Melbourne with Moreland City and the Greater Dandenong City region in the southeast Melbourne also show higher percentages. This scenario is quite a reverse of the scenario for overall rates of housing stress. Typically, this may be because most of the low-income households are purchasing their house in the outer regions.

Moreover, the spatial distribution (see Figure E.7b) for the estimated numbers reveals that the higher numbers of households in housing stress are residing within the SLAs located in the greater middle of western and southeastern regions. In particular, *Brimbank (C), Keilor* in the western region, and *Dr. Dandenong (C), Bal* in the southeast region, are in the fifth quantile. On the other side, almost the whole inner Melbourne and surroundings are in lower quantiles. Some SLAs in the far Melton–Wyndham, northern outer including Yarra Rangers Shire, and southeastern outer Melbourne have also lower estimated numbers. Furthermore, the finding indicates that about 60% of SLAs in Melbourne have only up to 633 buyer households in housing stress, and most of them are located in these lower quantile regions.

8.3.2.2 Estimates for Public Renters

The spatial analyses of the estimates of housing stress for public renter households are given in Figure E.8. The map for percentage estimates shows a missing estimate at the SLA *Casey (C), south,* in southeastern outer Melbourne (since in this SLA there is no household living in public renting, so that makes the denominator of percentage zero and leads the estimate to be missing). Besides, the SLAs with higher percentage estimates are mainly centered within three parts of the Melbourne statistical division, and they are greater inner Melbourne (except *Port Phillip (C), west; Maribyrnong (C);*

Yarra (C), north; and *Banyule (C), Heidelberg),* southeastern outer and northern outer Melbourne. More specifically, the top five hot spots with the highest estimated percentages located in these regions are *Boroondara (C), Camberwell N; Melbourne (C), S'bank-D'lands; Whittlesea (C), north; Melton (S), east;* and *Melbourne (C), inner.*

Besides, the distribution pattern of the housing stress rates for Melbourne's SLAs is much clustered between 15% and 21%, and more than 60% of SLAs are within this range (see the alternative legend). It is worth mentioning that even though the rates for public renters look higher in Melbourne, the estimated number of households experiencing housing stress for public renters is very low compared to Sydney. For example, the SLA of *Melbourne (C), inner,* has the highest rate of 39.10% of households, but there are only 48 public renter households living at this SLA. As a result, the map for the estimated numbers shows the estimate in the second quantile. Furthermore, estimated numbers of housing stress are higher for the SLAs mostly located in inner western and northern regions of Melbourne. In contrast, the SLAs in all outer regions of Melbourne have very low estimated numbers, and more than 80% of SLAs have up to an estimated number of about 125 households (see Figure E.8b).

8.3.2.3 Estimates for Private Renters

The spatial distribution of housing stress estimates for private renter households is presented in Figure E.9. Almost the entire parts of inner to middle Melbourne including western and southern SSDs and some parts of eastern Melbourne have SLAs in the lower quantiles. More specifically, all SLAs (except *Melbourne (C), inner)* in inner Melbourne have the lowest percentage estimates, whereas most of the SLAs in Mornington Peninsula Shire, Greater Dandenong City, Yarra Ranges Shire (except *Yarra Ranges (S)—Seville),* Hume City region including northern middle to outer, in western Melbourne, and Melton–Wyndham have the estimates in higher quantiles.

Moreover, the map (b) in Figure E.9 demonstrates that SLAs in inner, midwestern, and southeastern parts in Melbourne clearly have higher numbers of households in housing stress for private renters. As observed in the earlier section, *Brimbank (C), Sunshine; Darebin (C), Preston; Frankston (C), west; Glen Eira (C), Caulfield; Gr. Dandenong (C) Bal; Maribyrnong (C); Melbourne (C), remainder;* and *Port Phillip (C), St Kilda,* are some of the higher-ranked SLAs for estimated numbers of households experiencing housing stress. However, there are also a few SLAs that have been noticed across those parts of Melbourne, which have smaller estimated numbers for lower quantiles. The entire outer parts of Melbourne, on the other hand, have shown comparatively very low estimated numbers for private renter households.

The findings reveal that the prevalence of housing stress ranging from 12.8% to 43.3% is more common in private renter households, compared to the other tenures' households, and most of the SLAs with higher prevalences are located around outer Melbourne regions. However, the estimated numbers of housing stress are lower for the SLAs in the outer regions. This result indicates that SLAs in the outlying parts of Melbourne have relatively smaller numbers of private renter households, but a significant number of them are experiencing housing stress. In other words, among the private renter households in outer regions who are experiencing housing stress, a considerably high number of these households may have lower income. And that is why the estimated rates are higher in these locations. Alternatively, the results show that most of the low-income households in private tenure seemingly residing within greater inner and middle Melbourne regions and proportions of housing stress are low for SLAs in these regions. Obviously, it is due to a greater representation of high-income households also living in those areas. Furthermore, the division of greater inner and outer regions of Melbourne is fairly apparent in the maps for both the percentage and number estimations. However, the scenarios are realistically converse to each other.

8.3.2.4 Estimates for Total Renters

Figure E.10 offers the spatial distributions of the estimates of housing stress for total renter households in Melbourne. The spatial map of percentage estimates for total renters shows coherent likeness with the map of percentage estimates for private renter households, while the map of estimated numbers for total renter households shows similarity to both the maps of estimated numbers for private renter households and the overall households. This means, in Melbourne where more private renter households are in housing stress, that the estimates in total renter as well as overall households are also more for those small areas. Specially, the estimated numbers of households experiencing housing stress are seemingly condensed into some areas in Melbourne (such as SLAs in southern to southeastern outer and western suburbs) for all tenure groups.

Moreover, the map of percentages estimation shows that the SLAs in inner Melbourne, outlying Yarra Ranges, Hume, Brimbank, Moreland, Gr. Dandenong, Frankston, and the Mornington P'sula coast have estimates in the highest quantile for total renter households. The results reveal that the small areas in Melbourne have much higher prevalences of renter households in housing stress with the highest estimate of 40.45% in central Yarra Ranges and the lowest estimate of 13.60% in west Port Phillip.

Additionally, the estimated numbers are also higher for some of the SLAs located in these regions (such as *Brimbank (C), Sunshine; Frankston (C), west; Gr. Dandenong (C), Dandenong;* and *Gr. Dandenong (C) Bal*) and across the central Melbourne including southern and northern middle. The SLAs *Glen Eira (C), Caulfield,* in southern part of Melbourne, and *Darebin (C), Preston,* in northern middle part of Melbourne, have the second highest count after the highest estimate of 2785 households for *Melbourne, Reminder.* On the other end, the outer SLAs *Cardinia (S), south; Nillumbik (S) Bal;* and *Manningham (C), east,* have estimates of less than 75 households in housing stress. The findings suggest that most of the outer regions in Melbourne statistical division have relatively small numbers of renter households with housing unaffordability.

Overall, the spatial distributions of housing stress estimates by tenures reveal that housing stress measures by both percentage and number estimates are somewhat prevailing in the private renter households and more generally in the renter households. The maps of percentage estimates point to the southern part of Mornington P'sula; the large outer southeastern (with Frankston and Dandenong) region; the Yarra Ranges Shire; the northern and southwestern suburbs including Craigieburn, Broadmeadows, and Whittlesea in the outer northwestern part of the city; inner area in the central Melbourne; and Melton city have a higher proportion of households in housing stress. These areas may perhaps experience an inadequate supply of affordable housing and somewhat low level of socioeconomic circumstances such as a high density of lower-income households. However, the outer parts of the Melbourne statistical division have SLAs mostly with less estimated numbers of households in housing stress.

8.4 Brisbane

Brisbane, the capital statistical division of Queensland, is divided into 12 SSDs, which encompass a comparatively larger number of 215 small-sized SLAs. It is noted that one SLA, *Moreton Island*, located in the southeast outer Brisbane is considered as missing since it did not converge at the reweighting process. The distributions of SLAs, households, and housing stress for various SSDs in Brisbane are provided in Table D.3. The results show that the two outer Brisbane SSDs consist of 92 SLAs in which 53 are in the northwest outer. However, Beaudesert Shire Part A has only one SLA with 1282 households (10.45%) in housing stress. An estimated number of 110,702 households are living in the northwest outer Brisbane of which 9,339 households (8.44%) are in housing stress, and this proportion is relatively the lowest for SSD level in Brisbane statistical division.

Among the other SSDs, Pine Rivers Shire and southeast inner Brisbane have also low percentage estimates. Moreover, nearly three quarters of housing stress households in Brisbane are residing within the SLAs located in northwest and southeast outer and inner Brisbane regions, Logan City, Caboolture Shire, and Ipswich City. Inner Brisbane (which consists of 18 SLAs) has a relatively low number of 4227 households in housing stress, but the estimated rate (12.37%) is higher. Three more SSDs such as Logan City, Redcliffe City, and Caboolture Shire have even much higher percentage (approximately 13.50%) of households in housing stress.

8.4.1 Housing Stress Estimates for Overall Households

Figure E.11 portrays the spatial distributions of the estimates of housing stress for overall households in Brisbane. As Brisbane is quite different (in terms of SLA's sizes and numbers) from the other capital cities in Australia, more variation is observed in the geographic maps. The map for percentage estimates shows that the rates of housing stress are higher for most of the SLAs located in far southeast Redland Shire, middle of the Logan City, west part of southeast outer Brisbane, inner city, and the far northeast of Caboolture Shire and Redcliffe. In particular, *Margate–Woody Point* (with 16.80%) in Redcliffe; *Caboolture (S), central* (with 16.90%) in Caboolture Shire; *Redland (S) Bal* (with 16.90%) in Redland Shire; *city, inner*, and *city, remainder* (with about 19.40%), in inner Brisbane, *St Lucia* (with 19.50%) in northwest inner Brisbane and *Marsden, Kingston*, and *Waterford West* (with about 17.10%); and Woodridge (with 20.60%) in Logan are the 10 hot spots' SLAs for housing stress. In contrast, most of the SLAs located in the mid-southeast inner, and outer regions, northwest of outlining inner to far outer, Pine Rivers Shire, and midwest of Caboolture Shire are seen in the lower quantiles. The households living in these regions may perhaps encompass a very high proportion of households with better income, and thus, most of the small areas have been placed into the lower quintiles.

Moreover, the spatial map for the estimated numbers shows that SLAs in Ipswich, Redcliffe, and Logan regions; in Beaudesert, Caboolture, and Redland Shire regions; and in southeast and northwest inner Brisbane have very high numbers of households experiencing housing stress, compared to the rest of Brisbane. The Pine Rivers Shire region in the far northwest of Brisbane also has a higher estimated number (see Figure E.11b). Besides, the peak 11 SLAs in the fifth quantile are *Bribie Island* (with 960 households) and *Caboolture (S), central; Deception Bay* and *Morayfield* (with an average of 1116 households) in Caboolture Shire; *Redcliffe–Scarborough* (with 1063 households) in Redcliffe City; *Beaudesert Shire* (with 1282 households), *Marsden* and *Browns Plains* (each with an average of 1013 households), and *Woodridge* (with 1345 households) in Logan City; and *Ipswich (C), east* (with 2114 households), and *Ipswich (C), central* (with 2933 households), in Ipswich City. But it is observed that the only SLA *Ipswich (C), central*, from Brisbane (as well as from the state of Queensland) has counted into the top 50 lists of SLAs with the highest numbers of housing stress households in Australia. That means, compared to all SLAs in Australia, Brisbane's SLAs have much lower numbers of households in housing stress. Typically, Brisbane SLAs are more homogeneous and that is why they have noticeably more estimates in the lowest figures.

In fact, Brisbane has many SLAs with the highest rates of housing stress, relative to all SLAs in Australia; it has many of the lowest as well. SLAs in Brisbane have estimated numbers ranging from 9 to 2933 households (and percentages ranging from 2.40% to 20.60%) in housing stress. More than 80%

of SLAs have less than 500 households in housing stress, and nearly 50% of these SLAs have less than 200 households in housing stress (see cumulative frequency legend of Figure E.11b). Conversely, the percentage rates are much higher for most of these SLAs. Hence, it is worth mentioning that although the numbers of households experiencing housing stress are very low in Brisbane SLAs, the small number of total households living in these SLAs leads them to higher rates (mostly between 5% and 15% as clearly suggested by the alternative cumulative legend in Figure E.11a).

8.4.2 Small Area Estimation by Households' Tenure Types

8.4.2.1 Estimates for Buyers

Figure E.12 provides the spatial distributions of the estimates of housing stress for buyer households in Brisbane. The map of percentage estimates shows obvious dissection of Brisbane into the greater mid (except five SLAs in Bowen hills, City inner, and Dutton perk) regions that have SLAs in lower quantiles and outsider regions (except Pine Rivers Shire and the north Ipswich) that have SLAs in higher quantiles. Specifically, there are two SLAs *Rochedale* and *Willawong* located in the southeast outer Brisbane that stand out with the highest estimated values of 20.30% and 35.00%, respectively. On the other side, most of the SLAs have very low values, and more than 80% of the SLAs have less than 10% of buyer households experiencing housing stress.

The map of the estimated numbers shows some similarity to those of the percentage estimates such as SLAs in the greater inner regions have low numbers of households in housing stress. Again, the Redland, Logan, Beaudesert, Pine Rivers, Caboolture, and Ipswich in the far external parts of Brisbane include the SLAs with higher housing stress counts. The respective cumulative frequency legend (in Figure E.12b) demonstrates that a half of the small areas in Brisbane have only up to 50 buyer households in housing stress. Furthermore, there are three SLAs (*Ipswich (C), east* with 699; *Ipswich (C), central* with 805; and *Beaudesert (S), Pt A,* with 885 households) with rather extreme estimated numbers, compared to most of the SLAs in Brisbane.

8.4.2.2 Estimates for Public Renters

Figure E.13 presents the spatial distributions of the estimates of housing stress for public renter households. It is reported that 32 SLAs across the Brisbane statistical division are in the "missing class" as the estimated total counts for public renter households are found to be zero in these small areas. There are seven SLAs with more than half of the households in housing stress, and among them two have 100%. However, each of these small areas has very few (ranging from 3 to 9) households living in public renting. In particular, *Riverhills* with four households in northwest outer and *Pallara–Heathwood–Larapinta* with three households in southeast outer Brisbane have reported 100% of estimates. Besides, the SLAs in higher quantiles are scattered in the inner and outer parts of Brisbane as well as in the Redland and southwest Ipswich.

Maps for the numbers of households in housing stress show somewhat reverse scenarios into at least several parts in Brisbane. For example, Redland, Logan, Ipswich, and Caboolture regions are apparently noticeable for high estimates. Most of the SLAs in inner city are observed in the lower quantiles, whereas small areas in the northwest and southeast inner Brisbane are in the upper quantiles. Typically, a significantly high number of SLAs in Brisbane have very low estimated numbers of public renter households experiencing housing stress. The alternative frequency legend (map (b) in Figure E.13) clearly shows that more than a quarter of SLAs have zero and about three quarters have less than 35 households in housing stress.

8.4.2.3 Estimates for Private Renters

Figure E.14 presents the geographic distributions of estimates for private renter households. The prevalence of housing stress is more common in households living within SLAs across the outlying regions of Brisbane. The five SLA hot spots with the highest rates are *Woodridge* in Logan; *Caboolture (S), Hinterland,* and *Bribie Island* in Caboolture; *Redland (S) Bal* in Redland; and *Pinjarra Hills* near the Ipswich city. On the other hand, except for *inner city, St Lucia,* and *Dutton Park,* the entire inner and outer Brisbane are seemingly the domains of lower proportions of housing stress.

The map of estimated numbers shows that although inner to outer Brisbane has some SLAs in the highest quantile, their proportions are relatively low. This may be the result of not only a significantly high number of private renter households living in these SLAs but also their representation into the higher-income households for these areas. Moreover, there are two SLAs in Ipswich City (which are *Ipswich (C), East* with 1287 households, and *Ipswich (C), central* with 1939 households) that have the upmost estimated numbers. Also, these two SLAs have much higher percentage estimates (more than 30%), compared to most other SLAs (nearly 80%, see frequency legend in Figure E.14b) in Brisbane.

8.4.2.4 Estimates for the Total Renters

The maps in Figure E.15 display the scenarios of housing stress estimates for the aggregated renter households in Brisbane. In terms of the percentage measure, the division of the greater inner regions and outlining regions is fairly apparent. Once again, the far southeast to southwest (such as far Redland, Beaudesert, and Ipswich) and northwest (including *Bribie Island* in *Caboolture*) of Brisbane have SLAs mostly in the higher quantiles. In particular, almost all SLAs in Ipswich (except *Ipswich (C), north*) and Logan

(*Carbrook–Cornubia* and *Shailer Park*) cities, Redland (except *Thornlands* and *Redland Bay*), and Pine rivers and Caboolture (except *Caboolture (S), midwest*) shires have the highest percentages of estimates for total renter households. However, the middle Brisbane east to west and south to north (except few SLAs in inner city with *St Lucia, Rocklea, Anstead,* and *Pinjarra Hills*) including some northwest outer regions have most of the SLAs in the lower quantiles. The small areas located in these parts of Brisbane presumably have better socioeconomic conditions and could have much higher-income households in renting tenures.

Moreover, the map for estimated numbers virtually shows consistency with the map of estimated numbers for the private renter households. The findings support the results of private renters that are more frequent in housing stress, and the estimates for this group greatly influence the overall housing stress estimates.

In general, the spatial distributions of housing stress estimates for the Brisbane statistical division demonstrate that the rates of housing stress are much higher due to a typical small-size population at various SLAs across Brisbane, particularly SLAs located in inner city and far outer regions. The estimated numbers of households experiencing housing stress are very low (specifically for buyer and public renter households) at SLA level in Brisbane, compared to some other major capital cities. Finally, although the micro-level scenarios within various maps for buyer, public, and private renters households moderately vary with their distributional patterns (clearly observable in frequency legend), lots of SLAs in Brisbane have the highest percentage estimates but much lower estimated numbers in relation to both Brisbane itself and the entire Australia.

8.5 Perth

Perth, the capital statistical division of Western Australia, is divided into five SSDs, which includes 37 SLAs. It is noted that one SLA *Perth (C), inner*, is considered as missing since it did not converge at the *re*weighting process. The distributions of SLAs, households, and housing stress for various SSDs in Perth are provided in Table D.4. The results reveal that the central metropolitan area consists of the highest numbers of SLAs, but it has the lowest estimates in both the numbers (4322 households) and percentage (9.10%) of housing stress. Besides, more than half of the households (about 53.27%) living in Perth statistical division are located within the SLAs geographically located in north and southeast metropolitan areas. Only the north metro SSD contains the highest numbers of 16,090 households in housing stress. The southeast and west metropolitans also have higher estimated numbers. Moreover, although the percentage estimates slightly vary with the SSDs, households experiencing housing stress are more prevalent in the southeast metropolitan, compared to all other regions in Perth.

8.5.1 Housing Stress Estimates for Overall Households

Figure E.16 provides the spatial distributions of the estimates of housing stress for overall households in Perth. The map for percentage estimates reveals that the housing stress rates are quite low at the SLAs of Perth ranging from 5.40% to 13.80%. The fluctuation within rates for various small areas is also stable throughout the distribution. The SLAs in the highest quantile are scattered across south and northwest *Wanneroo* and *Stirling* in north, *Fremantle* and *Kwinana* in west and *Victoria Park* in southeast, and the *remainder* of central in Perth metropolitan areas. In addition, *Rockingham, Armadale,* and *Mosman Park* in the south, *Subiaco* in inner, and SLAs in Belmont–Bayswater region in the southeast have higher rates of households in housing stress. In contrast, most of the SLAs in east and west and south and north outlying Perth have fallen in lower quantiles. Additionally, *Cottesloe (T), Peppermint Grove (S), Cambridge (T),* and *Nedlands (C)* all are located in the central metropolitan area and have the lowest percentages of households in housing stress.

Moreover, the spatial map of the estimated numbers reveals that the entire central region and outer east, southeast, and north side of Perth have lower numbers of households in housing stress. On the other hand, most of the SLAs in north, southeast, and west metropolitans including far *Cockburn* and *Rockingham* islands regions have very high estimated numbers. The results also demonstrate that the majority of SLAs in Perth have more than 1000 households experiencing housing stress. The cumulative frequency legend clearly indicates the four SLAs with more than 3000 households in housing stress. These SLAs are *Gosnells (C), Rockingham (C), Swan (C),* and *Stirling (C), central,* and among them the last one stands out at an estimate of 5294 households.

It has been observed previously that all of these four SLAs and *Canning (C)* (a southeast metro SLA that has 2930 households in housing stress) are listed in the set of top 50 SLAs with the highest estimated numbers of housing stress households in Australia. Hence, compared to the SLAs within some other cities in Australia, although Perth as a whole has an estimated lower percentage of households in housing stress, it has many SLAs with much higher counts, and at least a few of them have been placed in significantly uppermost levels.

8.5.2 Small Area Estimation by Households' Tenure Types

8.5.2.1 Estimates for Buyers

Figure E.17 displays the spatial distributions of the estimates of housing stress for buyer households in Perth. The map of percentage estimates clearly indicates the isolation of the greater central (from north to southeast) region, with a low proportion of households in housing stress. Buyer households living in these areas could have better economic conditions such as higher-level incomes, compared to other small areas in Perth. For a specific example, inner *Fremantle* in southwest metropolitan has 66 households being in purchasing tenure and none of them experiencing housing stress. Moreover, southeastern *Stirling* and east *Fremantle* with the entire part of central (such as *Peppermint Grove; Cottesloe; Mosman Park; Claremont; Perth (C), remainder; Subiaco; Nedlands; Cambridge;* and *Vincent*) have fairly low percentage estimates. On the other hand, most of the small areas located in the outer regions of Perth show higher proportions of buyer households experiencing housing stress. More specifically, SLAs in the highest quantile located around the outer regions are central *Stirling*, south as well as northeast and northwest *Wanneroo*, southeast *Armadale* and *Gosnells*, and southwest *Kwinana* and *Swan*. It is likely that most of the households who are purchasing their house in the outer regions of Perth may have a lower-level household income.

Moreover, the spatial map for the estimated numbers (in Figure E.17b) shows a somewhat similar scenario with the percentage estimates. Although changes into colors are quite apparent, still most of the outer regions' SLAs (except *Serpentine–Jarrahdale* in southeast and *Bassendean* in east) are in the higher quantiles, whereas SLAs in the greater inner area are in the bottom quantile. Besides, the spatial distribution also reveals that 40% of the SLAs in Perth have 0 to 211 buyer households in housing stress. In contrast, nearly half of the buyer households experiencing housing stress are residing within just seven SLAs in Perth. These highly estimated SLAs are south *Wanneroo*, south *Joondalup, Cockburn, Rockingham,* central *Stirling, Gosnells,* and *Swan*. It is further remarkable that one-fifth of the total housing stress counts for buyer households are living in the last two SLAs, and they stand together at the right end of the cumulative legend. These findings may suggest that a significant number of buyer households in Perth are experiencing housing stress, and they are mostly located within some SLAs located outside the greater inner region.

8.5.2.2 Estimates for Public Renters

Figure E.18 depicts the spatial distributions of the estimates of housing stress for public renter households. The map for percentage estimates shows a missing estimate at the SLA *Peppermint Grove (S)* in the central metropolitan area of Perth. Given that there is zero household living in public renting in this SLA, the denominator of percentage holds zero and that leaves the estimate missing. The geographic distribution clearly separates (at least visually) the statistical division into two parts: mid-southeast to far north part (except *Claremont*) is in the higher quantiles and outer east to far southwest part is in the lower quantiles. In particular, most of the SLAs with higher percentage estimates are mainly centered within the first part such as inner *Fremantle*, north and south *Joondalup*, southeastern *Stirling*, and central *Cottesloe, Subiaco*, and *Vincent*. On the other hand, most of the SLAs with the lower percentage rates are located within the second part such as outer southeast *Serpentine–Jarrahdale, Armadale* and *Gosnells*, southwest *Kwinana*, and eastern *Kalamunda, Mundaring*, and *Bassendean*. It is also clear from the cumulative legend that the pattern of the distribution of housing stress rates for Perth's SLAs is mostly clustered between 14% and 19%. It suggests that quite high proportions of public renter households in Perth are experiencing housing stress. Moreover, there are two SLAs that have shown rationally isolated estimations, which are 4.00% for *Serpentine–Jarrahdale (S)* and 36.40% for *Fremantle (C), inner*.

Although the rates for public renter households look somewhat high in Perth, the estimated number of households experiencing housing stress for this group is quite low compared to some other capital cities such as Sydney, Melbourne, and Brisbane. For example, the SLA of *Fremantle (C), inner*, has the highest estimated rate, but there are only eight public renter households experiencing housing stress. Moreover, estimated numbers of housing stress are higher for the SLAs mostly located in the middle northeast to southwest parts with SLAs in outer central metro. The hot spots' areas for estimated numbers are Belmont, Cockburn, Fremantle (remainder), Stirling (central and costal), Canning, and Swan. In contrast, the SLAs in the rest of the Perth including inner regions have very low estimated numbers, and more than 60% of the SLAs have up to an estimated number lower than a hundred households (see in Figure E.18b).

8.5.2.3 Estimates for Private Renters

The spatial maps of the housing stress estimates for private renter households are presented in Figure E.19. The results reveal that the estimated rates of housing stress for SLAs range from 14.60% to 36.90%. Sixty percent of SLAs in Perth have more than a quarter (about 27%) of private renter households experiencing housing stress. Essentially, these rates are much higher, compared to the estimates for SLAs in some other major cities. Besides, the distribution of percentage estimates shows that almost the entire parts of central metropolitan to north Joondalup consist of SLAs in the lowest quantile. However, the range of this quintile is from 14.60% to 24.20%, again relatively high and most of the small areas herein are closer to the upper bound (see cumulative frequency legend). On the other side, most of the SLAs in the southwest to east region (other than *Serpentine* area), far-east region, and few SLAs in the north region fall into higher quantiles. For instance, some hot spots located in these regions are *Cockburn, Rockingham, Fremantle, Swan, Bassendean*, central *Stirling, Mundaring, Kwinana, Armadale*, and *Wanneroo*.

Moreover, the map in Figure E.19b demonstrates that SLAs in inner southeast, far southwest, and mid-north to far eastern parts in Perth clearly have higher numbers of households in housing stress for private renter households. In particular, the small areas *Bayswater, Canning, Rockingham*, and central *Stirling* have the greatest estimated counts of households experiencing housing unaffordability. It has been observed in the earlier analysis that these four SLAs are found in the top 50 SLAs lists for the highest estimated numbers of private renter households experiencing housing stress (where central Stirling from north metropolitan was at the fourth place with 3588 households in housing stress). However, most of the SLAs in the inner parts of Perth, on the other hand, have shown comparatively smaller estimated numbers to lower quantiles. Moreover, the outer Perth regions such as Serpentine–Jarrahdale, Mundaring, and northeast Wanneroo have also low counts of private renter households in housing stress.

Like some other capital cities, the incidence of housing stress is more common in private renter households compared to the other housing tenures, but the estimates for almost all small areas are much higher in Perth. On the other hand, the estimated numbers are lower for the SLAs in the central outer north, south, and east regions (see Figure E.19b). The findings also suggest that although the numbers are smaller for these regions, some SLAs in other parts (inner southeast and middle north to far eastern) of Perth have significantly high numbers of households in housing unaffordability. The proportions of housing stress are also relatively higher for small areas in these regions of Perth. It may be a result of a greater representation of low-income households living in those areas.

8.5.2.4 Estimates for the Total Renters

Figure E.20 presents the spatial distributions of the estimates of housing stress for total renter households in Perth. The maps show reasonable similarity to the maps of the estimates for private renter households. Obviously, the percentage estimates are highest at the SLAs *Kwinana*, central *Stirling, Rockingham*, inner *Fremantle, Mundaring*, south *Wanneroo*, and Armadale. These small areas are located across the east, northeast, southwest, and east metropolitan suburbs. Additionally, the total estimated numbers are also higher for these regions including the SLAs *Bassendean, Cockburn*, and *Swan*.

In terms of percentage statistics, *Armadale* shows the highest estimate of 32.50%, while on the other side, a small area in central *Cottesloe* shows the lowest estimate of 14.80%. In addition, the central *Stirling* stands alone with the largest estimated numbers of 4011 households in housing stress, while, on the other side, *Peppermint Grove* located in central metro has the smallest estimate of just 23 households. However, the maps of percentage and number estimates evidently show various scenarios for housing stress by renting tenures. The findings suggest that predominately the central metropolitan areas have low level of estimates with some other SLAs in the outer regions.

8.6 Adelaide

Adelaide, the capital statistical division of South Australia, is divided into four SSDs, which includes 55 SLAs. The distributions of SLAs, households, and housing stress for various SSDs in Adelaide are presented in Table D.5. It is noted that there is one small area (*Unincop* in western Adelaide) that did not converge at the *re*weighting process, and, thus, it is considered as a missing SLA.

The results reveal that most of the households (about 60.12%) in Adelaide are living in north and south parts of the city. In particular, northern Adelaide consists of 17 SLAs with the residence of 131,597 households, and among them 15,626 households experience housing stress. This SSD also shows the highest percentage rate, slightly higher than the rate of 11.59% for western Adelaide.

Moreover, although eastern suburbs have a relatively higher number of households living in them, both the estimated number and percentage of households experiencing housing stress are the lowest for this region. This may be because a significantly small number of households living in this SSD have household income within the bottom two quintiles of the income distribution. In other words, there may be a high proportion of households with better income.

8.6.1 Housing Stress Estimates for Overall Households

Figure E.21 provides the spatial distributions of the estimates of housing stress for overall households in Adelaide. The map for the percentage estimates shows quite plain divisions of the east part with the west part of Adelaide. The SLAs in the highest quantile are scattered across the west region. For example, SLAs with higher prevalences of housing stress are located within suburbs Playfold, Enfield of Port Adelaide, Salisbury, Onkaparinga, and Charles Sturt. More specifically, two SLAs such as *Playford (C), Elizabeth,* and *Playford (C), west central,* located in the middle of northern Adelaide have the highest percentage values.

In contrast, SLAs in Onkaparinga reservoir (including south Marion and hills) sides, Mitcham hills, Burnside, Adelaide hills, and Tea Tree Gully hills (including Playford hills) located across the east have lower prevalence of housing stress. Another small area *Walkerville (M)* in inner Adelaide also shows a much low percentage estimate of households in housing stress, compared to the estimates of other SLAs in the inner region.

Moreover, the alternative cumulative frequency legend points out that the two SLAs in Playfold area set together alone at the right top end with about 19% of households in housing stress. Also they make the length of the fifth quantile quite long, while other quantiles show consistency in lengths. Except for these two, most of the SLAs in Adelaide statistical division have smaller rates from about 5% to 15%. Thus, compared to the SLAs within some other capital cities in Australia, most of the SLAs in Adelaide show better places in terms of housing affordability to the low-income households.

The spatial map for the estimated numbers of housing stress households also rather splits the city into the outlying southeast and northern west parts in lower quantiles and the north inner and southern west parts in higher quantiles (see Figure E.21b). Almost all SLAs around the outlying Adelaide (except coast of Onkaparinga, north Holdfast Bay, coast of Charles Sturt, and Port Adelaide) and SLA of *Walkerville (M)* in inner Adelaide have low esti-mated numbers of households in housing stress. It is noted that Walkerville also shows a lower percentage estimate. This finding supports the existence of better socioeconomic indicators in Walkerville. In other words, most of the households residing in this small area represent a high-income group.

At the other end, east *Port Enfield*, central *Marion, West Torrens, Salisbury,* and *Playford–Elizabeth* have the highest estimated numbers, and among them *Playford–Elizabeth* stands alone with the highest number of 1949 households in housing stress (see the cumulative legend). The ratio estimate for this SLA is also the highest with 19.30%, which means a significant number of low-income households are residing in Playford–Elizabeth area. However, most parts of Adelaide are seemingly affordable to relatively low-income households.

8.6.2 Small Area Estimation by Households' Tenure Types

8.6.2.1 Estimates for Buyers

Figure E.22 displays the spatial distributions of the estimates of housing stress for buyer households in Adelaide. The graph for the percentage esti-mates indicates that all of the SLAs in eastern Adelaide have very low per-centages of buyer households in housing stress. Also, some small areas in the northern part of southern Adelaide, coasts of western SSD, and Playford hills of northern SSD show lower rates. In contrast, the outer southern part and the northern west part of Adelaide show higher proportions of buyer households experiencing housing stress. More specifically, SLAs in the highest quantile located around these regions are coasts of Onkaparinga with Hackham, central and inner north of Salisbury, west and west central of Playford with *Elizabeth*, and *Park* as well as *Port* of Enfield in Port Adelaide. Moreover, other small areas located in these suburbs also show higher percentages of house-holds in housing stress. Typically, this may be because most of the low-earner households are purchasing their house in south and/or northwest regions. Moreover, the pattern of cumulative frequency reveals that the percentage estimates for buyers are consistently distributed across SLAs in Adelaide, and mostly between 4.70% and 13.60%. There are three SLAs, namely, *Playford (C), west central; Playford (C), Elizabeth;* and *Port Adelaide Enfield (C), park*, with the highest estimates of 15.60%, 15.70%, and 17.50%, respectively.

The spatial map for the estimated numbers reveals that most parts of Adelaide from southern hills to northern hills and from southeast Burnside to west Playford have fairly low numbers of households in housing stress. About three quarters of small areas in Adelaide statistical division have estimates of only up to 300 households (see legend in Figure E.22b), whereas the higher estimated numbers are found for the SLAs located in middle northern suburbs and central west to far coast of southern suburbs. Additionally, it is remarkable that few SLAs located in Onkaparinga (such as *south coast* and *Morphett*), Salisbury (such as *central* and *inner north*), and Playford (such as *Elizabeth*) regions have higher estimates for both in counts and percentages. This finding suggests that significant numbers of buyer households who are experiencing housing stress in Adelaide live within these small areas. Also, it may be that the overall socioeconomic indicators (such as household income) for those few SLAs are relatively poor in Adelaide.

8.6.2.2 Estimates for Public Renters

Figure E.23 provides the spatial distributions of the estimates of housing stress for public renter households. The map for percentage estimates shows a missing estimate at the SLA *Playford (C), hills*, in the outing east of northern Adelaide, since in this SLA there is zero household living in public renting that assigns the denominator of the percentage estimate zero and then leads to the estimate being missing. In addition, although the map illustrates that many SLAs across the west part of Adelaide are in the lower quantiles, most of them have higher percentage estimates. Besides, the overall distributive pattern of the estimates is quite different (see cumulative frequency legend). For instance, there are two small areas (*Adelaide Hills (DC)—ranges* and *central*) in the first quantile that have a zero estimate, while another eight have estimates from 12.10% to 14.80% (and this fact is not apparent on the usual legend of a choropleth map). Additionally, the density of the estimates for SLAs in the second and third (even though several in the fourth) quantiles is remarkably close, but all are much higher. Nonetheless, the estimate of 33.30% for the SLA *Tea Tree Gully (C), hills*, seems to be an outlier because there are only three households in public renting and just one of them is experiencing housing stress. Similarly, *Onkaparinga (C), hills*, and *Mitcham (C), hills*, in southern Adelaide have small numbers of households living in public renting with a significant proportion of them in housing stress.

The map for estimated numbers displays that the SLAs within higher quantiles are mostly located in southern west coast, west and central Marion, Port Adelaide (Enfield region), Salisbury, and Playford suburbs. As observed in the earlier text, *Playford (C), Elizabeth; Marion (C), central; Marion (C), north; Port Adelaide Enfield (C), park*; and *Charles Sturt (C), northeast*, are some of the higher-ranked SLAs for estimated numbers of households experiencing housing stress (see Figure E.23b). In contrast, the SLAs in all southeast and northern west outer regions have very low estimated numbers. Eighty percent of SLAs in Adelaide statistical division have up to an estimated number of 159 public renter households in housing stress. Further, it is worth indicating that even though the rates for public renter households appear higher in Adelaide, the estimated number of households experiencing housing stress is very low compared to Sydney and Melbourne but much higher compared to other capitals.

8.6.2.3 Estimates for Private Renters

The spatial distributions of housing stress estimates for private renter households are presented in Figure E.24. The map for percentage estimates indicates much higher rates ranging from 22.90% to 55.50%. Although all of the SLAs in Adelaide have much higher prevalences of housing stress for private renters, some SLAs with the highest rates are *Port Adelaide Enfield (C), park* (with 51.80%); *Playford (C), west central* (with 52.40%); and *Playford (C), Elizabeth* (with 55.50%). Few other SLAs in these regions, Salisbury, and north-coast Onkaparinga are also in the top quantile. Moreover, the map (b) in Figure E.24 demonstrates that SLAs in the greater inner and southwestern Adelaide clearly have higher numbers of households in housing stress for private renters. However, a few SLAs have been noticed across those parts (such as *Walkerville* and *Port Adelaide Enfield*), which have smaller estimated numbers to lower quantiles. The entire southeast to northwest outer regions of Adelaide, on the other hand, have shown comparatively very low estimated numbers for private renter households.

The findings reveal that the prevalence of housing stress is more common in private renter households compared to the other tenures, and most of the SLAs with much higher prevalences are located around western inner and northern west Adelaide regions, whereas the estimated numbers of housing stress are lower for the SLAs in these regions (except *Playford (C)—Elizabeth*). Typically, *Playford (C), Elizabeth,* is the hot spot for both counts and percentage estimates for private renters. However, there is no SLA from the entire Adelaide in the top 50 lists with the highest estimated numbers for private renter households. This result indicates that although percentage estimates for public renters are significantly higher within small areas in Adelaide, the estimated numbers are lower compared to the estimates in some other major cities such as Sydney, Melbourne, and Perth.

8.6.2.4 Estimates for the Total Renters

Figure E.25 displays the distributions of the estimates of housing stress for the total renter households in Adelaide. In terms of the percentage estimates, the SLAs in mid-northern to outer west (such as northwest *Salisbury*, central and west *Playford* including *Elizabeth* and *Gawler*), inner eastern (such as west and east *Campbelltown*), inner southern hills of *Mitcham*, and south coast of *Onkaparinga* have the highest percentages of estimates for total renters.

However, the southern west coast to middle northern west and hill ranges of Adelaide (including some inner western suburbs) have most of the SLAs in the lower quantiles. Although the map represents mixing of the SLAs in different quantiles across the four SSDs, typically SLAs in Adelaide have high proportions of renter households in housing stress.

Moreover, the map for estimated numbers virtually shows consistency with the map of the estimated numbers for the private renter households as well as the total households (see Figures E.24b and E.25b). The finding supports the results that private renter households in Adelaide are more frequently in housing stress, and the estimated numbers for this group greatly influence the estimated numbers for total renters and, finally, the overall housing stress counts.

Overall, the spatial distributions of housing stress estimates for the Adelaide statistical division demonstrate that the rates of housing stress for the public as well as private renter households are much higher at various SLAs across Adelaide, particularly at the SLAs within northern west, inner western, and southern coastal regions. The estimated numbers of households experiencing housing stress are somewhat lower (specifically for buyer and public renter households) at many SLAs located in eastern Adelaide and outlining southeast to north hills regions, compared to some other major capital cities. It is obvious that the micro-level scenarios within various maps for buyer, public, and private renters households vary with the distributional patterns of the estimates; a significant number of SLAs in Adelaide have an overall lower percentage level of housing stress, as well as a much lower-level estimated number in relation to both Adelaide itself and the SLAs in other major capital cities.

8.7 Canberra

Canberra, the national capital statistical division of Australia, is divided into eight SSDs, which encompass a number of 109 small-sized SLAs. Table D.6 offers the distributions of SLAs, households, and housing stress for various SSDs in Canberra. The results show that there are 14 SLAs across Canberra (such as 5 within south and 3 within North Canberra) that did not converge at the *re*weighting process, and, thus, they are considered as missing. The estimates of 95 SLAs are taken for the analysis and distributed on the geographic maps. Besides, the table reveals that the Belconnen and Tuggeranong SSDs consist of 44 SLAs in which two SLAs (*Belconnen, SSD Bal,* and *Tuggeranong, SSD Bal*) are missing. More than half of the households in the ACT (about 51.09%) are residing in these two SSDs, and about 6.75% of them are experiencing housing stress.

Moreover, the remainder region, that is, *ACT, Bal,* has only one SLA with 77 households and, of them, 6 households (7.79%) are in housing stress.

An estimated number of 16,046 households are living in North Canberra, of which 8.31% are experiencing housing stress, and this proportion is highest, compared to all SSDs within the Canberra statistical division.

Apparently, compared to most other major capital cities in Australia, the Canberra regions have fairly low estimated percentages and numbers of households in housing stress. This could be because most of the families living in Canberra may have much higher household income with a better job security.

8.7.1 Housing Stress Estimates for Overall Households

Figure E.26 depicts the spatial distributions of the estimates of housing stress for overall households in Canberra. In terms of SLAs sizes and numbers of missing areas in Canberra, the spatial maps are very different from the other capital cities in Australia. For instance, in the figure, most of the outer Canberra SLAs seem to be missing. The map for percentage estimates of housing stress shows that the rates are higher for SLAs located within the town center region in North Canberra, Belconnen, Gungahlin-Hall, and Tuggeranong. In particular, *Reid, Duntroon, Acton,* and *city* in North Canberra; *Page, Scullin, Charnwood,* and *Belconnen Town Centre* in Belconnen; *Gungahlin* and *Hall* in Gungahlin-Hall; and *Gilmore* in Tuggeranong are remarkable hot spots' SLAs for housing stress.

Among the SLAs in Canberra, *Belconnen Town Centre, Acton,* and *city* have the highest estimates of 12.80%, 13.80%, and 23.20%, respectively. The frequency legend indicates that the SLAs of city in North Canberra stand alone with the peak rate in Canberra. In contrast, most of the SLAs located in South Canberra (except *Symonston*), Weston Creek–Stromlo (except *Waramanga*), Woden Valley (other than *Lyons* and *Mawson*), and north outer part of Belconnen are seen in the lower quantiles.

The geographic map for the estimated numbers shows that SLAs in Tuggeranong, North Canberra, and Belconnen have higher numbers of households experiencing housing stress compared to the rest of Canberra. The Ngunnawal and Palmerston areas in far north and Narrabundah in the southeast of Canberra also have higher estimated numbers (see Figure E.26(b)). More specifically, the top 10 SLAs with the highest number of households in housing stress are *Ainslie, Watson, O'Connor,* and *Lyneham* (with an average of 150 households) located in North Canberra; *Kaleen* and *Belconnen Town Centre* (with an average of 165 households) located in Belconnen area; *Wanniassa, Gordon,* and *Kambah* (with an average of 240 households of which *Kambah* alone has the highest of 343 households) located in Tuggeranong area; and *Ngunnawal* (with 253 households) in Gungahlin-Hall. It is observed that the estimated total number of households experiencing housing stress in the entire Canberra region is lower than the estimated number for just one SLA (*Canterbury (C)*) in Sydney, and much lower compared to most of the other capital cities.

8.7.2 Small Area Estimation by Households' Tenure Types

8.7.2.1 Estimates for Buyers

Figure E.27 presents the spatial distributions of the estimates of housing stress for buyer households in Canberra. The map of percentage estimates reveals that most of the SLAs in Tuggeranong (other than *Fadden, Macarthur, Oxley,* and *Bonython*), Gungahlin-Hall, outer South Canberra, Belconnen, and some SLAs *Acton, City,* and *Reid* in North Canberra are in higher quantiles. In inner South Canberra, Woden Valley to Weston creek (except *Lyons* and *Rivett*) and middle Belconnen to the upper North Canberra suburbs have SLAs with lower proportions of households in housing stress. Although nearly all SLAs (about 90%; see the cumulative legend) in Canberra demonstrate estimates between 1% and 6%, *Symonston* in South Canberra shows an extremely high rate of 60%. It is noted that like few other SLAs in Canberra, the small area *Symonston* has only 15 households purchasing their home, but most of them may have low household income.

Moreover, the other map of estimated numbers shows rather diverse scenarios (see Figure E.27b). For example, the greater parts of inner to middle (including both south and west, Woden Valley, and Weston creek) Canberra are in the lower quantiles, whereas most of the outer Canberra regions such as Gungahlin, the northeast to northwest part of Belconnen, and Tuggeranong have relatively high numbers of households experiencing housing stress. However, the results reveal that, like the percentage estimates, the estimated numbers of buyer households in housing stress in Canberra are very low (ranging from 0 to 110 households) and only two SLAs, namely, *Kambah* and *Ngunnawal*, have counts of just over a hundred households experiencing housing unaffordability.

8.7.2.2 Estimates for Public Renters

Figure E.28 offers the spatial distributions of the estimates of housing stress for public renter households. It is reported that further six SLAs are in the missing category as the estimated total counts for public renter households are found to be zero for these SLAs. Three SLAs, *Isaacs, City,* and *Chapman,* have prevalence of housing stress ranging from 50% to 100%. However, each of these small areas has very few households living in public renting tenure. For example, *Chapman* located in Weston creek region has just three households living in public renting and all of them are observed in housing stress. Besides, other SLAs in higher quantiles are scattered throughout Canberra, mostly in Belconnen, South Canberra, and Tuggeranong regions.

The cumulative legend for the map of estimated numbers points out that there are 11 SLAs (such as *Duntroon, remainder of ACT, Acton, Macarthur, Bruce, Pialligo, O'Malley, Barton, Harrison, Weetangera,* and *Fadden*) in Canberra where none of the households living in public renting are experiencing housing stress. Most of the SLAs in the northwestern Canberra and inner Belconnen are observed in the higher quantiles. In fact, all SLAs in Canberra have very low estimated numbers of households experiencing housing stress for the public renter category. This is not only because of the households' representation in higher income but may also be due to the recipients of ACT governments' renting assistance through the CHC affordable housing scheme.

8.7.2.3 Estimates for Private Renters

Figure E.29 represents the spatial distributions of estimates for private renter households. The map of the percentage estimates reveals that housing stress is more frequent in households living within SLAs mostly located across Belconnen, inner city, and Tuggeranong. The far north *Hall*, middle *Downer* and *Hackett*, and south *Symonston* also have higher percentage estimates. The seven hot spots' small areas with the highest rates are *Higgins, Scullin, Charnwood, Gilmore, City, Acton,* and *Hall*. On the other hand, the entire North Canberra (except only *Symonston*) and most of Gungahlin are regions with a lower proportion of private renter households in housing stress.

Moreover, the numbers of households experiencing housing stress are much lower for all SLAs in Canberra. Eighty percent of SLAs have less than or only up to a count of 60 households. The map (b) in Figure E.29 indicates that SLAs in Belconnen town, inner and north city regions, and Gungahlin are mostly in the higher quantiles (as does the *Narrabundah, Mawson, Kambah,* and *Gordon*). However, there are only three SLAs in Canberra that have more than 100 private renter households in housing stress, which are *Belconnen Town Centre* (with 111 households), *Ngunnawal* (with 119 households), and *Kambah* (with 141 households).

8.7.2.4 Estimates for the Total Renters

The maps of the estimates for total renters clearly reveal that percentage estimates are higher at SLAs within middle Belconnen, inner city, outer parts of North and South Canberra, and Tuggeranong (see Figure E.30), whereas estimated numbers are higher in middle Belconnen and North Canberra, as well as Gungahlin and northeast of Tuggeranong. Besides, a significantly high number of SLAs have estimated rates from about 14% to 22%, but most of these small areas have less than 100 households in housing stress. In terms of percentage measures, *Hall* shows an extreme estimate of 42.90%, while it has just only 21 renter households of which 9 are in public and 12 in private renting tenure. On the other side, the SLA of *Kambah* has the highest estimated numbers of 237 housing stress households for total renters, even though it shows a relatively lower estimate of 19.47%.

Overall, the national capital Canberra benefits from better socioeconomic conditions and could have much higher-income households living in rental tenures. The findings report that both of the maps for the estimates of total renter households exhibit fairly similar patterns of the scenarios in the

maps for the estimates of overall households. Hence, the estimates of housing stress for renter households substantially influence the overall results
of housing stress. In addition, although the percentage estimates for many
SLAs in Canberra are relatively much higher, their corresponding estimated
numbers are much lower. Canberra's higher-income levels and possibly also
the ACT government's initiation of supports such as the CHC affordable
housing assistance, rather than any difference in housing costs, probably
hold down the estimated number. Regardless of fairly mixed-looking spatial
distributions of housing stress for different tenures' households, typically
the majority of households experiencing housing stress are residing in the
inner north, part of Belconnen, Gungahlin, Narrabundah, and Tuggeranong
regions including *Gordon* and *Kambah*.

8.8 Hobart

Hobart, the capital statistical division of Tasmania, is the one SSD called
"Greater Hobart," which encompasses eight SLAs. Table D.7 offers the distributions of SLAs, households, and housing stress for the Greater Hobart. It is
noted that there is one SLA (*Hobart (C)—inner*) in Hobart or Greater Hobart
that did not converge at the *re*weighting process, and, thus, it is considered
as missing. The results show that Greater Hobart is the residence of 76,361
households of which 7,856 households are experiencing housing stress. The
rate of housing stress for Hobart is 10.29%, which is much higher than the
rates for Canberra and Darwin. The estimates of seven SLAs are taken for
the analysis and distributed on the geographic maps.

8.8.1 Housing Stress Estimates for Overall Households

Figure E.31 portrays the spatial distribution of the estimates of housing stress
for overall households in Hobart. In terms of SLAs' sizes and missing area,
the spatial maps for Hobart are very different from the other capital cities in
Australia. For instance, in the figure, a small area of very little size within
eastern central of the west region seems to be missing and this is *Hobart
(C)—inner*. However, all other SLAs are larger in size.

    The map for percentage estimates of housing stress shows rather com-
mon rates ranging from 8.40% to 11.90% for SLAs across Hobart. The SLA
of *Clarence* in the east region of Hobart has the lowest estimate. Moreover,
Kingborough in the south of the west region has an estimate of 8.90%.

    Small areas *Derwent Valley* located in the far northwest and *remainder* of
central Hobart are found to be in lower quantiles with the rates of 11.20%
and 10.90%, respectively, while the other part *Glenorchy* (in the middle of the

west region), *Brighton* (in northwest), and *Sorell* (located in outer east) have an average of 11.67% of households in housing stress.

Moreover, the map for the estimated number of households in housing stress reveals that SLAs in central of the west region have the highest counts. In particular, more than half of the households who are experiencing housing stress within the Hobart statistical division are residing in *Glenorchy* (with 1968 households) and *Hobart—remainder* (with 2116 households). The small areas *Clarence* and *Kingborough* have also much higher estimated numbers compared to the rest of the SLAs in Hobart. More explicitly, *Clarence* has 1549 households and *Kingborough* has 910 households in housing stress, whereas SLAs *Sorell*, *Brighton*, and *Derwent Valley* have an average of 438 estimated numbers (altogether 1313 households), while *Derwent Valley* has just 262 households in housing unaffordability.

8.8.2 Small Area Estimation by Households' Tenure Types

8.8.2.1 Estimates for Buyers

Figure E.32 presents the spatial distributions of the estimates of housing stress for buyer households in Hobart. The map shows that all small areas in Hobart have low levels of percentage estimates ranging from 6.10% to 12.50%. The SLAs in far north and outer east show relatively higher values, and, on the other hand, SLAs in central and south have lower estimates. Specifically, buyer households residing in *Hobart, remainder,* have the lowest prevalence of housing stress. This may be because most of the buyers living there could have reasonably higher household income. In contrast, a significant number of households purchasing homes in far north and east SLAs (such as *Brighton* and *Sorell*) may have low level of incomes.

It is remarkable that these two SLAs have also large numbers of buyer households. As a result, its corresponding rates are lower and that means a significant number of buyers in these areas are from the class of better earning households. In general, the findings indicate that, like the percentage estimates, the numbers of buyer households in housing stress at the small area level in Hobart are also at a moderate level, where only two small areas have over 500 families experiencing housing unaffordability.

8.8.2.2 Estimates for Public Renters

Figure E.33 represents the geographic distribution of housing stress estimates for public renter households in Hobart. SLAs in Hobart have percentage estimates ranging from 12.40% to 17.30%. The map reveals that the prevalence of housing stress is quite high in the west region of the city. Specifically, *central* Hobart and *Kingborough* have an average of 17.20% public renters in housing stress. In contrast, *Brighton* and *Clarence* areas have much lower rates of 12.40% and 13.5%, respectively.

Besides, the map for estimated numbers indicates that SLAs in Hobart have comparatively less households in housing unaffordability ranging from 12 to 240 households. The estimated number is higher for *Clarence* (of 171 households) and the highest for *Glenorchy* (of 240 households). Conversely, the estimated number is lower for *Kingborough* (of 50 households) as well as *Derwent Valley* (of 33 households) and the lowest for *Sorell* (of just 12 households). Overall for the public renter households, almost all SLAs in Hobart show relatively small figures in the estimates, compared to the estimates for many SLAs in the other cities.

8.8.2.3 Estimates for Private Renters

The spatial maps of the estimates of housing stress for private renter households are presented in Figure E.34. The results reveal estimated rates of housing stress ranging from 29.90% to 40.70% in Hobart. Essentially, these rates are much higher compared to the estimates for SLAs in some other major cities. About 40% of private renter households in the northwest *Derwent Valley* and fat east *Sorell* suburbs are experiencing housing unaffordability. It is usual that a number of private renters living in the outer suburbs could have low household income, and thus, significantly high proportions of them are in housing stress. On the other hand, the rest of the SLAs in Hobart are seemingly the regions with lower proportion of housing stress (but they still have much high rates, compared to many SLAs in other cities).

Moreover, the estimated numbers of private renter households experiencing housing stress are also high for few SLAs in Hobart. For example, the SLAs in the central part of Hobart such as *Glenorchy* and *Hobart, remainder,* have, respectively, estimated numbers of 1089 and 1664 households in housing stress. Another SLA *Clarence* has also a noticeable number of 777 households experiencing housing unaffordability. Other SLAs in Hobart have low estimates, that is, *Derwent Valley* has 130 households, *Brighton* has 172 households, and *Sorell* has 250 households. In terms of the total private renter households living in Hobart, both the number and percentage estimates are significantly higher for the two SLAs *Glenorchy* and *Hobart—remainder*.

8.8.2.4 Estimates for the Total Renters

Figure E.35 presents the spatial maps of housing stress estimates for total renter households in Hobart. The maps show possible similarity to the maps of the estimates for private as well as public renter households. For instance, small area *Sorell* located in the outer east is in the high quantile for percentage estimates for both the private and total renter households. That means the estimates for the private renter greatly influence the percentage estimates for total renter households in *Sorell* area. Moreover, estimates for the public renters have influence on the total estimates in central and south areas.

However, the map looks rather similar for the estimated numbers to the map for private renter households. The results reveal that for total renter households, SLAs in Hobart have percentage estimates ranging from 19.90% to 37.40% and estimated numbers ranging from 163 to 1773 households. These data point to the fact that although greater Hobart consists of only several SLAs, they show a considerable variation in housing stress estimates. The percentage estimates are higher at the SLAs *Hobart, remainder, Kingborough,* and *Sorell.* More specifically, in terms of percentage measure, *Sorell* shows the largest value of 37.38%, while it has only a total of 262 renter households in housing stress. On the other side, the small area in central *Hobart—remainder—*shows the largest estimated numbers of 1773 households (with a rate of 28.62%) experiencing housing unaffordability. Hence, the findings suggest that presumably the central region in Hobart seems unaffordable mostly for renter families with a low household income. Furthermore, relative to overall households, Hobart has a far higher incidence of housing stress (particularly for private renter households) than most of the other capitals.

8.9 Darwin

Darwin, the capital statistical division of Northern Territory, is divided into three SSDs, which includes 41 SLAs. It is noted that four SLAs in Darwin such as the *city, inner, Lee Point–Leanyer swamp,* and *Winnellie* in Darwin City and *East Arm* in Palmerston–East Arm are considered as missing since they did not converge at the *re*weighting process. The distributions of SLAs, households, and housing stress for various SSDs in Darwin are provided in Table D.8. The results reveal that the SSD of Darwin City consists of the highest numbers of 30 SLAs, and most of which are relatively small in size (except the *city, remainder,* and *Lee Point–Leanyer swamp*). The outer east Litchfield Shire, on the other hand, has only two large-sized SLAs.

Moreover, 22,121 households (about 64.80% of the total households in Darwin statistical division) are residing in the central city areas, and 9.09% of them are experiencing housing stress. More than 7000 households are living within nine SLAs geographically located in the southeast part of Darwin known as Palmerston–East Arm. This SSD has 826 households in housing stress with a high proportional rate of 11.17%. Moreover, Litchfield Shire in the outer east has a total of 4620 households of which 7.25% are in housing stress. As the population size of Darwin statistical division is small, the overall estimated number of households in housing stress is also very small. However, the percentage estimate is fairly high and close to the estimates for other capital cities (except Canberra).

8.9.1 Housing Stress Estimates for Overall Households

Figure E.36 presents the geographic distributions of the estimates of housing stress for overall households in Darwin statistical division. The map of percentage estimates demonstrates that the housing stress rates are quite low at the SLAs of Darwin ranging from 3.20% to 14.90%. The variation within rates for small areas into the five quantiles is somewhat normal throughout the distribution. The SLAs in the highest quantile are scattered across northern suburbs (such as *Narrows, Coconut Grove,* and *Nightcliff*) of Darwin City and central suburbs (such as *Gray, Moulden, Bakewell,* and *Woodroffe*) of southeast SSD Palmerston–East Arm. The largest proportional rate of 14.90% was observed for both the *Narrows* and *Moulden*. Some other SLAs located in these regions, for example, *Fannie Bay, Millner, Alawa, Nakara, Wagaman,* and *Driver,* have also higher rates of households in housing stress. In contrast, the rest of the small areas in Palmerston–East Arm are in lower quantiles with some SLAs from the northern east (including the *city—remainder*) and west parts of the city. The far inland outlining SLA *Litchfield—part B*—is also in the lowest quantile with an estimate of 7.10% of households in housing stress.

The spatial map of the estimated numbers shows that SLAs in the northeast middle part of Darwin City have lower estimated numbers of households in housing stress; for example, *Bayview–Woolner* with the *Gardens; Narrows; city, remainder; Jingili; Wanguri;* and *Brinkin* all have the estimated numbers in the bottom quantile (as does the *Litchfield—Pt A*). On the other hand, several SLAs in the central Palmerston–East Arm region, northwest Nightcliff–Coconut Grove areas, and the greater outlying Litchfield have counts for the top quantile. Nevertheless, the estimated numbers of housing stress for all SLAs in Darwin are quite low and range from 19 to 316 households.

It is noticeable that when almost all small areas in Darwin have counts between 21 and 151 households with less variation in the distribution pattern, the two SLAs in Litchfield Shire, *Litchfield, Pt A,* and *Litchfield, Pt B,* have the lowest and highest counts, respectively. It is also clear from the cumulative frequency legend that one SLA stands alone at the right end, and it is found to be the *Litchfield—Pt B*. The reason behind this is a comparatively huge number of 4436 households residing in *Litchfield, Pt B,* while only 184 households are residing in *Litchfield, Pt A*.

However, the estimate for *Litchfield, Pt A,* shows relatively less variation like some other SLAs in Darwin as they also contain a small number of households. Furthermore, in terms of housing stress measures, the hot spots' SLAs in Darwin are *Nightcliff* and *Coconut Grove* in city west and *Gray, Bakewell,* and *Moulden* in Palmerston–East Arm, as they all are observed in the highest quantile for both the estimated numbers and percentages.

8.9.2 Small Area Estimation by Households' Tenure Types

8.9.2.1 Estimates for Buyers

Figure E.37 displays the spatial maps of the estimates of housing stress estimates for buyer households in Darwin. The map of percentage estimates
reveals that the greater central city region (except *City—remainder*) and most
of the SLAs in north Darwin have low proportions of households in housing stress. For example, the *Gardens, Stuart Park, Bayview–Woolner, Fannie Bay,
Ludmilla, Moil,* and *Nightcliff* all have shown less than 5% of buyer households in housing stress. In addition, the small area *Gunn–Palmerston City* in
the southeast has a low rate. Buyer households living in these areas could
have better economic conditions such as high level of income, compared to
other small areas in Darwin. On the other side, the entire Litchfield Shire
shows higher percentage estimates with most of the SLAs in the central
Palmerston–East Arm. There are few SLAs (such as *Wagaman, Nakara, Anula,*
and *Karama*) in the north Darwin have also higher proportions of buyers in
housing stress. For a particular remark, *Litchfield (S), Pt A,* has the highest
estimate of 14.50%; *city, remainder,* has the second highest of 11.10%; and all
other SLAs have estimates of up to 10.30%. Although the overall percentage
estimates of housing stress for buyer households are at low levels in Darwin,
it is likely that most of the households purchasing their house in Litchfield
Shire and in the *city, remainder,* may have a poor level of household income.

Moreover, the spatial map for the estimated numbers shows that the central
city to middle of Darwin are in the bottom quantile, whereas outer Litchfield,
central Palmerston, and the north of city parts are in higher quantiles. The
results also reveal that most of the SLAs (about 92%) in Darwin have less
than 50 buyer households with housing unaffordability. In contrast, there
are only three small areas such as *Karama, Bakewell,* and *Litchfield—Pt B—*
with estimated numbers of more than 50. Nearly one-third of the estimated
households are residing within these SLAs, and about a quarter (206 households) is in *Litchfield—Pt B.* This finding illustrates that a significantly high
number of households (about 65.19%, as the overall estimated number is 316
households) experiencing housing stress in *Litchfield—Pt B—*are the buyer
households. Perhaps, this is because households who have relatively low
income are purchasing homes in this outer region of Darwin.

8.9.2.2 Estimates for Public Renters

Figure E.38 depicts the spatial distributions of the estimates of housing stress
for public renter households. The map for percentage estimate shows a further missing estimate at the SLA *Brinkin* in the mid-northwest of Darwin
City. Given that there is no household living in public renting at this small
area, the denominator of percentage measure holds zero and that leaves the

estimate missing. Besides, the estimates of other SLAs in Darwin range from 0% to 61.50%. There are five small areas (which are *city, remainder, Stuart Park, Bayview–Woolner, Durack,* and *Litchfield, Pt A*) across Darwin that have almost a thousand (979) public renter households residing in those areas, but none of these has found to be in housing stress. Perhaps, one reason for this result could be that all of these households are the recipients of generous housing assistances as well as other social welfares. Moreover, housing stress rates for public renter households are comparatively very high at some SLAs in Darwin. For example, small areas *Marrara, Nakara,* and *Gunn–Palmerston City* have the estimates of 42.90%, 58.80%, and 61.50%. The variations within rates for SLAs in the bottom and top quantiles are also apparent into the cumulative frequency legend.

Although the rates for some SLAs look quite high in Darwin, the estimated numbers of households experiencing housing stress for these SLAs are very low. For example, the SLA with the highest percentage has only eight households experiencing housing stress. Moreover, estimated numbers of housing stress are higher for the SLAs mostly located in the middle of Palmerston–East Arm and the west and north parts of the city. Small areas *Moulden, Woodroffe,* and *Gray* in Palmerston region are the hot spots of estimated numbers for public renter households experiencing housing unaffordability.

8.9.2.3 Estimates for Private Renters

Figure E.39 provides spatial maps of the estimates of housing stress for private renter households. The results reveal that the prevalence of housing stress in Darwin's SLAs ranges from 3.60% to 30.30%. Housing stress is more common within some small areas scattered across the northwest city region and central Palmerston. The two SLAs in Litchfield Shire have also high rates.

Moreover, the estimated numbers of housing stress in Darwin's SLAs range from 8 to 117 households. The map (b) in Figure E.39 demonstrates that SLAs in outlining Litchfield Shire, northwest suburbs in Darwin City clearly have higher numbers of private renter households in housing stress. In particular, the small areas *Litchfield, Pt B, Stuart Park,* and *Nightcliff* have estimates of more than 100 households. However, most of the SLAs in Darwin, on the other hand, have estimated numbers less than 50 households for the private tenure.

8.9.2.4 Estimates for the Total Renters

Figure E.40 represents the spatial distributions of the estimates of housing stress for total renters in Darwin. The results for percentage estimates demonstrate that the housing stress rates for the total renter households are relatively lower for various SLAs in Darwin and range from 3.30% to 27.00%. Clearly, the percentage estimates are highest at the SLAs located across the west, mid-northwest part of Darwin City, and inner Litchfield Shire. In particular, *Fannie Bay, Ludmilla, Alawa, Narrows, Nakara,* and *Litchfield, Pt A,* within those regions have the highest estimates (see Figure E.40a). However, the total estimated numbers are lower for most of these SLAs.

In terms of percentage measure, *Litchfield—Pt A*—shows the utmost value among the estimates for small areas in Darwin, while it has the lowest value of 10 households in estimated numbers. In addition, *Nightcliff* has the highest estimated counts of 134 renter households in housing stress, while it shows the prevalence rate of 21.14%. Overall, the findings suggest that almost all SLAs in Darwin have very small numbers of households in housing stress and their proportional rates are also found to be quite reasonable, as the total number of households for most of the SLAs is relatively low in the Darwin statistical division.

8.10 Concluding Remarks

The results of housing stress estimates for the major capital cities in Australia have been thoroughly discussed in this chapter. It has observed that almost two-thirds of households with housing unaffordability are residing in Sydney, Melbourne, Brisbane, Perth, Adelaide, Canberra, Hobart, and Darwin. Among these capital cities, Sydney and Melbourne have about 40% of all housing stress households in Australia, and Sydney alone has one-third of the total estimate of housing stress for these eight cities. Besides, the distributive patterns of the housing stress estimates varied with cities, SLAs within a city, and by tenure types. For example, Canberra has much lower estimated numbers and percentages of households in housing stress, compared to all other cities. On the other hand, Sydney has much higher estimates. Housing stress estimates for the overall households in different cities are largely determined by renter households, specifically by the private renter households. Moreover, housing stress estimates show somewhat mixed increasing trends for some capital cities in Australia.

Spatial distributions of the estimates in Sydney have pointed out the SLAs in Canterbury–Bankstown, Fairfield–Liverpool, Blacktown, and Gosford–Wyong as the hot spots for housing stress. However, geographically half of Sydney has SLAs with low estimates of housing stress, and most of these small areas are located across northern Sydney as well as in the central and outer Western Sydney, which characteristically have better socioeconomic conditions. Additionally, in Melbourne, most of the SLAs with significantly high estimates of housing stress are located in western Melbourne, greater mid to north Melbourne, middle-southern Melbourne, Hume City, and central Melbourne regions. In contrast, most of the outer Melbourne regions and parts of the inner and middle Melbourne are seemingly much affordable places for relatively low-income households. The findings have demonstrated that some of the typical reasons for a high increase in housing stress estimates in different SLAs in Sydney and Melbourne are perhaps related to increase in populations, housing costs, interest rates, income risk, etc., and a relative decrease in social housing stock, supply of housing in the market, and per capita income of households in these areas. Further, spatial distributions of the estimates in Brisbane have revealed rather different results. Although the majority of SLAs across Brisbane have very small counts of households in housing stress, the percentage rates are much higher for most of these SLAs. Some hot spots for housing stress estimates in Brisbane are located in Ipswich, Redcliffe, Logan, Beaudesert, Caboolture, and Redland Shire regions within southeast and northwest inner Brisbane.

The findings for Perth have shown a relatively lower percentage of households in housing stress. But this capital has many SLAs with much higher estimated numbers, and at least few of them (such as Canning, Gosnells, Rockingham, Swan, and central Stirling) have significantly highest estimates. In addition, small areas in Adelaide are with a low level of housing stress estimates. That means most of the Adelaide regions are affordable places for low-income households. It is noted that Adelaide is somewhat a low-population-growing capital city, and supply of housing and housing costs are relatively low at many SLAs in Adelaide. However, the regions of Playfold, Port Adelaide, Salisbury, and Charles Sturt have significantly large estimates for the renter households, especially the private renter households in housing stress.

The spatial analyses of the estimates for Canberra have shown that the SLAs in inner North Canberra, Belconnen, Gungahlin, Narrabundah, and Tuggeranong including Gordon and Kambah were the hot spots for housing stress. However, almost all small areas in Canberra have a very low level of housing unaffordability estimates, compared to SLAs in all other cities. This is because the Canberra region has better socioeconomic conditions with significantly larger numbers of households with higher income. Besides, the housing-related policies including new land development strategies for adequate supply of housing and the CHC affordable housing program for the ACT seem to be functioning well for its community.

Furthermore, the findings for SLAs in Hobart have shown that private renters in Hobart are more common in housing stress, compared to most other capital cities. The central part of Hobart is quite unaffordable for families with low household income. In addition, most of the SLAs in Darwin have a very low estimated number of households experiencing housing stress, except only few SLAs: Marrara, Nakara, and Gunn–Palmerston City have relatively high prevalences for the public renters. Finally, it is noticeable that numbers of households residing in these two cities are quite low, compared to the other capital cities, and at SLA level, almost all of the estimates are significantly low.

9

Validation and Measure of Statistical Reliability

9.1 Introduction

One of the important issues relevant to the spatial microsimulation modeling approach of small area estimation is the difficulty in validating the model outputs and measuring the statistical reliability of the estimates. This is because the microsimulation modeling technology (MMT) approach simulates a spatially disaggregated synthetic micropopulation data set that previously did not exist and uses this data set to produce the small area estimates. Validation processes in MMT typically check the model outputs in a range of systematic ways to reveal errors in the outputs. Without examining such errors from the model, a modeler or researcher cannot claim the accuracy of the estimates produced by the model. Measures of statistical reliability can also be used to assess the stability of the model estimates.

Validation and creation of the statistical reliability for MMT estimates are an integral part of the overall development and application of the model. However, there has not been much research on testing the statistical significance of these model estimates and deriving estimates of how reliable these MMT outputs may be. Therefore, this chapter is designed to offer new validation techniques for testing the accuracy of the model outputs and showing where the model works well and where it functions less well. It also illustrates the creation of a statistical reliability measure such as a confidence interval (CI) for the MMT estimates of housing stress at the statistical local area (SLA) level in Australia.

An outline of the remainder of the chapter is as follows. Section 9.2 reviews some validation techniques from the literature. Section 9.3 describes the theories and results of two new validation methods used in the MMT for small area housing stress estimation. Section 9.4 creates the statistical reliability measure of the MMT estimates. Finally, Section 9.5 provides a concluding summary of the chapter.

9.2 Some Validation Methods in the Literature

There are a few techniques utilized in the literature for validating the MMT outputs. Most of these techniques are based on synthetically simulated microdata and known figures available from the actual population (Rahman 2014). As at small area levels the estimated statistics are typically unavailable from another source, some researchers have suggested reaggregating the small area estimates up to larger levels where reliable statistics are available to compare the results. Many researchers have also tried to utilize a number of alternative methods to determine the accuracy of their model estimates. The overall process of validation often involves using multiple modes of validation to reveal a better picture about the accuracy.

In general, techniques used by researchers in different countries for validation include basic comparisons (Taylor et al. 2004; Edwards and Clarke 2009); reaggregation up to actual comparable data sets (Ballas et al. 2001; Hynes et al. 2006); the total absolute error (TAE) measure and/or relative errors measures (Williamson et al. 1998; Kelly 2004; Hynes et al. 2006; Tanton et al. 2007; Rahman 2009a; van Leeuwen et al. 2009); the regression approach for the slope of best fit line (Ballas et al. 2005a; Edwards and Clarke 2009); the standard error about identity (Ballas et al. 2007; Tanton and Vidyattama 2010); the general-level equal variance t-test (Edwards and Clarke 2009); the Z-score measure, which is based on the difference between the relative size of the category in the synthetic and actual populations (Williamson et al. 1998; Voas and Williamson 2000; Hynes et al. 2006; Rahman et al. 2013; Rahman and Harding 2014); and the Z-score-based chi-squared (χ^2) measure, where the Z-scores are for cellular fit and chi-squared measures are for tabular fit (Hynes et al. 2006). Some of these validation methods are outlined in the following.

Taylor et al. (2004) have used three types of comparisons for validating the housing unaffordability estimates from an MMT in Australia. In the first type, the authors have performed an assessment of how close the model outputs were to subpopulation counts data from the Australian Bureau of Statistics (ABS) that were used in the reweighting process. Explicitly, in this type of validation, the small area counts of households in each of the relevant subpopulation groups such as tenure type, household age group, and income group were estimated by the GREGWT reweighting procedure and compared with respective census benchmarks used in the reweighting process. Their results confirm an approximate linear relationship between the model estimates and the census benchmarks with very low differences across the small areas.

Additionally, in the second type of validation, they used a comparison of model estimates against ABS census data that had not been used in the GREGWT reweighting stage. The authors demonstrate that comparison against such alternative census data from ABS was a useful means of validation as it was used by other researchers (e.g., see Percival et al. 2002) in Australia.

The third form of validation authors have used is expert validation; in this instance, they have asked people who know the areas to provide formal assessments of how well the output accorded with their knowledge of the actual small area figures. Fundamentally, it was the feedback given by the major stakeholders that accounted for the validation.

Edwards and Clarke (2009) illustrate the four key analyses used to validate small area estimates of health data generated by an MMT in the United Kingdom. In particular, the individual-level simulated micropopulation data are first compared with the census counterparts for all of the census-type variables: that is, for both the constraint variables used in the simulation and the nonconstraint variables that were not used as input variables in the simulation. In some way, this initial stage analysis is consistent with the first type of validation used in Taylor et al. (2004).

At the next step of the analysis, the error values for each of the census-type variables used in the simulation processes were measured by aggregating the differences between distributions of each actual constraint variable plus other census variables that are synthesized (nonconstraint variables) and the synthetic microdata at the small area level. The purpose is to check the fitting of the combination of synthetic microdata to the census benchmarks, which have been commonly employed by a number of researchers (e.g., see Williamson et al. 1998; Kelly 2004; Chin and Harding 2006; Tanton et al. 2007; van Leeuwen et al. 2009; Rahman et al. 2010b; Rahman n.d., among many others). The fit of a combination of individuals to known small area constraints is evaluated by the TAE, the aggregated absolute difference between simulation (observed) and census (expected) counts for each variable in each small area.

The third kind of validation, a regression analysis of microsimulated data against actual data, is undertaken to better understand the fit of the simulation. The coefficient of determination, or the R^2 statistic, is an indicator that ranges in value from 0 to 1 and reveals how closely the simulated values for the regression trend line fit the actual census data. A trend line is most reliable when its R^2 value is at or close to 1 and it reveals the closeness to the desired 45° line of the quadrangle. Several authors have used the approach or derivatives from this. One of the derivatives is the standard error around identity, which is the error around a 45° line (Ballas et al. 2005b,c, 2007; Tanton and Vidyattama 2010).

It is noteworthy that this regression analysis type of validation does not give the researcher any information about the fit of the microsimulated data to the ideal situation (where $y = x$ and the simulated data are the same as the true statistics). Rather, the approach expresses the fit of the data to the best fit 45° line through those data. Indeed, the coefficient of determination measure can provide numeric information about the precision of the overall MMT but not exactly about the accuracy of the small area estimates produced by the model.

Nevertheless, in the final means of their analysis, Edwards and Clarke (2009) have performed a two-tailed t-test with an equal variance assumption for statistical comparison of the two data sets to establish whether there are any statistically significant differences between the synthetic microsimulated data and real populations for different variables (used as constraints and nonconstraints). The results show that most of the synthetic spatial microdata simulate nonconstraint variables very poorly, that is, there exists a significant difference between the microsimulated data and the expected or real figures. Nevertheless, most of the constraint variables used in the simulation generate synthetic spatial microdata very well and show no significant difference between the simulated and expected populations. It is noted that the equal variance assumption may not be feasible for all variables (such as for nonconstraint variables) used in the calibration process. Additionally, the test cannot provide information about the statistical significance and reliability measure of a particular small area estimate, except for a comparison at the general level.

As referenced earlier, the Z-score measure has been employed by several researchers to check the cellular fit of microsimulated synthetic data sets. The Z-score is based on the difference between the relative size of the category in the synthetic micropopulation and actual population data sets. The Z-score has been obtained by assessing the standard normal proportions of the differences between the relative size of a specific category in the microsimulated synthetic population and the available census population (Rahman 2015). An adjustment to the measure can be considered when dealing with an actual count of zero in the census table by using the corresponding synthetic microdata count (Williamson et al. 1998). Nevertheless, when dealing with possible zero counts for any cells in the tables or to avoid division by zero, Hynes et al. (2006) used a random component that has been determined by "one over two times the sum of all the elements in the census table." The random component has to be added if the actual figure is larger than the corresponding synthetic data and subtracted if the actual figure is smaller than the corresponding synthetic data. While the Z-score measure has known statistical properties and wide acceptance as a valid measure of fit, there are a range of limitations of the Z-score-based microdata evaluation process, and these have been discussed in Voas and Williamson (2000).

Hynes et al. (2006) use a Z-score-based chi-squared measure to validate a farm-level spatial microsimulation model for Ireland along with some of the validation techniques mentioned earlier. The chi-squared measure only checks whether the synthetic data set is deemed to fit or not to fit with the actual tables. Clearly, the process can examine an overall tabular form of fitting of the synthetic microdata but not the statistical significance of individual estimates. There is, however, a lack of literature on using such a technique as a standard validation methodology that can test the statistical significance of the MMT estimates and can create its statistical reliability measure by means of a statistical procedure.

9.3 New Approaches to Validating Housing Stress Estimation

This section introduces two novel approaches to validate SLA-level housing stress estimates produced by the spatial microsimulation model. The first approach measures the standard normal Z-statistic for performing a systematic statistical hypothesis test of the MMT estimates and the second approach calculates an absolute standardized residual estimate (ASRE) for further analysis.

9.3.1 Statistical Significance Test of the MMT Estimates

Statistical hypothesis tests are particularly useful when the researchers need to make a decision on an estimate related to some policy issue. For example, suppose the media in a country is reporting that the housing costs of a particular small area within a big city are very high. Is this high housing cost a social problem in that area? How many households and which groups of people are mostly suffering from these high housing costs? A social researcher wishes to investigate the issue by socioeconomic modeling, and the research obtains an estimate of 37% of households living in that area and experiencing housing stress. However, the census or administrative data in that particular area show that 35% of households are in housing stress. Hence, the researcher needs to establish not only the statistical reliability of their own estimate but also how it is relative to the other existing estimate to claim that high housing costs in that particular small area are affecting the living of a significantly large proportion of households who may or may not necessarily be in poverty.

There are four basic steps necessary for conducting a systematic statistical hypothesis test: (1) formulate the null (H_0) and alternative (H_A) hypotheses, (2) compute the relevant test statistic, (3) calculate the resulting p-value for the test statistic, and (4) either reject or do not reject H_0 (as per the p-value by comparing with a specified level of significance) and interpret the results. These steps are performed in the subsequent text for testing the statistical significance of the MMT estimates.

Let us consider the following mathematical notations:

$\hat{p}_{ij}^m$ is an estimated proportion of households who are in the interested category m (such as *housing stress* in this case) for jth data in ith small area.

P_{i0}^m is the true proportion of households who are in the interested category m (housing stress) for the population in ith small area.

n_{ij} is an estimated size for jth data in ith small area.

The following null and alternative hypotheses can be defined to check the statistical significance of the small area housing stress estimates produced by the MMT:

$H_{i0} : \hat{p}_{ij}^m = P_{i0}^m$ (model estimate in ith small area is equal to the true value)

$H_{iA} : \hat{p}_{ij}^m \neq P_{i0}^m$ (model estimate is not statistically equal to the true value from census)

Hence, for testing these hypotheses, a Z-statistic is defined as (see Rahman et al. 2010b)

$$Z_i = \frac{\hat{p}_{ij}^m - P_{i0}^m}{\sqrt{\dfrac{P_{i0}^m(1 - P_{i0}^m)}{\sum_j n_{ij}}}}; \quad \forall i. \tag{9.1}$$

In terms of random variable properties, the values of the Z-statistic have the standard normal distribution (see DeVeaux et al. 2004; Gupta and Kapoor 2008), that is, the Z-statistic follows the normal distribution with a mean zero and a variance one. Once the value of the Z-statistic is obtained, the corresponding probability statistic can be estimated by observing the p-value in a standard normal probability curve or reference table. Additionally, since the form of the alternative hypothesis implies a two-sided statistical test, the pooled estimate of the p-value for the two-tailed test is $2P$ (where P is the rejection area on the standard normal curve (SNC) in one tail for a given value of the Z-statistic).

Figure 9.1 illustrates the standard normal distribution of the test statistic Z with the rejection and nonrejection regions of the null hypothesis for two sets of fixed critical values. The critical values −1.96 and 1.96 correspond to a 5% significance level or type I error under H_{i0}, whereas the critical values −2.58 and 2.58 correspond to a 1% significance level under H_{i0}.

The rejection regions are the areas marked with oblique lines under the two tails of the probability curve. Any estimate of the test statistic lying either below −1.96 or above 1.96 results in rejecting H_{i0} for a 5% level of significance. However, there are values between −2.58 and −1.96 or 1.96 and 2.58 that advocate not rejecting H_{i0} at a 1% (and even lower) level of significance. For example, three hypothetical values of the Z-statistic, −2.83, 0.75, and 2.47, result in rejecting H_{i0} at both 5 and 1% significance levels, not rejecting H_{i0} at both levels, and rejecting H_{i0} at 5% level but not rejecting at 1% and/or lower levels, respectively.

The decision rule of the test is based on the estimated p-value that is also sometimes known as the "observed significance level" of the test. As the p-value is the probability of getting a test statistic at least as large as the value actually observed by assuming the H_{i0} is true, no preset value of significance level is required. That means in this regard a p-value is an exact

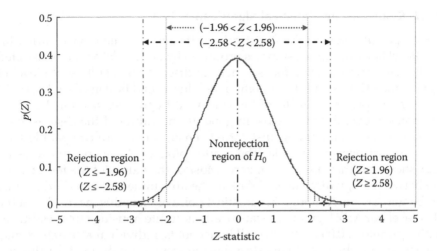

FIGURE 9.1
The standard normal distribution of Z-statistic for a two-sided statistical hypothesis test.

value of type I error (rejecting a true H_{i0}) of the statistical test. In fact, using the p-value approach, the null hypothesis could be rejected for a p-value < 0.0001. Often, it may be dependent on the clients' or researchers' choice.

Table 9.1 gives the area and p-value under the probability curve for some important values of the Z-statistic. It is obvious that more than 50% of the nonrejection region is encompassed by the values of the Z-statistic between −0.68 and 0.68. In addition, about 99.99% of the area under the probability curve of the Z-statistic is cut off by the critical values of ± 4. Hence, the decision rule indicates that any estimate of the test statistic lying either below −4 or above 4 results in rejecting H_{i0} for a 0.01% significance level (i.e., for p-value = 0.0001).

TABLE 9.1

Area Property of the Curve for Some Important Values of the **Z-Statistic**

Area Property of Z-Statistic	Area under the Curve[a]	p-Value	Decision Rule
$P(-0.68 < Z < 0.68)$	0.5034 [50.34]	0.4966	Reject H_{i0}: if p-value is less than the
$P(-1.00 < Z < 1.00)$	0.6826 [68.26]	0.3174	standard values (i.e., 0.05, 0.01, 0.001, or
$P(-1.96 < Z < 1.96)$	**0.95 [95]**	**0.05**	0.0001) for different choice of level of
$P(-2.00 < Z < 2.00)$	0.9544 [95.44]	0.0456	significance.
$P(-2.58 < Z < 2.58)$	**0.99 [99]**	**0.01**	
$P(-3.00 < Z < 3.00)$	0.9973 [99.73]	0.0027	
$P(-3.50 < Z < 3.50)$	0.9996 [99.96]	0.0004	Accept H_{i0}: otherwise.
$P(-4.00 < Z < 4.00)$	**0.9999 [99.99]**	**0.0001**	

[a] Percentage value is within the square brackets.
Note: The bold values represent the standard cut-off figures.

9.3.2 Results of the Statistical Significance Test

The empirical results of the statistical significance test mentioned earlier for SLA-level housing stress estimates produced by our model are demonstrated in the following text. The focus is here to discuss the results at 5% level of significance, that is, the cutoff point of *p*-value would be considered as 0.05.

Figure 9.2 presents results of the statistical hypothesis test for the MMT estimates of overall households. In general, the values of the Z-statistic for housing stress estimates against actual census values are compatible with the standard normal variable. The scatter plots of Z-statistic measures by SLA show that they are close to zero. Most of the values of the test statistic are between ± 2, which cover 95.44% of the area under the standard normal probability curve. Although in statistical point of view this means that at the SLA level our MMT produces statistically accurate housing stress estimates with a *p*-value of 0.05 for the overall households, individual researchers may have their own choice of *p*-value to judge the accuracy of the model estimates.

More specifically, any value of Z-statistic within the standard cutoff points of −1.96 and +1.96 has an estimated *p*-value ≥ 0.05, which means we can accept the null hypothesis and determine that the corresponding model estimates

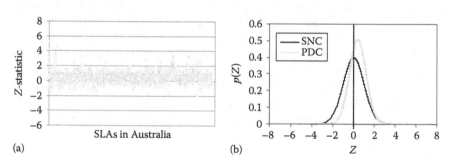

(a)   (b)

Statistical Decision about the Model Estimates	At *p*-Value = 0.05		At *p*-Value = 0.01		At *p*-Value = 0.0001	
	SLA (#)	%	SLA (#)	%	SLA (#)	%
Inaccurate (reject H_{i0})	47	3.7	20	1.6	5	0.4
Accurate (accept H_{i0})	1231	96.3	1258	98.4	1273	99.6
Total	1278	100	1278	100	1278	100

(c)

FIGURE 9.2
Results of the hypothesis test at statistical local area (SLA)–level housing stress estimates for overall households. (a) Z-statistic estimation by SLAs. (b) Probability curve of Z-statistic. (c) A tabular report on statistical decision.

are statistically accurate at the reference p-value of 0.05 or at 5% level of significance. However, there are some values of the test statistic that are located outside of the cutoffs mentioned earlier. This result demonstrates the MMT estimates of housing stress are statistically inaccurate for several SLAs.

It can also be seen from the probability curve that the MMT estimates are somewhat overestimated for many SLAs as the "probability density curve" (PDC) of the Z-statistic is shifted to the right with a slight leptokurtic in shape (refer to Figure 9.2b). That is, under the standard normal distribution property, the Z-statistic would be much higher on the right side of the "SNC" than on the left side. The curve also indicates that there are more small areas with inaccurate results in the right tail. Although to some extent the model yields overestimates of housing stress for many SLAs in Australia, it should be noted that most of them show no statistically significant difference to the actual census figures.

The adjoining tabular report on the statistical significance test reveals that our model has produced accurate small area estimates for 1231 (96.3%) SLAs in reference to the p-value of 0.05, for 1258 (98.4%) SLAs in reference to the p-value of 0.01, and for 1273 (99.6%) SLAs in reference to the p-value of 0.0001, respectively. These findings confirm that the MMT for small area housing stress estimation has performed very well by producing statistically acceptable housing stress estimates at SLA level across Australia.

The Z-statistic and its corresponding p-value for SLAs with statistically inaccurate results are presented in Table 9.2. There are five SLAs in particular that reveal a significantly large difference between the MMT estimates of housing stress and actual census values. In particular, the findings demonstrate that the housing stress estimates for overall households of the SLAs, namely, *Melbourne (C)—Remainder* in Melbourne; *Waverley (A)*, *Warringah (A)*, and *Woollahra (A)* in Sydney; and *Petermann-Simpson* in the Alice Springs, are inaccurate by any conventional level of p-value. As the values of the Z-statistic for these SLAs are <-4 or more than $+4$, they have estimated p-values of 0.0000 that indicate the existence of significantly large differences between the microsimulated synthetic data and actual census values.

Among those SLAs with Z-statistic values that are outliers, four have shown overestimates of housing stress by the simulated data, while one has revealed an underestimate. To get an idea of why a significantly large difference arises for each of these estimates, we may check some microlevel results for an SLA (such as *Waverley (A)* in Sydney) along with its geographic characteristics. For instance, in Waverley, 9,479 private renter households are simulated, of which 1,788 households are estimated to be in housing stress (out of a total SLA-level simulated population is 22,807) against 9,203 private renter households with 1,189 households in housing stress from the actual census (the total SLA-level census population is 22,809). It is noted that although the total numbers for simulated and actual populations are approximately equal, the simulated data for households experiencing housing stress are different

TABLE 9.2

Statistical Local Areas Show Insignificant Z-Statistic with p-Value <0.05

SLA Name	ID	Z-Statistic	p-Value
Melbourne (C)—Remainder	205054608	−5.03	0.0000
City—Remainder	705051138	−2.05	0.0404
Fannie Bay	705051028	1.99	0.0466
Larrakeyah	705051044	1.99	0.0466
Armadale (C)	505250210	1.99	0.0466
Narrogin (S)	520056510	2.00	0.0456
Broome (S)	545100980	2.01	0.0444
Gosnells (C)	505253780	2.04	0.0414
Glen Eira (C)—Caulfield	205652311	2.04	0.0414
Kingston (C)—North	205653431	2.05	0.0404
Stonnington (C)—Malvern	205656352	2.06	0.0394
Sydney (C)—East	105057204	2.07	0.0384
East Pilbara (S)	540053220	2.10	0.0358
Mermaid Wtrs-Clear Is. Wtrs	307103562	2.13	0.0332
Alice Springs (T)—Ross	710400207	2.17	0.0300
Cockburn (C)	505201820	2.18	0.0292
North Sydney (A)	105555950	2.21	0.0272
Rockingham (C)	505207490	2.22	0.0264
Kalamunda (S)	505104200	2.22	0.0264
Swan (C)	505108050	2.23	0.0258
Belmont (C)	505250490	2.28	0.0226
Port Phillip (C)—St Kilda	205055901	2.34	0.0192
Bayswater (C)	505100420	2.36	0.0182
Victoria Park (T)	505258510	2.37	0.0178
Holdfast Bay (C)—North	405202601	2.37	0.0178
Aspley	305071034	2.45	0.0142
Stuart Park	705051104	2.46	0.0138
Alice Springs (T)—Charles	710400201	2.54	0.0110
Katherine (T)	710302200	2.71	0.0068
Pittwater (A)	105656370	2.73	0.0064
Mackay (C)—Pt A	340054762	2.75	0.0060
Stirling (C)—Coastal	505157915	2.80	0.0052
Manly (A)	105655150	2.90	0.0038
Melville (C)	505205320	2.95	0.0034
City—Inner	305011143	2.96	0.0030
Sutherland Shire (A)—East	105157151	3.04	0.0024
Bayside (C)—Brighton	205650911	3.11	0.0018
Wyndham-East Kimberley (S)	545059520	3.11	0.0018
Joondalup (C)—South	505154174	3.13	0.0017
Marrickville (A)	105055200	3.19	0.0014

(Continued)

TABLE 9.2 (*Continued*)

Statistical Local Areas Show Insignificant Z-Statistic with *p*-Value <0.05

SLA Name	ID	Z-Statistic	*p*-Value
Ku-ring-gai (A)	105604500	3.26	0.0012
Mosman (A)	105555350	3.56	0.0004
Stirling (C)—Central	505157914	3.62	0.0003
Waverley (A)	105108050	4.18	0.0000
Warringah (A)	105658000	4.50	0.0000
Woollahra (A)	105108500	5.17	0.0000
Petermann-Simpson	710403009	7.41	0.0000

from the census data. Like some other SLAs that showed overestimates of housing stress, Waverley experiences a particularly low poverty rate with low actual percentage of private renter households in housing stress. In addition, it is in the residential catchment for many tourists (as it is close to the Bondi Beach) and/or part-time working households, perhaps suggesting a possibility of skewing the estimate.

Another possibility could be a very high rate of rent in that area (the median weekly rent in Waverley is estimated as 301 AUD compared to 190 AUD in Australia), which may affect a considerable number of households having income just above the average (median weekly household income in Waverley is estimated as 1179 AUD compared with 1027 AUD in Australia).

On the other hand, the SLA *Melbourne (C)—Remainder* has a remarkably low estimate of housing stress in the simulated population. The reason for this is not clear as the poverty rate in this small area is fairly high. Perhaps there is something geographically unusual about this small area, for instance, it may be the residential catchment for some universities in Melbourne. Potential localized features including group housing for students (especially for international students*) or casual workers can decrease the estimate of housing stress below the actual census figure.

Nevertheless, there are some SLAs with statistically inaccurate estimates of housing stress at the reference level of *p*-value of 0.05 (see Table 9.2, SLAs with Z-statistic ≥2.58). Some of these SLAs are located within major capital cities and may have different geographical features, socioeconomic characteristics, and household attributes that can affect the simulation results significantly compared to the census. For example, in Sydney, the Northern Beaches SLA *Manly (A)* is an area with housing costs being among the highest in Australia. It has well-known educational and religious institutions with a wide range of recreational opportunities including the Sydney Harbour, Pacific Ocean, and Manly Beach. The demand for housing in such an attractive area may have an impact on housing costs for many households in this

---

* For example, refer Tsutsumi and O'Connor (2005).

area, and therefore their housing costs most likely went up from the 30% threshold that may lead to an overestimate of housing stress.

The noncapital city SLAs that were not estimated well, such as *Katherine (T)* in the Northern Territory, are strategic growing areas in rural Australia. Economic growth in this SLA results from the flow-on effects of providing regional support services to major national projects such as mining developments, defense construction, culture and heritage conservation, forestry and horticultural trials, and a transport and logistics hub servicing the Australasia Railway. However, an inadequate supply of housing may increase housing expenses for lower-income households and may skew the MMT estimate.

Figure 9.3a through d presents results of the statistical significance test for MMT estimates of various households by tenure type. It is noticeable that at SLA level the model produces very good estimates for buyer and total renter households. In particular, for buyer households, our model shows statistically acceptable estimates for about 99% of SLAs (with a Z-statistic between −1.96 and +1.96) at the reference level of significance or p-value of 0.05. There are only three SLAs such as *Le Hunte (DC)* in South Australia, *Joondalup (C)— South* in Perth, and *Ku-ring-gai (A)* in Sydney that indicate inaccurate results with Z-statistic values 2.8, 3.0, and 3.6, respectively. Nevertheless, the PDC clearly reveals that the distribution of the test statistic has a leptokurtic shape having the highest peak at $Z = 0.5$. Additionally, for total renters, the MMT for housing stress estimation demonstrates accurate results for nearly 96.8% SLAs at p-value of 0.05. The shape of the PDC is more consistent with the SNC, and the peak is slightly higher around zero (see Figure 9.3d).

The findings for public renter households reveal a clear underestimation of housing stress by the MMT as remarkably large numbers of estimates are below zero, with a considerable number less than or equal to the left-sided critical value of −1.96 for p-value of 0.05. That is, the left tail of the PDC has most of the values of the Z-statistic that confirms inaccurate estimates of housing stress or the existence of a statistically significant difference between MMT estimates and the actual census values. The PDC also shows a notable platykurtic shape in nature and is shifted toward the left of the SNC.

In contrast, the results for private renter households demonstrate an overestimation of housing stress. The appropriate graph in Figure 9.3c shows that compared to the standard curve the PDC is shifted to the right with a very slight (perhaps ignorable) platykurtic shape. A large number of SLAs have values for the Z-statistic indicating unacceptable housing stress estimates in the right tail of the probability distribution. Thus, relative to the estimates for public renter households, the model produces much better results for the private renter households.

Further statistical results from the MMT estimates of housing stress for different tenure households are given in Table 9.3. It is noted that the total

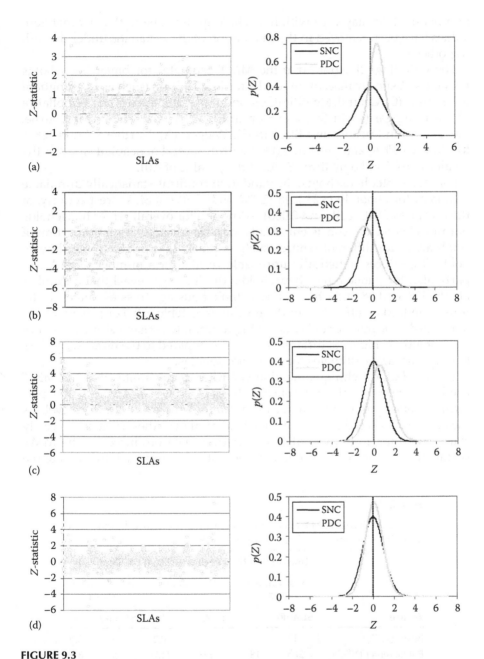

FIGURE 9.3
Results of the hypothesis test for different households by tenure types. (a) Plots of Z-statistic by statistical local areas (SLAs) and the probability curve for buyer households. (b) Plots of Z-statistic by SLAs and the probability curve for public renter households. (c) Plots of Z-statistic by SLAs and the probability curve for private renter households. (d) Plots of Z-statistic by SLAs and the probability curve for total renter households.

number of SLAs may vary with household groups due to the fact that zero values have been removed in the calculation of the Z-statistic under the null hypothesis.

These findings illustrate that the MMT estimates for buyer households provide statistically inaccurate results for just 13 SLAs (such as Rockingham (C), Stirling (C), Central, Melville (C), Joondalup (C), South in Perth; Woollahra (A), Ku-ring-gai (A) in Sydney; Stonnington (C), Malvern in Melbourne; Larrakeyah in Darwin; and Blackall (S), Walgett (A), Northern Areas (DC), Mackay (C), Pt A, and Le Hunte (DC) within several noncapital cities) at the p-value of 0.05 and only three SLAs at the p-value of 0.01.

For total renter households, the validation confirms statistically unreliable estimates for about 40 (3.2%), 16 (1.3%), and 5 (0.4%) SLAs, respectively, at the p-values of 0.05, 0.01, and 0.0001. Although for overall renter households our model generates consistent results for a significantly high proportion of small areas, for different tenures such as public and private renter households, it produces statistically unreliable results for a number of SLAs. In particular, for public renter households, the findings reveal that a number of 205, 128, and 38 SLAs have unacceptable housing stress estimates at the three standard cutoff points of the p-value (see Table 9.3). For private renter households, the number of SLAs having inaccurate housing stress estimates is substantially higher at different p-values compared to the number of SLAs having inaccurate estimates for buyer households.

The validation technique mentioned earlier demonstrates that the MMT used to calculate small area housing stress estimates in Australia has performed well in producing statistically accurate estimates at the SLA level. As the approach is based on a systematic statistical hypothesis test, it can easily clarify whether an individual small area estimate produced by the MMT is statistically reasonable or not. Additionally, this approach to validate the

TABLE 9.3

Report on Statistical Decision for Inaccurate Microsimulation Modeling Technology Estimates by Tenure

	Inaccurate MMT Estimate (i.e., H_{i0} Is Rejected)					
	At p-Value = 0.05		At p-Value = 0.01		At p-Value = 0.0001	
Tenure	SLA (#)	%	SLA (#)	%	SLA (#)	%
Buyer (1257)[a]	13	1.0	3	0.2	0	0.0
Public renter (1097)	205	18.7	128	11.7	38	3.5
Private renter (1244)	136	10.9	81	6.5	14	1.1
Total renter (1265)	40	3.2	16	1.3	5	0.4

[a] Numbers of total SLAs with a determinable Z-statistic under the null hypothesis. This can differ by tenure due to zero values, which causes a division by zero.

estimates can not only check whether the estimate for an area is different from a reliable figure but also identify and describe features of the small areas that show unacceptable results.

9.3.3 Absolute Standardized Residual Estimate Analysis

This approach to validation is very straightforward. First, we have to calculate an ASRE for a small area (in this case, SLA-level housing stress estimation) and then analyze the values of the ASRE to make a decision about the accuracy. The complete procedure involves several steps that are outlined here:

1. Estimate the absolute difference between an observed total and its respective actual total of households for the sample in a small area.
2. Obtain the average empirical mean squared error (AEMSE) of the estimates with the actual data for a large region, which consists of all the relevant small areas.
3. Calculate an estimate of the absolute standardized residual by dividing the value of the absolute difference by the square root of the AEMSE.
4. Analyze the ASRE to reach a decision about the model outputs.

The mathematical formulas for the ASRE use the following standard notations:

$\hat{Y}_{ij}$ is an observed household total in the jth data in the ith small area.

Y_{ij} is the total households in the jth population in the ith small area.

m_r is the number of small areas in a rth region and $r > i$.

The ASRE can be defined as

$$\text{ASRE} = \left(\frac{\delta_{ij}}{\sqrt{\text{AEMSE}}} \right) \tag{9.2}$$

where

$$\delta_{ij} = \left| Y_{ij} - \hat{Y}_{ij} \right|$$

$$\text{AEMSE} = \frac{1}{m_r} \sum_m \left(Y_{ij} - \hat{Y}_{ij} \right)^2$$

(e.g., see Gomez-Rubio et al. 2008).

The decision criterion for this validation technique is also precisely defined. When the value of *ASRE* is close to zero for a small area, then the corresponding estimate of households produced by the model is acceptable, which means the performance of the model estimate is good. On the other hand, if the *ASRE* value is ≥2, then it is usually considered as large (Field 2000) and that unexplained errors exist in the model estimates and/or the microsimulated data sets. For an explicit example, the SLA of *Campbelltown (C)—North* in Sydney has an ASRE value of 0.012 that confirms the model estimate of housing stress (i.e., 3180 households) is approximately close to the census figure (i.e., 3183 households). In contrast, the SLA of *Ipswich (C)— East* in Brisbane has an ASRE value of 3.395 that demonstrates the model estimate of housing stress (i.e., 2114 households) is significantly apart from the census figure (i.e., 1293 households). It means the model has produced inaccurate estimates of housing stress for *Ipswich (C)—East* due to some unexplained errors.

9.3.4 Results from the ASRE Analysis

The ASRE analysis for estimates of housing stress demonstrates very similar results to the statistical significance test. In particular, the values of ASRE for the total number of households in housing stress confirm that for about 94.3% of SLAs (1205 of 1278 SLAs) in Australia, our MMT for housing stress estimation determined very accurate housing stress estimates (see Figure 9.4). This result is approximated to the finding from the statistical significant test at the 5% level of significance or p-value 0.05.

There are 73 (5.7%) SLAs with an ASRE ≥ 2, and most of them are located in major capital cities and coastal centers such as Wollongong, Newcastle, Coffs Harbour, Tweed-Heads, Gold Coast, Hervey Bay, and Mackay. It is noted that these 73 SLAs with considerably large values of ASRE include almost all SLAs that have demonstrated inaccurate small area housing stress estimates using the significance test (see Table 9.3).

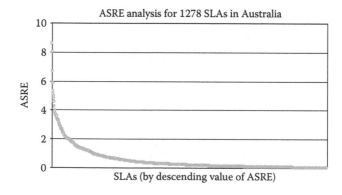

FIGURE 9.4
Absolute standardized residual estimate analysis for MMT estimates of total households in housing stress.

Looking at the specific results from the ASRE analysis in different cities, Figure 9.5 shows the results for eight key cities in Australia. The ASRE for 64 SLAs in Sydney illustrates that there are four SLAs such as *Marrickville (A), Waverley (A), Sutherland Shire (A)—East,* and Warringah (A) that have ASRE values greater than 2, confirming unreasonable estimates of housing stress from the model.

In addition, six SLAs, namely, Melbourne (C), Remainder; Port Phillip (C), St Kilda; Maribyrnong (C); Moonee Valley (C), Essendon; Darebin (C), Preston;

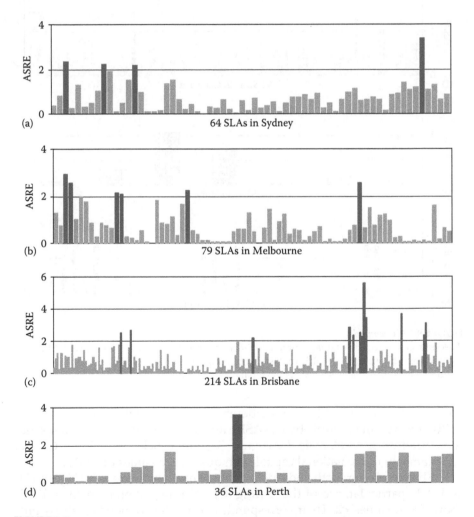

FIGURE 9.5
Absolute standardized residual estimate (ASRE) analyses for statistical local areas (SLAs) in eight major capital cities. (a) ASRE analysis for SLAs in Sydney. (b) ASRE analysis for SLAs in Melbourne. (c) ASRE analysis for SLAs in Brisbane. (d) ASRE analysis for SLAs in Perth.
(Continued)

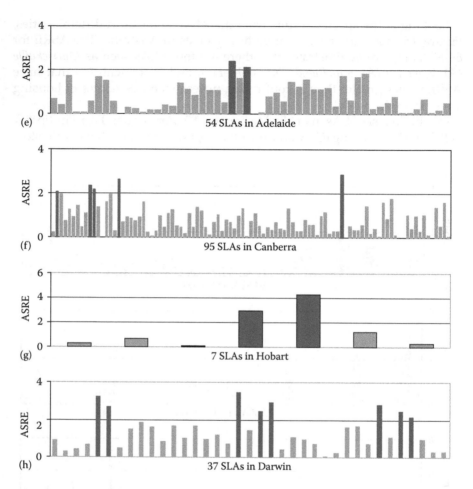

FIGURE 9.5 (*Continued*)
Absolute standardized residual estimate (ASRE) analyses for statistical local areas (SLAs) in eight major capital cities. (e) ASRE analysis for SLAs in Adelaide. (f) ASRE analysis for SLAs in Canberra. (g) ASRE analysis for SLAs in Hobart. (h) ASRE analysis for SLAs in Darwin.

and Glen Eira (C), Caulfield in Melbourne, have moderately large values of ASRE (red bars in Figure 9.5b). The ASRE analysis for Canberra illustrates that there are five SLAs (Ainslie, Lyneham, O'Connor, Belconnen Town Centre, and Kambah) in the national capital with quite large values of ASRE.

The bar graphs for Perth and Adelaide show much better results (Figure 9.5b and c). In particular, except the Stirling (C)—Central, all other SLAs in Perth show *ASRE* values <2. Their corresponding estimates of housing stress produced by the model are acceptable, which means the performance of the model is very good for SLAs in Perth. In addition, SLAs in Adelaide demonstrate very rational results in terms of ASRE, where only two SLAs such as West Torrens (C)—East and Adelaide (C) have large ASRE of 2.4 and 2.1, respectively.

The ASRE analysis for Brisbane shows somewhat different results. There are 12 SLAs in Brisbane in which our model generates unreasonable estimates of the number of households in housing stress according to this validation. It is noted that the total SLA number in Brisbane is 214 (excluding the missing SLA Moreton Island) as Brisbane has very small-sized SLAs. This could be due to a comparatively small number of households living in these small-sized SLAs in Brisbane and also the homogeneity of these areas in Brisbane. Overall, the graph demonstrates that although there are several SLAs with ASRE bigger than 2, the model produces statistically reliable results in Brisbane.

There is a cluster of SLAs with inaccurate results in Brisbane (Figure 9.5c). For instance, few SLAs in Ipswich show very high values of ASRE, which confirms that the MMT has produced statistically inaccurate results in this area. Ipswich is one of the very first growing regions in Brisbane, and the population characteristics are quite different. In particular, a significantly large number of working population families (about 60%) are technicians and trades workers, community and personal service workers, clerical and administrative workers, and laborers who usually may have lower income. But the housing cost for this area is relatively high. These factors were not included as part of the model benchmarks; as a result, our model simulates significantly high estimates of housing stress for the region.

The result for the city of Hobart shows significantly large (>2) ASRE values for the SLAs *Glenorchy (C)* and *Hobart (C)—Remainder*. It seems that for these two SLAs the model overestimates the number of households experiencing housing stress. Typically, the supply of housing in these areas is fairly inadequate with a growing housing demand. Also, households' low income may potentially increase the figure of housing stress. Moreover, the ASRE analysis for Darwin highlights that there are eight SLAs (see Figure 9.5h), namely, *Coconut Grove, Fannie Bay, Nightcliff, Rapid Creek, Stuart Park, Gray, Moulden*, and *Woodroffe*, that have estimates of ASRE > 2. Like Brisbane and Canberra, the capital of the Northern Territory encompasses fairly small-sized SLAs. Most of these SLAs are with a comparatively small number of households. Also, less housing supply in the marker for these small areas in Darwin and low households' income may perhaps be the two reasons for overestimating the numbers of housing stress toward a remarkably greater value of ASRE.

9.4 Measure of Statistical Reliability of the MMT Estimates

This section is designed to calculate a statistical reliability measure, *CI*, for the SLA-level housing stress estimates produced by the MMT for small area estimation.

9.4.1 Confidence Interval Estimation

A CI for a model estimate is an estimated range of numeric values in which the true value of the estimate falls based a given probability or level of confidence. The range of the CI can be defined by the following simple relationship:

SLA-level MMT estimate ± margin of error

The margin of error (ME) is frequently defined as the *radius* of a CI for a particular estimate given by the model. Essentially, there are two specific things associated with calculating the ME. These are

1. A critical value
2. A standard error of the model estimate

Determining the ME at various levels of confidence (such as 95%, 99%, or 99.99%) is quite easy when the model estimates follow a conventional probability distribution and the standard errors of the estimates are measurable. It is obvious that values of the Z-statistic for the SLA-level housing stress estimates follow the standard normal distribution and standard errors of the estimates are available using the methods outlined earlier. The ME for the small area housing stress estimates can be defined as

$$ME = z_{(p)} \times SQRT\left[\frac{p_{ij}^m(1-p_{ij}^m)}{\sum_j n_{ij}}\right], \tag{9.3}$$

where
$z_{(p)}$ is the critical value based on the confidence limit chosen by the researcher
SQRT stands for the square root
All others are usual notations used in Section 9.3 (Section 9.3.1)

Hence, an indicative CI measure for the estimate of housing stress can be mathematically expressed as

$$CI = \hat{p}_{ij}^m \pm ME \tag{9.4}$$

(e.g., see Mason and Lind 1996).
Now, by using the expression of *ME* from Equation 9.3, we have

$$CI = \hat{p}_{ij}^m \pm z_{(p)} \times \sqrt{\frac{\hat{p}_{ij}^m(1-\hat{p}_{ij}^m)}{\sum_j n_{ij}}} \tag{9.5}$$

As the critical value for CI estimation relies on the researchers or clients level of confidence, this arbitrary value can be selected from the set of standard levels of significance, that is, $p = 0.05, 0.01$, or 0.0001. Let us consider the fixed level of confidence at 95% for the analysis. In that case, the corresponding level of significance value is $p = 0.05$. Now, since the percentage estimates for the SLA-level housing stress follow approximately a standard normal distribution, at the 99% confidence limit the respective critical value for estimating the CI is given as (Cochran 1993) $z_{(p = 0.05)} = 1.96$.

Therefore, by using this critical value in Equation 9.5, the CI for the MMT estimate for ith SLA in Australia can be stated by the following formulas:

$$
\begin{aligned}
CI &= \hat{p}_{ij}^m \pm z_{(p=0.05)} \times \sqrt{\frac{\hat{p}_{ij}^m (1 - \hat{p}_{ij}^m)}{\sum_j n_{ij}}} \\
&= \hat{p}_{ij}^m \pm 1.96 \times \sqrt{\frac{\hat{p}_{ij}^m (1 - \hat{p}_{ij}^m)}{\sum_j n_{ij}}}.
\end{aligned}
\tag{9.6}
$$

To establish an interval of statistical reliability such as CIs for the MMT estimates are important. Although a single estimate of statistical reliability is often of interest to the researcher, the CI of the model estimate eventually gives the researcher a range of values within which the unknown true figure of the population parameter (in this case, the true percentage or proportion of households in housing stress) may lie. Therefore, the researcher can declare with some level of confidence that the unknown true figure of the MMT estimate for a particular SLA lies in that particular range/interval.

It is noted that a CI must always be qualified by a specific confidence level that is usually articulated as a percentage of confidence to the researcher. Additionally, a CI consists of all possible real values of interest about the population within a particular range. The end points of the CI are therefore known as confidence bounds, where the highest and lowest values are referred to as the upper bound (U) and the lower bound (L), respectively. Of course, for any values in an interval, the difference between the actual data for the interest of population and the model estimate may be statistically insignificant at the selected significance value (Cox and Hinkley 2000, pp. 214–233), for example, at $p = 0.05$ or at the 5% level in our case.

Researchers also use the CI to describe the amount of uncertainty associated with the model estimate of an unknown population parameter. The width of the CI gives some idea about the uncertainty. Typically, a narrow interval may indicate less uncertainty with better precision for the estimate, whereas a very wide interval may indicate more uncertainty about the parameter estimation.

9.4.2 Results from the Estimates of Confidence Intervals

This section presents the empirical results for the proposed method to estimate CI. At first, the CIs are created for the estimates of housing stress for
every SLA across Australia. General discussions of the results are given in
the following text. Subsequently, the specific results for SLAs in some major
capital cities are illustrated.

The 95% CI obtained using the mathematical equation (9.6), for overall household level housing stress data created by the MMT for all 1278
SLAs across Australia, are shown in Figure 9.6. The point estimates (PEs)
of housing stress for each of these SLAs are also in the figure. This clearly
shows that although there is considerable variability in the CI measures
by the SLAs and the PEs, a significantly large number of SLAs have fairly
narrow CIs.

In particular, based on the difference between the upper and lower
bounds (i.e., the length) of the 95% CI, a total of 1223 SLAs have differences of just 2%–20% points; among them, 800 SLAs have narrow differences up to only 10% points. Additionally, only about 4% (55 SLAs)
of small areas in the analysis have relatively wide limits, that is, from
>20% points to <42% points. From this information, one can infer that our
MMT has produced housing stress estimates for about 96% of SLAs with

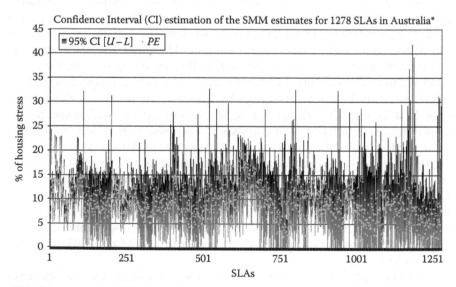

* U is the upper bound of CI; L is the lower bound of CI; and PE is the point estimate.

FIGURE 9.6
Measure of confidence intervals of the statistical local area–level housing stress estimates for
overall households.

a statistical reliability at the 95% level according to the confidence limits of up to 20% points.

However, there are 47 SLAs with the CI measures ranging from 20% to 30% points, which may be considered as moderately wide. There are also 8 SLAs that have CI measures showing a wider length of 30% points or more, and among them, only one SLA (*Acton* in Canberra) has the highest length of CI of 41.9% points. It is worthwhile mentioning that most of these SLAs that show a relatively wider length of CIs typically also have a small number of households, and they are geographically located outside the major cities (except a few in Brisbane and Canberra).

Specific results for eight major capital cities are given in Figure 9.7a through h. It can be seen that the confidence bounds produced by the proposed method are reasonably narrow for almost all SLAs in Sydney, Melbourne, Perth (except *Peppermint Grove (S)* and *Fremantle (C)—Inner*), Adelaide, and Hobart. Usually, these capital cities have larger and more populous SLAs than SLAs in the other three capital cities (Brisbane, Canberra, and Darwin). For the more populous SLAs, the populations of the model simulates are closer to the accurate census data, as the sample sizes for those SLAs from the survey data are also large. This means that with a reasonable number of households simulated by the model, there should be a sufficiently small CI, which means a better statistical reliability measure. Nevertheless, the two SLAs in Perth such as *Peppermint Grove (S)* and *Fremantle (C)—Inner* have relatively small numbers for the microsimulated data that may cause the much wider CI measures.

Rather, different results are found in the SLAs of Brisbane, Canberra, and Darwin (see Figure 9.7c, f, and h). These three capital cities have not only large numbers of small-sized SLAs, but also many of them have fewer households. It is apparent from the figure that several SLAs in each of these three cities show significantly bigger CI measures. This is because all of these SLAs have very small household populations. For instance, *Pinjarra Hills* in Brisbane, *Acton* in Canberra, and *Narrows* and *Litchfield (S)—Pt A* in Darwin have rather wide CIs as their synthetic populations are too small to create adequate precision for their small area housing stress estimates.

As the approach uses the normal probability distribution to quantify the CI (appropriately given the evidence that the model estimates values of the Z-statistic that follow approximately the standard normal distribution), where less populous areas have a wider CI, particular caution must be exercised in interpreting the statistical reliability of these areas. In the case of excessive distance between the lower and upper bounds, a researcher should assess the synthetic data set to ensure it can create more valid results before making a decision. Another option is to consider the results for those few SLAs that are ignorable as the uncertainty associated with it tends to have only a very negligible effect on the overall results.

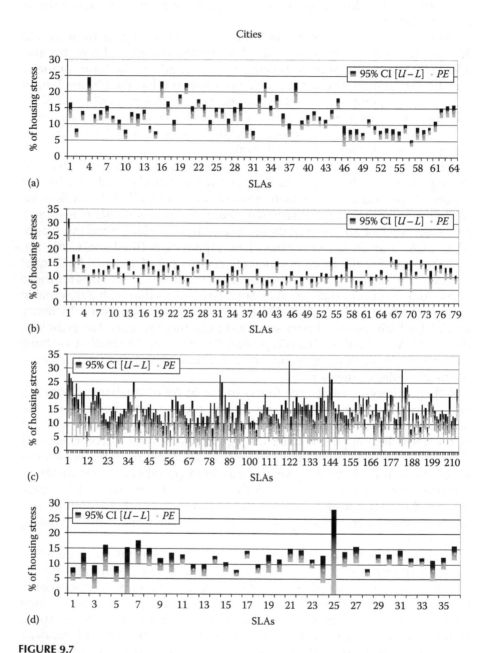

FIGURE 9.7
Confidence interval (CI) measures for estimates of housing stress for major capital. (a) CI estimation for statistical local areas (SLAs) in Sydney. (b) CI estimation for SLAs in Melbourne. (c) CI estimation for SLAs in Brisbane. (d) CI estimation for SLAs in Perth.   (*Continued*)

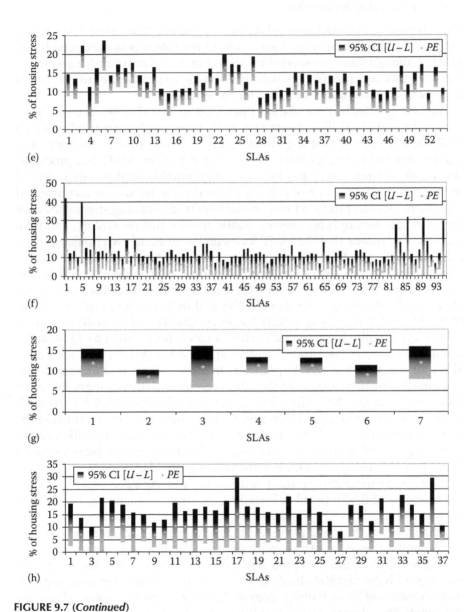

FIGURE 9.7 (*Continued*)
Confidence interval (CI) measures for estimates of housing stress for major capital. (e) CI estimation for SLAs in Adelaide. (f) CI estimation for SLAs in Canberra. (g) CI estimation for SLAs in Hobart. (h) CI estimation for SLAs in Darwin.

9.5 Concluding Remarks

This chapter has offered two validation techniques that create tests of statistical reliability for SLA-level estimates of housing stress produced by an MMT. Although there are several validation methods commonly used in the literature by many researchers, none of them use a statistical significance test and a statistical reliability measure for the estimates.

The first approach of validation discussed in this chapter is based on a systematic statistical hypothesis test, which can easily reveal whether an individual SLA-level housing stress estimate is statistically accurate. The results from the technique have demonstrated that our MMT has produced statistically accurate estimates for a very high number of SLAs (about 96.3%) in reference to the p-value of 0.05. Additionally, the technique has revealed that the model has produced reasonable results for housing stress estimates by household tenure type (buyers, public renters, private renters, and total renters). This means that the MMT for estimates of small area housing stress has performed well by producing statistically acceptable housing stress estimates at SLA level across Australia. The findings also suggest that this new proposed approach to validate the MMTs can not only check the significance of an SLA-level estimate but also identify and describe the possible features of the SLAs that may have unacceptable results. The SLAs with Z-statistic values significantly bigger/smaller than a cutoff point (such as ±1.96 at 5% level) demonstrate inaccurate housing stress estimates for the respective SLAs by rejecting the null hypothesis. In such a case, researchers would examine the microlevel data for these SLAs along with their geographic attributes. For instance, the finding for the Northern Territory SLA *Katherine (T)* (with Z-statistic value of 2.71) has revealed that the model overestimated the housing stress data, and this noncapital but strategically economic growth area is experiencing an inadequate supply of housing that perhaps increases demand and housing costs for low-income households.

The second validation approach outlined in this chapter is the ASRE analysis, where an ASRE for the SLA-level housing stress estimate has been calculated and then analyzed using standard cutoff criteria for making a decision. Normally, when the value of ASRE is close to 0 or <2 for an SLA, then the corresponding housing stress estimate from the model is acceptable, otherwise not. The results from the ASRE analysis show somewhat consistent findings with the first statistical significance test, at least in terms of approximate numbers of SLAs having accurate housing stress estimates. Moreover, the results have demonstrated that there are a number of SLAs with considerably large values of ASRE, and most of them are geographically located in various capital cities, including Melbourne, Brisbane, Canberra, and Darwin, as well as major coastal centers in the east part of Australia.

An alternative approach to estimate CIs has been proposed in this chapter to create a measure of statistical reliability of the MMT estimates. In general, the empirical results for the 95% CI estimation have revealed that a total of 1223 SLAs estimates illustrate reasonably narrow CIs with a fairly good level of precision. There are also some SLAs showing relatively wider CIs. These SLAs are often less populous and mostly located outside of the major urban areas or located within capital cities that consist of SLAs with smaller sizes. In particular, the measures of CIs for several SLAs in Brisbane and Canberra have shown rather uncertain results due to a too small size of the population. Finally, for such SLAs with very wide CI measures, an optimality check can be performed before reaching any decision about the statistical reliability.

10

Conclusions and Computing Codes

10.1 Introduction

This final chapter draws this book to a close by reflecting on the study's key results and summarizes the basic limitations of research. Furthermore, some potential areas of future research are outlined, followed by some concluding statements regarding the most significant contributions of this book.

10.2 Summary of Major Findings

The original purpose of this book was to provide an extensive and comprehensive state-of-the-art presentation of all of the existing methodologies in small area estimation and then to develop a novel and more robust methodology called microsimulation modeling technology (MMT) for small area estimation with its real-world applications. These aims were targeted by particularly focusing on methodological developments by bridging classic ideas and conventional statistical small area estimation concepts with the latest Bayesian prediction–based computational method for spatial microdata simulation and ultramodern MMT-based geographical small area estimation theories, and then on the role that small area estimation and microsimulation modeling can play in supporting rational decision-making process and policy analysis along with overcoming the lacuna in the validation techniques and measures of the statistical reliability of the MMT-based small area estimates. To achieve this, a range of objectives (listed in Chapter 1) were established and sought to be undertaken. Each of these objectives is addressed in this section with regard to the major findings of this book to evaluate whether they were successfully attained.

Chapter 2 provides adequate evidence that small area estimation plays a vital role in various real-world practices by producing reliable and accurate small area estimates. These days, most of the developed nations are utilizing small area estimation methodology as an essential means to support the process of knowledgeable and effective decision-making and policy analysis at local or regional levels. The results show that, overall, small area methodology covers a range of simple direct estimation techniques to very complex indirect model-based estimations. The review of direct small area estimation has revealed that although the theories and formulas of direct estimators are very simple and straightforward, they can be used only when enough sample data are available at a small area level. Eventually, the findings reveal a significant deficiency of direct methods in terms of statistical inferences.

Chapter 3 covered two objectives on the statistical approaches of indirect small area estimation, where a thorough review is provided on a range of indirect model-based statistical approaches for small area estimation. The overall methodologies for indirect statistical estimation include two types of statistical models, commonly known as the implicit and explicit models, and a range of statistical techniques for the estimation of small area models. The implicit model-based methods can generate three types of estimators, namely, the *synthetic, composite,* and *demographic estimators.* Although these indirect estimators can be obtained by borrowing strength from available auxiliary data, with a confirmation of better precision over the direct estimators, with regard to the theory, advantages, drawbacks, and applications, they are relatively easier to calculate but typically biased and limited for applications with the nature and suitability of the applicable data.

Explicit models include the *area level, unit level,* and *general linear mixed models.* In the area-level models, information on the response variable is available only at the small area level; for unit-level models, data are available at the unit or respondent level; and in the general linear mixed models, data are available at the area and/or unit levels. All of these models are studied by researchers in different fields of statistics such as in social and economic statistics, agricultural statistics, and government statistics. For making statistical inferences of explicit model-based estimators, three statistical procedures have been used that include the (empirical-) best linear unbiased prediction (E-BLUP), empirical Bayes (EB), and hierarchical Bayes (HB).

Findings of the review suggest that the E-BLUP approach is applicable to the linear mixed models that are usually designed for data with continuous variables, but not applicable for binary or count data. This is a considerable limitation of the E-BLUP approach. In principle, the EB approach is considered as a frequentist approach and does not depend on a prior distribution of the model parameters, whereas the HB approach essentially uses a prior distribution of model parameters and can handle complex problems using Markov Chain Monte Carlo (MCMC). Even though all of these statistical approaches use data from different sources to obtain the small area estimators, they do not involve generating a base microdata file at small area level, which is a significant resource for further analyses.

The concepts and methodologies for the geographic approaches of indirect small area estimation have been extensively discussed in Chapter 4, which manages objectives 5 and 6, in addition to objective 1. It is noticeable in the literature that until recently most of the research on small area estimation has been conducted using statistical model-based estimation only. However, spatial microsimulation model–based geographic approaches are emerging of late as a very useful alternative means of small area estimation. The key challenge for indirect small area estimation techniques, which is highlighted continuously throughout the literature, is being able to produce reliable synthetic spatial microdata. The appraisal of methodologies has revealed that the two reweighting methods—combinatorial optimization (CO) and GREGWT—are commonly used tools to produce small area microdata.

The CO technique uses an intelligent searching algorithm, that is, *simulated annealing*, which selects an appropriate set of households from survey microdata that best fit to the benchmark constraints by minimizing the total absolute error/distance with respect to the *Metropolis Criterion*. The new weights give the actual household units, which are the best representative combination. Thus, CO is a selection process to reach an appropriate combination of sample units rather than calibrating the sampling design weights to a set of new weights. On the other hand, the GREGWT technique utilizes a truncated chi-squared distance function and generates a set of new weights by minimizing the total distance with respect to some constraint functions. The minimization tool of Lagrange multipliers has been used in the GREGWT process to minimize the distance function and is based on the Newton–Raphson iterative process. The results show that sets of new weights can vary substantially with changing values of the vector of difference between the benchmark totals and sample-based estimated totals. In fact, the chi-squared distance measures show smoother fluctuations than the absolute distance measures.

The findings of the comparison between the CO and GREGWT reveal that they are using quite different iterative algorithms and their properties vary considerably. The CO routine has a tendency to include fewer households but give them higher weights, and, conversely, the GREGWT routine has a tendency to select many households but give them smaller weights. The overall performances are fairly similar for both the reweighting techniques from the standpoint of use in spatial microsimulation modeling. However, in the context of available data in Australia, the GREGWT algorithm is capable of producing good results relative to the CO technique. Furthermore, like other methods, the GREGWT technique has some limitations. For instance, the GREGWT reweighting algorithm does not meet "accuracy criterion" for few small areas where the sample size is too small or especially zero, and it also takes more CPU run time compared to the CO.

Ultimately, the spatial microsimulation model–based geographic approaches to indirect small area estimation techniques are robust, in the sense that further aggregation or disaggregation is possible on the basis of choice of spatial scales or domains. As well, since the spatial microsimulation framework uses a list-based approach to microdata representation, it is possible to use the microdata file for further analysis and updating. Furthermore, by linking spatial microsimulation models with static micro-simulation models, it is also possible to assess small area effects of policy changes. In contrast, the traditional statistical model-based approaches do not have such utility.

Chapter 5 was concerned with objective 7 that is to develop an alternative Bayesian prediction–based microdata simulation technique and also linked to objectives 5 and 6. As the review of different methods to small area micro-data simulation for small area estimation seeks further developments, this chapter introduced a new methodology for generating the small area micro-data toward the spatial microsimulation modeling approaches of small area estimation. The new approach is based on the Bayesian prediction theory and can simulate complete scenarios of the whole population (i.e., observed units plus unobserved units within the population) in each small area. As a result, the process can produce more accurate and statistically reliable small area estimates, compared to the estimates from the other reweight-ing techniques. More explicitly, the theorical findings revealed that the pro-posed methodology is able to produce variance estimations of the small area estimates and the Bayes credible region or confidence interval (CI) for the estimates to confirm their statistical reliability. However, the reweighting methods such as the GREGWT- and CO-based estimates customarily cannot confirm such types of measures of reliability.

Furthermore, the Bayesian prediction–based microdata simulation is a probabilistic approach, which is quite different from the deterministic approach used in GREGWT and the intelligent searching tool *simulated annealing* used in the CO reweighting. Since the Bayesian joint posterior prob-abilities of the parameters for the observed sample units and unobserved population units are estimated through the MCMC method, the proposed microdata simulation methodology is somewhat linked to a chain Monte Carlo sampling, but rather different from the multiple imputation technique. The basic computation process of the new approach is mainly associated with a prediction distribution of the unobserved population units, given the sam-ple units derived from a robust model. Nonetheless, the new approach can use the generalized regression model operated in the GREGWT algorithm to link observed units in the sample and unobserved units in the population. In contrast, from the view point of the CO reweighting, it uses the MCMC simu-lation with a posterior density-based iterative algorithm. Although it seems that to find out an appropriate linking model is somewhat a challenging task to deal with, the main advantages of this proposed new approach are that it can not only simulate the complete scenarios of the small area population, but also create the measures of statistical reliability of all small area estimates.

In Chapter 6, a detailed description of the construction procedure for the spatial microsimulation model used to calculate small area housing stress estimates is presented. This chapter principally covers research objec-tive 8, which was to describe the construction of an MMT-based spatial microsimulation model for small area estimation, by revisiting research objectives 6 and 1. First, background information about various data sources and issues related to the datasets is discussed. Most of the data is sourced from the Australian Bureau of Statistics (ABS), which provides reliable census and survey data. As the individual- and/or household-level raw survey microdata are generally not available at small geographic scales due to the confidentiality agreement with the respondent, some of the data files such as the Survey of Income and Housing–Confidentialised Unit Records Files (SIH-CURF) used in this book are supplied by the ABS on special request and under specific conditions.

The specification of the spatial microsimulation model for small area housing stress estimation confirmed that it has two computational stages: the first stage simulates statistical local area (SLA)-level households' synthetic weights by the GREGWT algorithm, and the second stage calculates the ultimate small area housing stress estimates after producing the small area synthetic microdata using the synthetic weights. Essentially, the five groups of files named the *general model file, unit records data files, benchmarks files, auxiliary data files,* and the *GREGWT program file* are required for executing the first stage of the model to generate synthetic weights for small areas in Australia. Then using this *synthetic weights file* and other data files such as the *SIH-CURF files* and the *consumers price index file,* the model runs the second-stage computation through a *SAS program file* to achieve spatially disaggregated microdata and determines the small area housing stress estimates.

The chapter has also discussed the basic concepts of housing stress estimation. The findings demonstrate that housing stress is measured by two quantities: household's income and its expenditure on housing, where the cutoffs of housing costs can vary by the circumstance, income level, as well as the location of the household. Although there are three common measures: "30 only rule," "30/40 rule," and "30/(40–10) rule," the 30/40 rule-based measure is the most common rule exercised by researchers as this definition is accountable and valid for the use in various socioeconomic policy analyses related to the housing stress issue. Additionally, a comparison of three definitions promotes the 30/40 rule-based measure as being more consistent and appropriate than the others in terms of policy significance as well as the statistical view. For instance, the findings have revealed that one of the major policy significance of the 30/40 rule-based housing stress definition is that this rule is widely used as the basis for determining household eligibility for entry to public rental housing and/or receipt of rent assistance. Moreover, compared to the other two variants of housing stress estimation, the coefficient of variation measure for the "30/40 rule"-based variant has the lowest value and has also a more rational pattern toward the usual normal distribution curve, which statistically indicates that the "30/40 rule" is a more consistent and appropriate measure.

By deriving and discussing various empirical results of the methodologies using Australian data, Chapter 7 covers objective 9 of this book. The findings reveal that housing stress estimation (defined by the 30/40 rule) in Australia varies significantly with geography and by tenure types of households. One of the key findings using outputs from the spatial microsimulation model was that in 2006 around one in ten Australian households were experiencing housing stress, with considerably large numbers of these households residing in the east coast states of New South Wales, Victoria, and Queensland. When looking at the housing stress at a higher geographic disaggregation, the findings from the model outputs have revealed that households experiencing housing stress were mostly residents of the Sydney, Melbourne, Brisbane, Perth, Adelaide, Gold Coast, Hunter, Illawarra, Mid-North Coast statistical divisions and some other statistical divisions located across coastal centers of New South Wales and Queensland. The Canberra, Hobart, and Darwin statistical divisions all have relatively low housing stress levels. Besides, breaking the geographies down further, we find greater heterogeneity in housing stress estimates, but still the households are concentrated in these main locations or spots. The areas with a high proportion of households living in housing stress were those concentrated in the outer fringes of capital cities along the east coast of Australia. Of particular interest was Newcastle, which has the largest estimated number of households in housing stress of all estimated statistical subdivisions in Australia.

Breaking the geographic scale down even further to one of the smallest and administratively helpful areas—SLA, we can really get at which small areas are suffering the most by housing stress. The findings have demonstrated that a large number of SLAs in the New South Wales coastal cities including Sydney were with the highest numbers of households in housing stress. Most of the SLAs in Melbourne, Adelaide, and Hobart were also with significantly higher estimates. Moreover, the rapidly growing mining areas around inland locations in different states have shown many SLAs with relatively higher estimates of housing stress. These could be because of a significant lack in the supply of housing within these quickly growing mining areas, which create a high demand of housing and then increase of housing costs for mainly low- and middle-income households. In contrast, a significantly large number of SLAs in Brisbane, Canberra, and Darwin have shown much lower number estimates of housing stress. This is probably because these SLAs not only are small in sizes, but also relatively have much smaller household populations. The results of the percentage estimates reveal somewhat opposite results to that of the count estimates, that is, many SLAs with very low estimated numbers across Australia showed very high rates of housing stress. Nonetheless, various SLAs in different capital cities indeed confirm significantly large values in housing stress for both counts as well as percentages.

When analyzing the estimates by household tenure, it is found that private renter households were more likely to be experiencing housing stress than those households who own or are purchasing their home. In most states' capital cities, the SLA-level housing stress estimates for private renter households are high and influence the estimates for total renters (public and private), and then consequently for the overall households living in housing stress within that area. A significant number of buyer households in Australia were also experiencing housing stress, and most of these households were the residents of SLAs located in various major cities. For example, around a quarter of the total buyer households in housing stress were located at several hotspots in Sydney, Melbourne, and Perth. Normally in these major city SLAs, the supply of housing is practically limited to low- and middle-income households with its typical growing trends in populations and house prices.

Chapter 8 is concerned with objective 10, which was to analyze small area estimates in capital cities. Using the spatial analysis technique, the chapter thoroughly discusses the results of housing stress estimates for the SLAs in several major capital cities to see the local level variation in housing stress. Almost two-thirds of households with housing stress were resided in Sydney, Melbourne, Brisbane, Perth, Adelaide, Hobart, Canberra, and Darwin. Among these capital cities, Sydney and Melbourne have about 40% of all housing stress households in Australia, and Sydney alone has one-third households compared to the total estimates from the eight capital cities. Nevertheless, housing stress estimates show rather mixed increasing trends for some capital cities in Australia.

Spatial analysis of the housing stress estimates in Sydney has indicated the SLAs in Canterbury-Bankstown, Fairfield-Liverpool, Blacktown, and Gosford-Wyong as the hotspots for housing stress due to a significant increase of housing costs with a very limited supply of housing for low- and middle-income households. However, SLAs in the rest of Sydney with better socioeconomic conditions have low housing stress estimates. In Melbourne, the outer Melbourne regions and parts of the inner and middle Melbourne appeared as much affordable places for relatively low-income households. The findings reveal that although the majority of SLAs across Brisbane have very small counts of households in housing stress, the percentage rates are much higher for most of these SLAs. Some hotspots for housing stress estimates in Brisbane were observed in Ipswich, Redcliffe, Logan, Beaudesert, Caboolture, and Redland shire regions with southeast and northwest inner Brisbane.

Findings for Perth have shown relatively lower percentage of households in housing stress. But this capital has many SLAs with much higher estimated numbers, and at least few of them (such as Canning, Gosnells, Rockingham, Swan, and central Stirling) have estimates significantly at the highest level. In addition, most of the Adelaide regions were affordable places for low-income households, except the regions of Playford, Port Adelaide, Salisbury and Charles Sturt, which demonstrated reasonably large estimates for the renter households, particularly for the private renter households. Besides, spatial analysis for Canberra has shown that almost all small areas were with very low level of housing stress estimates relative to other cities, as the Canberra region has much better socioeconomic conditions with a big proportion of households having higher income. The supply of housing including the commonwealth rent-assisted housing is also relatively better in this regional and low-population growth area. Furthermore, the results for Hobart and Darwin have shown only considerable variation in housing stress by SLAs for private renter and public renter households, respectively. The findings for SLAs in the central part of Hobart and some SLAs in Darwin such as Marrara, Nakara, and Gunn-Palmerston city have relatively high prevalences of housing stress. It is further noticeable from the analysis that numbers of households residing in these two cities are quite low, compared to the other capital cities, and almost all of the SLA-level estimates are significantly low in Darwin.

To fulfill further two objectives 11 and 12, Chapter 9 has offered two new types of validation techniques for checking the statistical significance of the SLA-level housing stress estimates produced by the spatial microsimulation model, along with an alternative approach to create the statistical reliability of the spatial microsimulation model estimates. Although there are several techniques available in the literature for validating spatial microsimulation models, none of these methods have been directly looked at on the statistical significance test of the model estimates and then have created any statistical reliability measure for the estimates. As a result, the logical and straightforward statistical inferences about the model estimates are yet unavailable.

The theories of the proposed validations are coordinated with the typical statistical procedures such as the Z-test and standarized residuals analysis. For example, in the first approach of validation, a systematic hypothesis test has been performed to check the statistical significance toward accuracy of an individual SLA-level housing stress estimate produced by the model. The empirical findings validate that the model has produced statistically accurate (at the level of significance value, $p = 0.05$) small area estimates of housing stress in considerably large numbers of SLAs across Australia for the overall households, buyers, public renters, private renters, and total renters. In a general view, findings also reveal the distributive patterns of the test results for each group of households. They have typically explored a degree of accuracy of the estimates compared with the standard normal distribution. It is obvious the test results of the housing stress estimates for overall households and the total renter households are highly consistent with the actual census figures. As the approach is based on a systematic statistical hypothesis test, it can easily clarify whether an individual small area estimate produced by the MMT is statistically reasonable or not. In such context, the key advantages of this validation technique are that it is scientific and straightforward and has standard index toward making statistical inference. Additionally, this approach to validating the small area estimates can identify and describe the possible features of SLAs, which particularly show inaccurate estimates. Besides this, the features of this validation technique are practicable to create the statistically reliability measure of the model estimates, while most of the other validation tools do not have such attributes.

The second validation technique called the *absolute standardized residual estimate* (ASRE) analysis has also some statistical properties for making statistical decision about the model estimates. Usually, when the value of ASRE is close to zero or less than 2 for an SLA, then the corresponding housing stress estimate from the model is acceptable, otherwise not. Although the cutoff point 2 is standard one to use, a modeler can choose alternative values as cutoffs if necessary for the clients' demand. The findings of the ASRE analysis also validate that our model has produced acceptable small area estimates for a significantly large number of SLAs throughout Australia. In other words, the analysis has demonstrated that only about 5% of the total SLAs in Australia showed an ASRE estimate more than or equal to the cutoff point 2, and most of these SLAs were located within major capital cities and coastal centers, including Wollongong, Newcastle, Coffs Harbour, Tweed-Heads, Gold Coast, Hervey Bay, Mackay, etc. In addition, the findings have revealed that almost all the SLAs with considerably large values of ASRE have also demonstrated inaccurate small area housing stress estimates by the statistical significance test.

Moreover, the chapter proposes an alternative means to create a measure of statistical reliability such as "confidence intervals" for the spatial microsimulation model estimates. The description of the methodology is given with a discussion of the outputs of CI measures for housing stress estimates produced by the model. The findings are indeed appreciable, since at the 95% CI level the measures of CIs for most of the small area housing stress estimates show fairly narrow length with a higher level of precision. More explicitly, the study has demonstrated that the CI estimations of 1223 SLAs have a length of less than 20% points, and among them about 800 SLAs have a further narrow CI length of up to 10% points. Accordingly, the uncertainty comes from the spatial microsimulation modeling—particularly, in the process of microdata simulation for these small area estimates seems low. However, the uncertainty associated with the wider-length CI measures perhaps comes from the bias of the model toward the small areas with small numbers of households or populations. The key advantage of this approach is that it is simple and very straightforward in terms of computational requirements. Eventually, the method is well functional in the situation where the spatial microsimulation model can effectively produce statistically accurate small area estimates.

Finally, the last objective of providing the overall concluding remarks and all computing codes has been covered in this conclusions chapter.

10.3 Limitations

It is evident from the earlier discussion that this research has achieved its principle aims by successfully fulfilling all the particular objectives, which have been set in the introduction chapter of this book. However, in all research studies, there are inevitably some limitations that need to be acknowledged. This research is no exception. In particular, the outcomes of this study were limited by the complex methodology of the research chosen, as well as by time restrictions and data availability. So far, these limitations have somehow proactive influence on the research results, as they involve factors that are relevant but have not been incorporated into the present study for one reason or another. Nonetheless, it is appropriate that the limitations are taken into consideration when interpreting these results and making related recommendations. The key limitations are reported in the following.

These analyses are based on spatially disaggregated simulated data, not "real" data. As a result, there perhaps is uncertainty associated with the small area estimate that is not assessable in many cases. Besides this, the validation against the ABS census figure is very practical and credible but may not necessarily be viewed as a comparison against absolutely a *true* figure. In fact, the ABS census figures are from the "census survey data." In theory, although census covers all populations in its data collection process, typically there may also be some sorts of errors associated with it in terms of possible errors in the process of data collection, inputs, processing, etc. However, worldwide, the census data are commonly considered as the most consistent values for the true figure.

The housing stress estimates produced by the model are at SLAs in Australia. However, there is considerable variation in the size and population of SLAs in different cities. In particular, the size and population of SLAs in Brisbane and Canberra are much smaller than those of SLAs in most other capital cities such as Sydney, Melbourne, Perth, and Adelaide. Thus, the results of the SLAs in Brisbane and Canberra are usually not comparable to the results of the SLAs in other capital cities.

The number of benchmarks used in the reweighting is limited by the size and nature of the SIH data used in the research. The number of benchmarks and classes used in the reweighting has implications of the resultant estimates. Typically, if a variable is included in a benchmark, then the estimate of that variable is reliable. Essentially, not all variables are included in the benchmarks for constructing the model, as ten benchmarks have been used in this research.

In addition, the GREGWT reweighting algorithm was not able to provide new weights by meeting the benchmarks for a number of SLAs, which are geographically located in the rural central desert regions of Australia, including most of the SLAs in central Western Australia, South Australia, and the Northern Territory. The analysis revealed that, compared to the total SLAs in Australia, these isolated small sets of non convergence SLAs have only about 0.4% households. That means, most of these non convergence SLAs have a population size of zero or too small to converge the algorithm toward generating new weights. As a result, the model was unable to determine small area housing stress estimates for these SLAs.

As highlighted in Chapter 6, there are limitations of the survey datasets used for empirical results. Mainly, the SIH only surveys people aged 15 years and older in private households. It does not include data against people younger than 15 years and does not cover the population resident in non private dwellings (such as hospitals, institutions, nursing homes, hotels, hostels, etc.) and dwellings in collection districts defined as very remote. However, these limitations were overcome by using the SIH linkage data file developed by the National Centre for Social and Economic Modelling (NATSEM) that covers the information about households living in non private dwelling.

Although the data collection and computing facilities for spatial microsimulation have advanced substantially over the last few years, this model is less strong in terms of behavior modeling and typically can only capture the one-direction impact of the policy on households, for example, it can assess the impact of the housing policy on households or individuals but overlooks the impact of households or individuals on the policy (Williamson 1999; Gilbert and Troitzsch 2005).

Moreover, microsimulation modeling techniques have been criticized for embodying more technical knowledge than theory (Halpin 1999). Indeed, the spatial microsimulation models are relatively more complex, have significant data handling issues as well as computing requirements, are costly to build and maintain and usually require a team of developers with a wide range of expertise and skills. Also, these types of models are limited by design, assumptions and reweighting or microdata simulation algorithms, and data requirements. The key is to make these explicit and then interpret the results within the model's limitations and capacities.

10.4 Areas of Further Studies

Some potential future research areas are discussed in this section.

First of all, the existing simulated spatially disaggregated microdata could be used to address other research questions, such as to see the SLA-level housing stress estimates for individuals or households by their age structure, labor force status, family composition, etc. in an individual city or state or across Australia. The data could be also useful to perform further spatial analysis in relation to housing stress or by upgrading the model with the latest available datasets.

The possible second area of research could do the policy analysis. By relating this spatial microsimulation model to the NATSEM static microsimulation model STINMOD, the future research can explore various scenarios of small area effects of policy changes, such as changes in government programs providing cash assistance to households with children and/or housing assistance.

The next possible significant research can perhaps use the new validation techniques to a further extent in terms of theoretical backgrounds and the application to other spatial microsimulation models. More importantly, to create and attach the CI measures for the small area estimation of other variables of interest.

The fourth option of future research could be checking the performance of the internal instrument of the model at various levels. For example, further research can check the effects of the model outputs toward better results by changing the benchmarks numbers, selection of the variables in benchmarks, and order of benchmarks.

Last but not least, the potential future research area is to make further advancement in the spatial microdata simulation methodologies. This could be done by implementing new types of distance functions in the GREGWT method instead of using the chi-squared distance function and changing the computing codes from SAS to other programs such as "R" or FORTRAN. On top, certainly, the future research will seek to compose the full computing codes for the new Bayesian prediction–based methodology toward spatial microdata simulation and then apply it to the real-world practices for producing empirical results.

10.5 Computing Codes and Programming

This section provides the SAS codes and programs for MMT-based small area estimation. There are two computational stages in MMT. The first stage of the model computes small area synthetic weights using reweighting techniques, and the second stage computes microdataset and the ultimate small area estimates.

10.5.1 The General Model File Codes

As discussed earlier, the creation of a detailed micropopulation dataset at fine small area levels is one of the major key objectives of MMT. To achieve this, small area synthetic weights need to be generated first by using different reweighting methods. The overall process of creation of small area synthetic weights is discussed in Chapter 6 (e.g., see Section 6.3.2 and Figure 6.1). The process starts by running the *general model file* in SAS language that contains the path to all input data files, the reweighting algorithms, and the link of a folder where all output files would be stored. Computing codes for the general model file are given here.

```
--------------------------------------------------------------------
/* SAS program to run reweighting algorithms for producing
   small area synthetic weights; By AZ, Aug 2013, © Authors */:
--------------------------------------------------------------------

options mautosource sasautos=("C:\Folder …\Algorithms","C:\
Folder …\SAS code") mprint;
libname sihlink "D:\Restricted survey datasets files paths)"
access=readonly;
libname Census "C:\Benchmarks";
libname Census2 "C:\Census Data";
data Census2.allhh06; *table 1 - all households;
            attrib Classdescription length = $ 50;
            set Census.allhh06 (drop=Classdescription);
            Classdescription = "OPDs";
run;
data lfagsx06; *table 2 - labor force by age by sex;
      set Census.lfagsx06;
        if ageP10_2 = 1 then ageP10_2=1; *0-14;
         else if ageP10_2 = 2 then ageP10_2=2; *15-24;
          else if ageP10_2 in (3,4) then ageP10_2=3; *25-54;
           else if ageP10_2 = 5 then ageP10_2=4; *55-64;
            else if ageP10_2 = 6 then ageP10_2=5; *65+;
             LFS_2_i = LFS_2;
          if LFS_2_i in (1,2) then LFS_2=1; *employed;
           else if LFS_2_i in (3,4,99) then LFS_2=2; *NILF;
run;
data lfagsx06_1;
   length Classdescription $ 50;
      set lfagsx06;
            attrib age_desc length = $ 5;
            if ageP10_2 = 1 then age_desc="0_14";
             else if ageP10_2 = 2 then age_desc="15_24";
              else if ageP10_2 = 3 then age_desc="25_54";
               else if ageP10_2 = 4 then age_desc="55_64";
                else if ageP10_2 = 5 then age_desc="65_ov";
            if sexp_2 =1 then sex_desc= "M";
             else if sexp_2 = 2 then sex_desc="F";
            if LFS_2 =1 then LFS_desc="Emp";
             else if LFS_2 =2 then LFS_desc="NILF";
               else if LFS_2 =0 then LFS_desc="NA";
         Classdescription=sex_desc||"_"||age_desc||"_"||LFS_desc;
run;
proc summary data=lfagsx06_1 nway sum;
      class sla_code sexp_2 ageP10_2 LFS_2;
      var total;
```

```
        id Classdescription;
        output out=Census2.lfagsx06new (drop=_TYPE_ _FREQ_) sum=;
run;
data llr06; *table 3: tenure type by weekly HH rent;
        set Census.llr06;
        length classdescription $ 50;
                rent_06_3_i = rent_06_3;
                if tenure_06_3=3 then do;
                if rent_06_3_i in (1,2,3) then rent_06_3=1;*0-139
                  public renters (PR);
                  else if rent_06_3_i in (4,5) then rent_06_3=2;
                    *140-224 and above PR;
                  else if rent_06_3_i in (6,7,8) then
                     rent_06_3=3; *225-449 and above PR;
                  else if rent_06_3_i in (9,10) then rent_06_3=4;
                    *450+ and above PR;
              if rent_06_3=1 then classdescription="Pub_0_139";
              else if rent_06_3=2 then classdescription="Pub_
                 140-224_ov";
                        else if rent_06_3=3 then classdescription=
                           "Pub225-449_140_ov";
                        else if rent_06_3=4 then classdescription=
                           "Pub450+_140_ov";
                end;
  if tenure_06_3=4 then tenure_06_3=3;
                if tenure_06_3=3 then do;
                    if rent_06_3_i in (1,2,3) then
                       rent_06_3=1; *0-139;
                      else if rent_06_3_i in (4,5) then
                         rent_06_3=2; *140-224;
                        else if rent_06_3_i in (6,7,8) then
                           rent_06_3=3; *225-349;
                          else if rent_06_3_i in (9,10) then
                             rent_06_3=4; *350-549+;
                            *else if rent_06_3_i in (10) then
                               rent_06_3=5; *550 and over;
  if rent_06_3=1 then classdescription="Rent_0_139";
                else if rent_06_3=2 then classdescription="Rent_
                   140-224";
                else if rent_06_3=3 then classdescription="Rent_
                   225-349";
                           else if rent_06_3=4 then
                              classdescription="Rent350-549";
                           else if rent_06_3=5 then
                              classdescription="Rent550+";
                  if rent_06_3=1 then classdescription="Priv_0_139";
                  else if rent_06_3=2 then classdescription=
                     "Priv_140_OV";
                  else if rent_06_3=3 then classdescription=
                     "Priv_225_449";
```

```
                        else if rent_06_3=4 then
                            classdescription="Priv_450_ov";
                end;
                if total=. then total=0;
run;
proc summary data=llr06 nway sum;
        class sla_code tenure_06_3 rent_06_3;
        var total;
        id classdescription;
        output out=census2.llr06 (drop=_TYPE_ _FREQ_) sum=;
run;
data census2.TENHHTYPE06; *table 4: tenure type by HH type;
        set census.TENHHTYPE06;
run;
data TENHHI06; *table 5: tenure type by weekly HH income;
        set census.TENHHI06;
        format hhincome_desc $16.;
        format tenure_desc $5.;
        hhincome_10_i = hhincome_5;
        if hhincome_10_i in (1,2,3) then hhincome_5=1; *0-249
          includes -ve income;
      else if hhincome_10_i in (4,5) then hhincome_5=2; *250-649;
          else if hhincome_10_i in (6,7) then hhincome_5=3;
             *650-999;
        else if hhincome_10_i in (8,9) then hhincome_5=4;
           *1000-1699;
          else if hhincome_10_i in (10,11) then hhincome_5=5;
           *1700 and over;
        if hhincome_5=1 then hhincome_desc="income below $250";
          else if hhincome_5=2 then hhincome_desc="income
            $250-$649";
            else if hhincome_5=3  then hhincome_desc="income
              $650-$999";
              else if hhincome_5=4 then hhincome_desc="income
                $1000-$1699";
                 else if hhincome_5=5 then hhincome_desc="income
                    $1700 over";
        if tenure_06_5 =1 then tenure_desc="O";
          else if tenure_06_5 =2 then tenure_desc="P";
            else if tenure_06_5 =3 then tenure_desc="RPub";
              else if tenure_06_5 =4 then tenure_desc="RPriv";
                else if tenure_06_5 =5 then tenure_desc="OTH";
    classdescription=tenure_desc||"_"||hhincome_desc;
run;
proc summary data=TENHHI06 nway sum;
        class sla_code tenure_06_5 hhincome_5;
        var total;
        id classdescription;
        output out=census2.TENHHI06 (drop=_TYPE_ _FREQ_) sum=;
run;
```

```
data census2.npd06 (drop=_type_ _freq_); *table 6: persons in
  non-private dwelling;
      set census.npd06;
      if total=. then total=0;
run;
data ttih06; *table 7: mortgage by weekly HH income;
      length classdescription $ 50;
      set census.ttih06 ;
      HHINCOME_7_i = HHINCOME_7;
       if HHINCOME_7_i in (1,2,3,4) then HHINCOME_7 = 1;
         *<$500 includes -ve/nil;
        else if HHINCOME_7_i in (5,6,7) then HHINCOME_7 = 2;
          *$500-999;
          else if HHINCOME_7_i >=8 then HHINCOME_7 = 3; *$1000+;
       if hhincome_7=1 then hhincome_desc="income below $500";
        else if hhincome_7=2 then hhincome_desc="income
          $500-999";
         else if hhincome_7=3  then hhincome_desc="income
           $1000+";
        MORT_06_7_i = MORT_06_7;
        if MORT_06_7_i in (1,2,3,4) then MORT_06_7 = 1;
          *Mortgage $1-949;
           else if MORT_06_7_i in (5,6) then MORT_06_7 = 3;
              *Mortgage $950-1999;
              else if MORT_06_7_i in (7,8) then MORT_06_7 = 4;
                *Mortgage $2000+;
              if MORT_06_7 = 1 then mort_desc = "Mortgage
                $1-949";
                else if  MORT_06_7 = 3 then mort_desc =
                  "Mortgage $950-1999";
                  else if MORT_06_7 = 4 then mort_desc =
                    "Mortgage $2000+";
                classdescription= COMPBL (mort_desc ||",
                  "||hhincome_desc);
                if total=. then total=0;
run;
proc summary data=ttih06 nway sum;
      class sla_code MORT_06_7 HHINCOME_7;
      var total;
      id classdescription;
      output out=census2.ttih06 (drop=_TYPE_ _FREQ_) sum=;
run;
data census2.Dshhtft06; *table 8: dwelling structure by HH
  composition;
    set census.Dshhtft06;
run;
data npers06; set census.npers06; *table 9: No.of persons in HH;
  NPERSONS_9_i = NPERSONS_9;
  if NPERSONS_9_i = 6 then NPERSONS_9 = 5;
```

```
     if NPERSONS_9_i = 6 then Classdescription= "Five+ persons";
run;
proc summary data=npers06 nway sum;
       class sla_code NPERSONS_9 ;
       var total;
       id classdescription;
       output out=census2.npers06 (drop=_type_ _freq_) sum=;
run;
data RENTHHINC06; *table 10: weekly rent by HH income;
       set census.RENTHHINC06;
        length hhincome_desc $ 20;
        length rent_desc  $ 15 ;
    hhincome_10_i = hhincome_10;
    if hhincome_10_i in (1) then hhincome_10=1; *nil or -ve
       income;
       if hhincome_10_i in (2,3,4,5) then hhincome_10=2;
          *income $1-649;
     else if hhincome_10_i in (6,7,8) then hhincome_10=3;
        *income $650-1399;
          else if hhincome_10_i >=9 then hhincome_10=4; *income
             $1400+;
       if hhincome_10=1 then hhincome_desc="nil or -ve income";
        else if hhincome_10=2 then hhincome_desc="income
           $1-649";
         else if hhincome_10=3 then hhincome_desc="income
            $650-1399";
             else if hhincome_10=4 then hhincome_desc="income
                $1400+";
       rent_06_10_i = rent_06_10;
       if rent_06_10_i in (1,2,3) then rent_06_10=1; *0 - 139;
        else if rent_06_10_i in (4,5) then rent_06_10=2;
           *140-224;
          else if rent_06_10_i in (6,7) then rent_06_10=3;
             *225-349;
            else if rent_06_10_i in (8,9) then rent_06_10=4;
               *350-549;
         else if rent_06_10_i = 10 then rent_06_10=5; *550 and
            over;
       if rent_06_10=1 then rent_desc="rent $0-139";
        else if rent_06_10=2 then rent_desc="rent $140-224";
          else if rent_06_10=3 then rent_desc="rent $225-349";
             else if rent_06_10=4 then rent_desc="rent $350-549";
               else if rent_06_10=5 then rent_desc="rent $550ov";
        classdescription=COMPBL (rent_desc||", "||hhincome_desc);
        if total=. then total=0;
run;
proc summary data=renthhinc06 nway sum;
       class sla_code rent_06_10 hhincome_10;
       var total;
```

```
        id classdescription;
        output out=census2.renthhinc06 (drop=_type_ _freq_)
           sum=;
run;
Data Input; *prepare input micro data;
 Set sihlink.sihlinkage;
run;
data input;
       set input;
HHTYPE_1=HHTYPE; *table 1;
 if HHTYPE= 0 then HHTYPE_1=0;
   else HHTYPE_1=1;
 if HHTYPE_1=1 then do; *table 2;
AGEP10_2=AGEP10;
           if ageP10_2 = 1 then ageP10_2=1  ; *10-14;
         else if ageP10_2 = 2 then ageP10_2=2  ; *15-24;
          else if ageP10_2 in (3,4) then ageP10_2=3  ; *25-54;
           else if ageP10_2 = 5 then ageP10_2=4  ; *55-64;
            else if ageP10_2 = 6 then ageP10_2=5  ; *65+;
SEXP_2=SEXP; LFS_2_i = LFS;
             if lfs_2_i=0 then lfs_2=0;
           else if LFS_2_i in (1,2) then LFS_2=1; *employed
             full time;
           else if LFS_2_i in (3,4,99) then LFS_2=2; *NILF;
           else if lfs_2=0 then LFS_2= 0;
end;
else if HHTYPE_1=0 then do;
             AGEP10_2=0;
             SEXP_2=0;
             LFS_2_i = 0;
end;
TENURE_06_3=TENURE_06; *table 3;
rent_06_3=rent_06;
if tenure_06_3=4 then tenure_06_3=3;
             if tenure_06_3=3 then do;
                if rent_06 in (1,2,3) then rent_06_3=1;
                  *0-139;
                  else if rent_06 in (4,5,6,7,8,9,10) then
                  rent_06_3=2; *140-OV224;
end;
TENURE_06_4=TENURE_06; *table 4;
HHTYPE_4=HHTYPE;
TENURE_06_5=TENURE_06; *table 5;
      if hhincome in (1,2,3) then hhincome_5=1; *0-249
         includes -ve income;
     else if hhincome in (4,5) then hhincome_5=2; *250-649;
         else if hhincome in (6,7) then hhincome_5=3; *650-999;
        else if hhincome in (8,9) then hhincome_5=4; *1000-1699;
        else if hhincome in (10,11) then hhincome_5=5; *1700
           and over;
```

```
             else if hhincome in (0) then hhincome_5=0;
                *0 instead of missing;
NPDTYPE_6=NPDTYPE; *table 6;
HHINCOME_7_i = HHINCOME; *table 7;
          if HHINCOME_7_i in (1,2,3,4) then HHINCOME_7 = 1;
            *<$500 includes -ve/nil;
            else if HHINCOME_7_i in (5,6,7) then HHINCOME_7 = 2;
              *$500-999;
              else if HHINCOME_7_i >=8 then HHINCOME_7 = 3; *$1000+;
          MORT_06_7_i = MORT_06;
          if mort_06_7_i = 0 then mort_06_7=0;
            else if MORT_06_7_i in (1,2,3,4) then MORT_06_7 = 1;
              *Mortgage $1-949;
              else if MORT_06_7_i in (5,6) then MORT_06_7 = 3;
                *Mortgage $950-1999;
                else if MORT_06_7_i in (7,8) then MORT_06_7 = 4;
                  *Mortgage $2000+;
dwstr_06_8 = dwstr_06; *table 8;
dcomp_06_8 = dcomp_06;
NPERSONS_9=NPERSONS; *table 9;
 if NPERSONS >=5 then NPERSONS_9 = 5;
hhincome_10_i = hhincome; *table 10;
if hhincome_10_i = 0 then hhincome_10=0;
else if hhincome_10_i in (1) then hhincome_10=1; *nil or -ve
  income;
   if hhincome_10_i in (2,3,4,5) then hhincome_10=2; *income
     $1-649;
     else if hhincome_10_i in (6,7,8) then hhincome_10=3;
       *income $650-1399;
         else if hhincome_10_i >=9 then hhincome_10=4; *income
           1400+;
       rent_06_10_i = rent_06;
       if rent_06_10_i=0 then rent_06_10=0; else /* NA Rent set
         to 0 */
       if rent_06_10_i in (1,2,3) then rent_06_10=1; *0 - 139;
        else if rent_06_10_i in (4,5) then rent_06_10=2;
          *140-224;
          else if rent_06_10_i in (6,7) then rent_06_10=3;
            *225- 349;
            else if rent_06_10_i in (8,9) then rent_06_10=4;
              *350 - 549;
          else if rent_06_10_i = 10 then rent_06_10=5; *550 and
            over;
run;

*====The standard sequence of tables used in the reweighting
  algorithms=====;

%macro Greg (state=);
%gregwt_algorithms_V2(
```

```
        unitdsn=Input /*Name of dataset that is coming in - see
          above dataset*/
            ,area_var=sla_code
            ,outfldr=D:\Output\Folder path
            ,BFLDR= C:\Census Data /*benchmark in file location*/
            ,ds_pop=C:\Benchmarks\ /*allhh06 /allSLAs*/
        ,GROUP=HH_ID
        ,UNIT= fam_id pers_id
          ,inweight=weight
        ,ID=_ALL_
,b2DSN=ALLHH06 /*Table 1. All household types*/
        ,b2CLASS=HHTYPE_1
        ,b2VAR=
        ,b2TOT=TOTAL
        ,b2GRP=HH_ID
        ,b2REPS=
,b1DSN=LFAGSX06new /*Table 2 - Age by sex by labor force
  status*/
        ,b1CLASS=SEXP_2 AGEP10_2 LFS_2
        ,b1VAR=
        ,b1TOT=TOTAL
        ,b1GRP=PERS_ID
        ,b1REPS=
,b3DSN=LLR06 /* Table 3 - Tenure by weekly household rent*/
        ,b3CLASS=TENURE_06_3 RENT_06_3
        ,b3VAR=
        ,b3TOT=TOTAL
        ,b3GRP=HH_ID
          ,b3REPS=
,b4DSN=TENHHTYPE06 /*Table 4 - Tenure by household type */
        ,b4CLASS=TENURE_06_4 HHTYPE_4
        ,b4VAR=
        ,b4TOT=TOTAL
        ,b4GRP=HH_ID
          ,b4REPS=
,b9DSN=TENHHI06 /*Table 5 - Tenure type by weekly household
  income;*/
        ,b9CLASS=TENURE_06_5 HHINCOME_5
        ,b9VAR=
        ,b9TOT=TOTAL
        ,b9GRP=HH_ID
        ,b9REPS=
,b8DSN=NPD06 /*Table 6 - Persons in non-private dwelling;*/
        ,b8CLASS=NPDTYPE_6
        ,b8VAR=
        ,b8TOT=TOTAL
        ,b8GRP=PERS_ID
        ,b8REPS=
```

```
,b7DSN=TTIH06 /*Table 7 - Monthly household mortgage by weekly
   household income;*/
         ,b7CLASS=MORT_06_7 HHINCOME_7
         ,b7VAR=
         ,b7TOT=TOTAL
         ,b7GRP=HH_ID
         ,b7REPS=
,b5DSN=DSHHTFT06 /*Table 8 - Dwelling structure by household
   family composition;*/
         ,b5CLASS=DWSTR_06_8 DCOMP_06_8
         ,b5VAR=
         ,b5TOT=TOTAL
         ,b5GRP=HH_ID
         ,b5REPS=
,b6DSN=NPERS06 /*Table 9 - Number of persons usually resident
   in household;*/
         ,b6CLASS=NPERSONS_9
         ,b6VAR=
         ,b6TOT=TOTAL
         ,b6GRP=HH_ID
         ,b6REPS=
,b10DSN=RENTHHINC06 /*Table 10 - Weekly household rent by
   weekly household income;*/
         ,b10CLASS=RENT_06_10 HHINCOME_10
         ,b10VAR=
         ,b10TOT=TOTAL
         ,b10GRP=HH_ID
         ,b10REPS=
,MAXITER=30 /*Number of iterations*/
         ,LOWER=0 /*minimum of zero means it won't return any
            negative values*/
         ,EPSILON=0.0001, log=0);
run;

%mend Greg;
%Greg (State=AUST); /*State AUST can be change NSW, SD, SSD,
   or particular SLAs)*/

options notes;
proc printto;
RUN;
```

10.5.2 SAS Programming for Reweighting Algorithms

Reweighting algorithms are essential tools to generate synthetic weights at the first-stage computations in spatial microsimulation models. A range of reweighting techniques are discussed in this book. Each of these methods is based on very sophisticated and lengthy computing encodings. SAS programs for the reweighting algorithms used in this book are as follows:

```
---------------------------------------------------------------------
GREGWT reweighting: Parameters for algorithms
---------------------------------------------------------------------
UNITDSN = Name of dataset to be weighted
OUTDSN  = Name for output weighted unit dataset
             Default _outdsn_
BENOUT  = Prefix for report datasets on benchmark
             convergence. First six characters only are used,
             followed by an integer suffix.
             There will be one such dataset for each benchmark
             dataset with suffixes 1,...
             Default value is benout (i.e. dataset names
             benout1,...)
BYOUT   = Name for report dataset on BY group convergence
             Default _byout_
EXTOUT  = Name for report dataset of extreme units
             Default _extout_
BY      = as for BY statement - optional
STRATUM = Stratum - optional
VARGRP  = Variance group - optional
             To get weighted residual variances, STRATUM and
             VARGRP must be specified and data must be sorted
             by BY STRATUM VARGRP
GROUP   = Level for weighting of groups rather than units.
             Optional
             In integrated weighting, specifies the grouping
             level at which the calibration is applied.
             The input weight for a group is that of the first
             unit in the group, unless OPTIONS=amean or
             OPTIONS=hmean.
             All units in a group will have the same final
             weight.
UNIT    = Other variables in the sort order below GROUP
             level.
             Optional. Used in integrated weighting: the list
                 BY STRATUM VARGRP GROUP UNIT must contain any
                 grouping variable used in the BnGRP macro
                 variables. For the use of these macro
                 variables see description below.
             Unit data MUST be sorted by &BY &STRATUM &VARGRP
                 &GROUP &UNIT
             *Although any of these can be left blank

INWEIGHT= Input weight (compulsory)
or  INWT=
```

REPID   = Gives the replicate identifier, a variable giving
             numbers 1 to m identifying the replicate a unit
             is in Use instead of setting up replicate input
             weights
             Optional
INREPWTS= Lists 2 or more replicate input weights - if these
             are provided, REPID is ignored. Optional
WEIGHT   = Name for output weight, default is &INWEIGHT
or   WT=
REPWTS   = Name for output replicate weights, default
             &INREPWTS

ID   = Output dataset will include variables used by the
             macro plus any additional variables named in ID.
             To get all variables, specify ID=_ALL_
PENALTY = Variable used to specify a weighted distance
             function - the distance contributed by this unit
             or group is multiplied by the PENALTY value for
             the unit, or the PENALTY value of the first unit
             in the group
             A high PENALTY value makes the weight less subject
             to change than that of other units
             Default is to use 1

Specifications for up to 30 datasets of benchmarks (n is 1-30):

BnDSN   = Name of nth dataset containing benchmarks
BnCLASS = Variables defining category for these benchmarks
BnVAR   = Variables on UNITDSN to be totaled to the
             benchmarks (default is to use weighted counts of
             units)
BnTOT   = Variables on BnDSN giving the benchmarks
BnREPS   = List of replicate benchmark totals
             - blank if this feature not used
             - otherwise lists <number of replicate weights>
             variables for each benchmark total
             e.g. B1TOT=psntot hhldtot, B1REPS=pt1-pt30
             ht1-ht30
BnRVAR   = List of replicate variables to be totaled to add to
             the replicate benchmarks
             - blank if this feature not used
             - otherwise lists <number of replicate weights>
             variables for each benchmark total
BnGRP   = Used in integrated weighting. Names a variable
             listed in &GROUP &UNITID that gives the grouping
             level at which this benchmark applies.
             Only the first record in the group is used in
             totals

```
                    e.g. if BnGRP=hhold, then, for this benchmark,
                    values from the first unit in the hhold are
                    assumed to be hhold records and are used in
                    totals.
LOWER      = Smallest value which weights can take or (if
                    followed by a % sign e.g. LOWER=50%) the smallest
                    percentage that weights can be multiplied by (but
                    if LOWER > 95% then 95% is used).
             The value may be an SAS expression involving
             variables available to the macro (can use ID to
             include them)
             e.g. LOWER=max(1,0.7*weight)
UPPER      = Largest value which weights can take or (if
                    followed by a % sign e.g. UPPER=200%) the largest
                    percentage that weights can be multiplied by (but
                    if UPPER < 105% then 105% is used).


OPTIONS = List of options - possible values are
             NOPRINT: turn off printing of output reports
             BADPRINT: only print benchmark and BY group data
                       where a benchmark was not met (also
                       defaults EXTNO to 5)
             NOLOG : turn off log information about BY groups
             NOTES : turn on notes (default is to turn most
                     notes to log off)
             EXP or EXPONENTIAL : Use exponential distance
                                  function instead of the
                                  default linear distance
             FIRSTWT: report on first weight (i.e. for
                      iteration 1) in addition to final
                      weight
             REPS   : attach replicate estimates to any table
                      produced (use names est_1-est_n or
                      names given by OUTREPS)
             UNIV   : print distribution of weights and weight
                      changes from PROC UNIVARIATE
             DEBUG  : do not delete intermediate datasets
             HMEAN  : Input weight for a group is the harmonic
                      mean of unit input weights
             AMEAN  : Input weight for a group is the
                      arithmetic mean of unit input weights
             Default input weight for a group (if neither
               HMEAN or AMEAN are specified) is the input
               weight from the first unit in the group

MAXITER = Maximum number of iterations in restricted version
             Default is 10
EPSILON = Convergence criterion: how closely must benchmarks
             be met, expressed as the discrepancy of estimate
             from benchmark divided by benchmark
```

Specification for an output table of estimates and variances:

```
OUT     = Name for output dataset containing table
CLASS   = Class variables defining categories for which
            estimates will be produced
          For ratio estimates, CLASS defines the categories
            to be used as denominators
          Estimates for totals across a class variable can
            be requested by giving the class variables
            prefixed by a #
          e.g. CLASS = state #sex gives estimates for states
            as well as for state by sex.
SUBCLASS= For ratio estimates, SUBCLASS defines the
            categories used as numerators
VAR     = List of continuous variables to be estimated for.
          For ratio estimates, these variables are used in
            the numerator only.
DENOM   = For mean or ratio estimates:
          - contains a list of variables used as
              denominators for the corresponding variables
              in VAR.
          - If DENOM lists fewer variables than VAR, the
              last variable in DENOM is used for the extra
              VAR members.
          - If more variables in DENOM than VAR the macro
              stops.
          - The keyword _one_ signifies using 1 as the
              denominator, giving estimates of mean
OUTGRP  = Used in integrated weighting. Names a variable
            listed in &GROUP &UNITID that gives the grouping
            level at which the table is being produced.
          Only the first record in the group is used in
            totals
          e.g. if OUTGRP=hhold then values from the first
            unit in the
          hhold are hhold records and are to be used in
            totals.

OUTREPS = Replicate estimates will be attached to the table
            if OUTREPS is given (in which case OUTREPS gives
            names for the variables).
NPREDICT= Number of predicted values to be attached to unit
            data
          on &OUTDSN. Predictions will be named hat_1-
            hat_&NPREDICT and (for level estimates) will
            correspond to the first &NPREDICT elements of
            the output tables.
          hat_n is the prediction under the regression model
            of the contribution of the unit to cell n of the
            output table.
```

If tables are not specified, NPREDICT has no
effect.
(Note that for a table of ratio estimates or means
the predictions of numerator and denominator
level estimates are produced)

WROUT   = Name for file of weighted residuals for tabulated
estimates
Default is to not produce this file

WRLIST   = List of variable names to contain the weighted
residuals.
Residuals are output in the same order as the
output tables.
WRLIST is only used if WROUT is specified.
Default is wr_1 i.e. only output the first residual

WRWEIGHT= True final weight for use in tabulations and
weighted residual calculations. This is only
required when the weighted residual calculations
are being re-done using a different benchmark
specification than that used for the original
weighting. This could be done to minimize
calculations (e.g. reduce to a post-stratified
ratio case), or because the original weighting
scheme is unknown.

TITLELOC= On output prints, the line at which GREGWT titles
should appear (using a title<n> statement)
leaving pre-existing titles on previous lines
intact
Default is the first line (i.e. a title statement)
NONE avoids printing any GREGWT titles

REPORTID= Unit identifiers to be used on extreme values
report (default uses as many as possible of the
variables listed in &BY &B1CLASS... &STRATUM
&VARGRP &GROUP &UNIT &B1VAR...)

MAXSPACE= Space available in kilobytes (roughly) for table
calculations (RAM, not hard drive space). Usually
leave at 500 - there is no advantage in
specifying more unless it is needed.
Program requires for each table category (in
bytes) total length of CLASS and SUBCLASS
variables + 8*(number of replicates + 1)
*(number of VAR and DENOM variables) + some
extra

REPWTMAX= Maximum number of replicate weights, default
5000 - used only with REPID, values over
REPWTMAX are considered invalid values of the
REPID variable

EXTNO   = Number of extreme values of each type to be printed

```
LINESIZE= Number of characters per line on output file.
          ONLY needed if the version of SAS does not support
          SYSFUNC or to fool the extreme values report into
          changing the number of id variables it prints (it
          allows 8 characters/variable).
RUNID   = Not used in this version
STEP    = Not used in this version
--------------------------------------------------------------------
© Authors
*---------------------------------------------------------------*
```

The reweighting algorithms have ten key consecutive parts in terms of their operation. Those parts can be named as (1) read initial variables and MACROS used by the main algorithms; (2) set up default values for macro variables; (3) initial creation of unit and BY datasets; (4) develop the list of variable names for tables; (5) start main calculations portion; (6) run calculations via each benchmark and then each *x* value for the small area; (7) aggregates calculations for benchmarks across table categories; (8) copy results into variables for outputs; (9) diagnostic reports on small area weights; and (10) delete all temporary files.

For convenience of the format and length of this *conclusions chapter*, the programming codes for part 1 are only included here. SAS programs for parts 2–10 are provided in Appendix F.

```
--------------------------------------------------------------------
%* PART 1: Initial variables reading MACROS used by the main
   algorithms *;

%MACRO GREGWT
(UNITDSN=,OUTDSN=,BENOUT=,BYOUT=,EXTOUT=,RUNID=,STEP=,
BY=,STRATUM=,VARGRP=, GROUP=,UNIT=,INWEIGHT=,INWT=,REPID=,
REPWTS=,INREPWTS=,WEIGHT=,WT=,ID=,PENALTY=,
B1DSN=,B1GRP=,B1CLASS=,B1VAR=,B1TOT=,B1REPS=,B1RVAR=,
B2DSN=,B2GRP=,B2CLASS=,B2VAR=,B2TOT=,B2REPS=,B2RVAR=,
B3DSN=,B3GRP=,B3CLASS=,B3VAR=,B3TOT=,B3REPS=,B3RVAR=,
B4DSN=,B4GRP=,B4CLASS=,B4VAR=,B4TOT=,B4REPS=,B4RVAR=,
B5DSN=,B5GRP=,B5CLASS=,B5VAR=,B5TOT=,B5REPS=,B5RVAR=,
B6DSN=,B6GRP=,B6CLASS=,B6VAR=,B6TOT=,B6REPS=,B6RVAR=,
B7DSN=,B7GRP=,B7CLASS=,B7VAR=,B7TOT=,B7REPS=,B7RVAR=,
B8DSN=,B8GRP=,B8CLASS=,B8VAR=,B8TOT=,B8REPS=,B8RVAR=,
B9DSN=,B9GRP=,B9CLASS=,B9VAR=,B9TOT=,B9REPS=,B9RVAR=,
B10DSN=,B10GRP=,B10CLASS=,B10VAR=,B10TOT=,B10REPS=,B10RVAR=,
B11DSN=,B11GRP=,B11CLASS=,B11VAR=,B11TOT=,B11REPS=,B11RVAR=,
B12DSN=,B12GRP=,B12CLASS=,B12VAR=,B12TOT=,B12REPS=,B12RVAR=,
B13DSN=,B13GRP=,B13CLASS=,B13VAR=,B13TOT=,B13REPS=,B13RVAR=,
B14DSN=,B14GRP=,B14CLASS=,B14VAR=,B14TOT=,B14REPS=,B14RVAR=,
B15DSN=,B15GRP=,B15CLASS=,B15VAR=,B15TOT=,B15REPS=,B15RVAR=,
B16DSN=,B16GRP=,B16CLASS=,B16VAR=,B16TOT=,B16REPS=,B16RVAR=,
```

```
B17DSN=,B17GRP=,B17CLASS=,B17VAR=,B17TOT=,B17REPS=,B17RVAR=,
B18DSN=,B18GRP=,B18CLASS=,B18VAR=,B18TOT=,B18REPS=,B18RVAR=,
B19DSN=,B19GRP=,B19CLASS=,B19VAR=,B19TOT=,B19REPS=,B19RVAR=,
B20DSN=,B20GRP=,B20CLASS=,B20VAR=,B20TOT=,B20REPS=,B20RVAR=,
B21DSN=,B21GRP=,B21CLASS=,B21VAR=,B21TOT=,B21REPS=,B21RVAR=,
B22DSN=,B22GRP=,B22CLASS=,B22VAR=,B22TOT=,B22REPS=,B22RVAR=,
B23DSN=,B23GRP=,B23CLASS=,B23VAR=,B23TOT=,B23REPS=,B23RVAR=,
B24DSN=,B24GRP=,B24CLASS=,B24VAR=,B24TOT=,B24REPS=,B24RVAR=,
B25DSN=,B25GRP=,B25CLASS=,B25VAR=,B25TOT=,B25REPS=,B25RVAR=,
B26DSN=,B26GRP=,B26CLASS=,B26VAR=,B26TOT=,B26REPS=,B26RVAR=,
B27DSN=,B27GRP=,B27CLASS=,B27VAR=,B27TOT=,B27REPS=,B27RVAR=,
B28DSN=,B28GRP=,B28CLASS=,B28VAR=,B28TOT=,B28REPS=,B28RVAR=,
B29DSN=,B29GRP=,B29CLASS=,B29VAR=,B29TOT=,B29REPS=,B29RVAR=,
B30DSN=,B30GRP=,B30CLASS=,B30VAR=,B30TOT=,B30REPS=,B30RVAR=,
CLASS=,SUBCLASS=,VAR=,DENOM=,OUT=,OUTREPS=,OUTGRP=,MAXSPACE=500,
OPTIONS =,TITLELOC=,REPORTID=,REPORTVS=,
MAXITER=,EPSILON =,LOWER=,UPPER=,REPWTMAX=200, EXTNO=,
NPREDICT=,WROUT=,WRLIST=,WRWEIGHT=,LINESIZE=);
%LOCAL XOPTIONS NOTESOFF MACID NUMALPHA DISTANCE CWDIST2
    LASTGRP
 LOCODE UPCODE ELAPTIME TEMP WORD I B CWFIRST
 WTDRES CK_TRASH GO_END EXTRA PRINTREP CWTIT CWTIT2 CWTIT3
 BADPRINT EXP NEWWT ORIGNOTE;
run; %*any preceding data step gives notes*;
%IF %LENGTH(&LINESIZE)=0 %THEN %DO ;
 %LET ORIGNOTE=%SYSFUNC(getoption(notes)) ;
 %LET LINESIZE=%SYSFUNC(getoption(linesize)) ;
%END ;
%ELSE %LET ORIGNOTE=NOTES ;
%LET XOPTIONS = %UPCASE(XXXXXXXX &OPTIONS X) ;
%IF %INDEX(&XOPTIONS,%STR( NOTES )) = 0
%THEN %LET NOTESOFF = %STR(options nonotes ;) ;
&NOTESOFF

data _null_ ;
  call symput("ELAPTIME",put(datetime(),32.)) ;
run;
%LET MACID=GREGWT ;
%LET NUMALPHA=123456789ABCDEFGHIJKLMNOPQRSTUVWXYZ ;
%IF %UPCASE(&TITLELOC)=_NONE_ %THEN %DO ;
 %LET CWTIT=* ; %LET CWTIT2=* ; %LET CWTIT3=* ;
%END ;
%ELSE %DO ;
 %LET CWTIT=title&TITLELOC ;
 %IF %BQUOTE(&TITLELOC)= %THEN %LET TITLELOC = 1 ;
 %ELSE %LET TITLELOC=%SUBSTR(&TITLELOC,1,1) ;
 %LET CWTIT2 = title%EVAL(&TITLELOC+1) ;
 %LET CWTIT3 = title%EVAL(&TITLELOC+2) ;
%END ;
```

```
%********** MACROS used by the main algorithms *******************;

%MACRO GREGPEXT(DATA=,EXTNO=,ID=,OPTIONS=,FIRST=
,TITLELOC=,WTFORMAT=) ;
 %LOCAL XOPTIONS CWFIRST ;
 run ; %* any preceding data step gives notes *;
 %LET XOPTIONS = %UPCASE(XXXXXXXX &OPTIONS X) ;
 %IF %INDEX(&XOPTIONS,%STR( NOTES )) = 0
 %THEN %LET NOTESOFF = %STR(options nonotes ;) ;
 &NOTESOFF
 %IF (&FIRST=1) OR (&FIRST=Y) %THEN %LET FIRST = 1 ;
 %ELSE %LET FIRST = 0 ;
 %IF %BQUOTE(&DATA)= %THEN %LET DATA=_extout_ ;
 %IF &EXTNO = %THEN %LET EXTNO = 5 ;
 %IF %BQUOTE(&WTFORMAT)= %THEN %LET WTFORMAT=6.1 ;
 %IF %UPCASE(&TITLELOC)=_NONE_ %THEN %DO ;
  %LET CWTIT=* ; %LET CWTIT2=* ; %LET CWTIT3=* ;
 %END ;
 %ELSE %DO ;
  %LET CWTIT=title&TITLELOC ;
  %IF %BQUOTE(&TITLELOC)= %THEN %LET TITLELOC = 1 ;
  %ELSE %LET TITLELOC=%SUBSTR(&TITLELOC,1,1) ;
  %LET CWTIT2 = title%EVAL(&TITLELOC+1) ;
  %LET CWTIT3 = title%EVAL(&TITLELOC+2) ;
 %END ;
 %IF &EXTNO > 0 %THEN %DO ;
  proc format ;
   value sevfmt
     -8  ="Negative weight"
     -7  ="Low weight, large change, adjusted:"
     -6  ="Low weight, large change:"
     -5  ="Low weight, adjusted:"
     -4  ="Low weight:"
    -3,3 ="Large change, adjusted:"
    -2,2 ="Adjusted:"
     -1  ="Large downward change:"
      1  ="Large upward change:"
      4  ="High weight:"
      5  ="High weight, adjusted:"
      6  ="High weight, large change:"
      7  ="High weight, large change, adjusted:" ;
  run ;

  data cwext(drop=_count_) ck_dsn(keep=severity _count_);
   set &DATA ;
   by severity ;
   if first.severity then _count_ = 0 ;
   _count_ + 1 ;
   if _count_ <= &EXTNO then output cwext ;
```

```
    if last.severity then output ck_dsn ;
  run ;
  data cwext ;
   merge cwext ck_dsn ;
   by severity ;
  run ;
  proc print data=cwext split='@' label ;
   &CWTIT "Extreme units (top &EXTNO in each type)" ;
   label _inwt_="input@weight"
   %IF &FIRST %THEN _regwt_="first@weight" ;
       _finwt_="final@weight"
       _wtrat_="final@/input"
       severity="type"
       _count_="number";
   format severity sevfmt. _finwt_ _inwt_
         %IF &FIRST %THEN _regwt_ ;
         &WTFORMAT _wtrat_ 5.3 ;
   by severity _count_ ;
   %IF &EXTNO>15 %THEN pageby _count_ ;;
   %IF %BQUOTE(&ID)^= %THEN id &ID ;;
   var _inwt_ _finwt_
   %IF &FIRST %THEN _regwt_ ;
       _wtrat_ ;
  run ;
  proc catalog c=formats et=format;
   delete sevfmt ;
  run ;
  quit ;
 %END ;
 options notes ;
%MEND GREGPEXT ;

%MACRO GREGPBEN(DATA=,BY=,ID=,OPTIONS=,FIRST=,TITLELOC=,TYPE=1);
 %LOCAL XOPTIONS CWRULER ;
 run ;
 %LET XOPTIONS = %UPCASE(XXXXXXXX &OPTIONS X) ;
 %IF %INDEX(&XOPTIONS,%STR( NOTES )) = 0
 %THEN %LET NOTESOFF = %STR(options nonotes ;) ;
 &NOTESOFF

 %IF (&FIRST=1) OR (&FIRST=Y) %THEN %LET FIRST = 1 ;
 %ELSE %LET FIRST = 0 ;
 %IF %BQUOTE(&DATA)= %THEN %LET DATA=benout1 ;
 %IF %UPCASE(&TITLELOC)=_NONE_ %THEN %DO ;
  %LET CWTIT=* ; %LET CWTIT2=* ; %LET CWTIT3=* ;
 %END ;
 %ELSE %DO ;
  %LET CWTIT=title&TITLELOC ;
  %IF %BQUOTE(&TITLELOC)= %THEN %LET TITLELOC = 1 ;
  %ELSE %LET TITLELOC=%SUBSTR(&TITLELOC,1,1) ;
```

```
   %LET CWTIT2 = title%EVAL(&TITLELOC+1) ;
   %LET CWTIT3 = title%EVAL(&TITLELOC+2) ;
 %END ;
 %LET BADPRINT = (%INDEX(&XOPTIONS,%STR( BADPRINT ))>0) ;

 data cwrep&B(drop=cwi) ;
   set &DATA ; %* Produce formatted versions of totals
     variables *;
   array cw_est{*} _bench_ _best_ _iest_ %IF &FIRST %THEN
     _rest_ ;;
   array cwxest{*} $ 7 cwbench cwbest cwiest %IF &FIRST %THEN
     cwrest ;;
   do cwi = 1 to dim(cw_est) ;
    if cwi > 1 &
     abs(cw_est{cwi} - _bench_) < 0.5 then cwxest{cwi} = ' .. ';
     else if (-9999999.5 <= cw_est{cwi} < -999.5)
          or (999.5 < cw_est{cwi} <= 9999999.5)
          then cwxest{cwi}=put(cw_est{cwi},7.) ;
     else if (-999.5 <= cw_est{cwi} < -0.995)
          or (0.995 < cw_est{cwi} <= 999.5)
          then cwxest{cwi}=put(cw_est{cwi},7.3) ;
     else cwxest{cwi}=put(cw_est{cwi},best7.) ;
   end ;
   if _bench_ <= 0.5*_iest_ then cwvs ='<<' ;
   else if _bench_ <= 0.9*_iest_ then cwvs ='< ' ;
   else if _iest_ <= 0.5*_bench_ then cwvs ='>>' ;
   else if _iest_ <= 0.9*_bench_ then cwvs ='> ' ;
   else cwvs =' ' ;
  run ;
%IF %BQUOTE(&TYPE) ^= 2 %THEN %DO ;
  proc print data=cwrep&B
%IF &BADPRINT %THEN (where=(cwbest^=' .. ')) ;
    label split='@';
    &CWTIT
"Estimate to benchmark &B comparison" ;
    &CWTIT2
".. indicates estimate = benchmark" ;
    %IF %BQUOTE(&BY)^= %THEN %STR(by &BY ;) ;
    %IF %BQUOTE(&ID)^= %THEN id &ID ;;
    label _count_="count"
          cwbench="bench@-mark"
          cwiest="input@weight"
  %IF &FIRST %THEN cwrest="first@weight" ;
          cwbest="final@weight"
          cwvs="vs" ;
    format _crit_ best6. ;
    var _count_ cwbench cwvs cwiest
        %IF &FIRST %THEN cwrest ;
        cwbest ;
  run ;
```

```
%END ;
%ELSE %DO ;
 %LET CWRULER=Rep. no.@_12345678901234567890123456789 0 ;
 %LET CWRULER=%SUBSTR(&CWRULER,1,10+&NWREPORT) ;
  proc print data=cwrep&B
   %IF &BADPRINT %THEN (where=(cwbest^='    ..  ')) ;
   label split="@";
      &CWTIT
"Convergence of replicates, benchmark &B" ;
      &CWTIT2
"   .=ok, I=impossible to meet, N=not converged   " ;
      %IF %BQUOTE(&BY)^= %THEN %STR(by &BY ;) ;
      %IF %BQUOTE(&ID)^= %THEN id &ID ;;
       label _report_="&CWRULER" ;
       var _report_ ;
  run ;
%END ;

%MEND GREGPBEN ;

%MACRO GREGPBY(DATA=,BYVARS=,OPTIONS=,GROUP=,TITLELOC=) ;
%LOCAL XOPTIONS ;
 run ;
 %LET XOPTIONS = %UPCASE(XXXXXXX &OPTIONS X) ;
 %IF %INDEX(&XOPTIONS,%STR( NOTES )) = 0
 %THEN %LET NOTESOFF = %STR(options nonotes ;) ;
 &NOTESOFF

 %IF %BQUOTE(&DATA)= %THEN %LET DATA=_byout_ ;
 %IF %UPCASE(&TITLELOC)=_NONE_ %THEN %DO ;
  %LET CWTIT=* ; %LET CWTIT2=* ; %LET CWTIT3=* ;
 %END ;
 %ELSE %DO ;
  %LET CWTIT=title&TITLELOC ;
  %IF %BQUOTE(&TITLELOC)= %THEN %LET TITLELOC = 1 ;
  %ELSE %LET TITLELOC=%SUBSTR(&TITLELOC,1,1) ;
  %LET CWTIT2 = title%EVAL(&TITLELOC+1) ;
  %LET CWTIT3 = title%EVAL(&TITLELOC+2) ;
 %END ;
 %LET BADPRINT = (%INDEX(&XOPTIONS,%STR( BADPRINT ))>0) ;
 proc print data=&DATA
%IF &BADPRINT %THEN (where=(_result_^='C')) ;
  label split='@' ; ;
  &CWTIT 'Report on overall convergence for BY groups' ;
  &CWTIT2 "C=met benchmarks, N=not converged, I=impossible" ;
  &CWTIT3 "R=met benchmarks but a replicate did not" ;
  label _iters_='iters@used'
    _result_='result@code' ;
  id &BYVARS _iters_ _result_ ;
```

```
  %IF %BQUOTE(&GROUP)^= %THEN %DO ;
   label
    _freq_='!----@non-@nils'
    _nilwt_=' unit@@nils'
    _negin_='count@-ve@input'
    _negwt_='----!@-ve@final' ;
   label
    _gfreq_='!----@non-@nils'
    _gnilwt_='group@@nils'
    _gnegin_='count@-ve@input'
    _gnegwt_='----!@-ve@final' ;
   var _freq_ _nilwt_ _negin_ _negwt_
    _gfreq_ _gnilwt_ _gnegin_ _gnegwt_ ;
   sum _freq_ _nilwt_ _negin_ _negwt_
    _gfreq_ _gnilwt_ _gnegin_ _gnegwt_ ;
  %END ;
  %ELSE %DO ;
   label _nilwt_='nils'
    _freq_='non-@nils'
    _negin_='-ve@input'
    _negwt_='-ve@final' ;
   var _freq_ _nilwt_ _negin_ _negwt_ ;
   sum _freq_ _nilwt_ _negin_ _negwt_ ;
  %END ;
 run ;

%MEND GREGPBY ;

%LOCAL LASTBY BYVARS PUTBY BYCODE NBYVARS
BYVAR1 BYVAR2 BYVAR3 BYVAR4 BYVAR5 BYVAR6 BYVAR7 BYVAR8 BYVAR9 ;
%MACRO HANDLEBY(BY=, PUTBLANK="all obs") ;
%LOCAL I WORD ;
%LET LASTBY = _last_ ;
%LET BYVARS = ;
%LET PUTBY = ;
%LET BYCODE = ;
%LET NBYVARS = 0 ;
%IF %QUOTE(&BY) ^= %THEN %DO ;
  %DO I = 1 %TO 20 ;
    %LET WORD = %SCAN(&BY,&I) ;
    %IF %LENGTH(&WORD) = 0 %THEN %LET I = 20 ;
    %ELSE %IF (%BQUOTE(&WORD) ^= DESCENDING)
             & (%BQUOTE(&WORD) ^= NOTSORTED) %THEN %DO ;
      %LET NBYVARS = &I ;
      %LET BYVAR&I = &WORD ;
      %LET LASTBY = LAST.&WORD ;
      %LET BYVARS = &BYVARS &WORD ;
      %LET PUTBY = %STR(&PUTBY "&WORD=" &WORD " ") ;
    %END ;
```

```
  %END ;
  %LET BYCODE = %STR(by &BY ;) ;
%END ;
%ELSE %LET PUTBY = &PUTBLANK ;
%MEND HANDLEBY ;

%LOCAL CWTESTPR ;
%MACRO TESTPR(LIST,NPRINTS=50) ;
  %LOCAL WORD ;
  %LET CWTESTPR = %EVAL(&CWTESTPR+1) ;
  cwpr&CWTESTPR + 1 ;
  drop cwpr&CWTESTPR ;
  if cwpr&CWTESTPR <= &NPRINTS then do ;
    %DO I = 1 %TO 50 ;
      %LET WORD=%SCAN(&LIST,&I,%STR( )) ;
      %IF %BQUOTE(&WORD)= %THEN %LET I = 50 ;
      %ELSE %DO ;
        cwtptemp = &WORD ;
        put "&WORD=" cwtptemp @ ;
      %END ;
    %END ;
    put ;
  end ;
%MEND ;
%LOCAL CWPRINT ;
%MACRO PR(DSN,TIT) ;
  %IF &CWPRINT %THEN %DO ;
    proc print data=&DSN ;
      %IF %BQUOTE(&TIT)= %THEN &CWTIT "&DSN" ;
      %ELSE &CWTIT "&DSN, &TIT" ;;
    run ;
  %END ;
%MEND PR ;

%MACRO CHECKERR(BOOLEAN,MESSAGE) ; %* Reports an error message
  if the BOOLEAN is true Sets the macro variable GO_END to 1
  to show error *;
  %IF (&BOOLEAN) %THEN %DO ;
    %LET GO_END =1;
    %IF %QUOTE(&MESSAGE)^= %THEN %DO ;
%PUT ************************************************************;
%PUT ERROR: &MESSAGE ;
%PUT ************************************************************;
    %END ;
  %END ;
%MEND CHECKERR ;

%MACRO CK_TRASH(DSN); %* for disposing of temporary datasets
  produced by macro *;
  %IF %BQUOTE(&DSN) ^= %THEN %DO ;
```

```
      %LET DSN = %UPCASE(&DSN) ;
      %IF %INDEX(%STR( &CK_TRASH ),%STR( &DSN )) = 0
      %THEN %LET CK_TRASH = &CK_TRASH &DSN ;
    %END ;
    %ELSE %IF %BQUOTE(&CK_TRASH)^= %THEN %DO ;
      proc datasets ddname=work nolist;
        delete &CK_TRASH ;
      run ;
      quit;
      %LET CK_TRASH = ;
    %END ;
%MEND CK_TRASH;
%******** End of macros used by the main algorithms *********;
```

NOTE: SAS programs for parts 2–10 are in Appendix F.

10.5.3 The Second-Stage Program File Codes

The SAS codes for the second-stage computation process of MMT to produce synthetic microdata and small area estimates are provided in the following. The model execution process of this stage for small area housing stress estimation is detailed in Chapter 6 (e.g., see Section 6.5.2 and Figure 6.3).

```
%let state=AUST;
%let datewt=090727; /* six digits date code from the
  reweighted file name */
libname wts&state. "C:\Folder …"; /* path to the outputs
  folder */
libname SIHCxxx "D:\Survey dataset file path"; /* for the
  latest survey */
libname SIHCxxx "D:\Survey dataset file path"; /* if old data
  are useable */
libname output "C:\Folder …"; /* path to the outputs folder */
libname libsas "C:\Folder …"; /* path to the outputs folder */
options mautosource sasautos=("C:\Folder …\Algorithms", "C:\
  Folder …\SAS code") mprint; /* same as to the 'general model
  file' */
%let outfldr = C:\Folder …; /* path to the outputs folder */

/*********SLA names *******/;
data slaname (keep=SLA_ID SLA_NAME);
    attrib SLA_ID label="SLA_ID";
      set slanames.aus_sla01 (rename=(SLA_MAIN=sla_id));
run;
```

```
/**********MICRODATA ***************/;
data SIH03HH;
       set SIHC0304.sih03bh;
       HH_ID=substr(ABSHID, 8)*10+3;
run;

data SIH05HH;
        set SIHC0506.sih05bh;
        HH_ID=substr(ABSHID, 9,5 )*10+5;
run;

data sih03hh;
       set sih03hh;
       if tenurecf = 1 then tenure03 = 1; /* Owner without
         mortgage */
       else if tenurecf = 2 then tenure03 = 2; /* Owner with
         mortgage */
       else if tenurecf in (0,4)then tenure03 = 5; /* Other */
       if ldlrdhcf = 2 and tenurecf = 3 then tenure03 =3;/*public
         renters*/
       else if ldlrdhcf ne 2 and tenurecf = 3 then tenure03 =
         4;/*private renters*/
run;

data sih05hh;
       set sih05hh;
               if tenurecf = 1 then tenure05 = 1; /* Owner
                 without mortgage */
               else if tenurecf = 2 then tenure05 = 2; /*
                 Owner with mortgage */
               else if tenurecf in (0,4)then tenure05 = 5;
                 /* Other */
               if ldlrdhcf = 2 and tenurecf = 3 then tenure05
                 =3;/*public renters*/
               else if ldlrdhcf ne 2 and tenurecf = 3 then
                 tenure05 = 4;/*private renters*/

/*keep variables of interest for calculation of housing
  stress; */

data SIH35( keep= HH_ID state
                         Adults
                         Kids
                         TotInc
                         DispInc
                         Rent
                         MortRep
                         RateGW
                         HCost
                         Tenure
```

```
                              AgeHhRef
                              FirstHome
                              MainIncSo
                              LifeCycle
                              HHWT )
                              ;
         set SIH03HH
             ( keep= HH_ID   statehbc
                             NOMEMHBC
                             NUMU15BC
                             INCTOTCH
                             DISPCH
                             WKRENTCH
                             TRPAY1CH
                             TRPAY2CH
                             TRPAY4CH
                             RATESCH
                             HCOSTSH
                             tenure03
                             AGERHBC
                             FSTHHCF
                             PSRCCH
                             LIFECYCH
                             SIHHHWT
                             )

             SIH05HH
             ( keep= HH_ID   statehbc
                             NOMEMHBC
                             NUMU15BC
                             INCTOTCH
                             DISPCH
                             WKRENTCH
                             TRPAY1CH
                             TRPAY2CH
                             TRPAY4CH
                             RATESCH
                             HCOSTSH
                             tenure05
                             AGERHBC
                             FSTHHCF
                             PSRCCH
                             LIFECYCH
                             SIHHHWT
                             )
                             ;
         Adults     =sum(NOMEMHBC);
         Kids       =sum(NUMU15BC);
         TotInc     =      INCTOTCH;
         DispInc    =      DISPCH;
```

```
        Rent         =         WKRENTCH;
        MortRep      =         sum (TRPAY1CH, TRPAY2CH, TRPAY4CH);
        RateGW       =         RATESCH;
        HCost        =         HCOSTSH;
        tenure       =sum (tenure03, tenure05);
        AgeHhRef     =sum(AGERHBC);
        FirstHome    =sum(FSTHHCF);
        MainIncSo    =  PSRCCH ;
        LifeCycle    = LIFECYCH;
        HHWT         =sum(SIHHHWT);
        state = statehbc;
        if DispInc > 0;   /* to exclude -ve or zero income HH*/;

run;

proc format;
        value state   1='NSW'
                      2='VIC'
                      3='QLD'
                      4='SA'
                      5='WA'
                      6='TAS'
                      7='NTACT';
        value tenure  1 = "Owner"
                      2 = "Buyer"
                      3 = "RenterPub"
                      4 = "RenterPriv"
                      5 = "Other" ;
run;

data SIH35;
        set SIH35;
                sih_yr=substr(hh_id,length(hh_id),1);
                if sih_yr=5 then SIHYR="SIH0506";
                else if sih_yr= 3 then SIHYR="SIH0304";
run;

/*factors to uprate income and house price; */

*%let CPI02to06=(155.5/139.5); *(=1.114695341) *Dec to Dec;
  *ABS 2007 6401.0 Consumer Price Index, Australia, TABLES 1
  and 2. CPI: All Groups, Index Numbers and Percentage
  Change
*%let CPI03to06=(155.5/142.8); * (=1.088935574) Dec to Dec;
  *these factors to uprate housing cost, rent, and mortgage;
*%let HCPI02to06=(132.9/114.2); * (=1.163747811) *dec to dec,
  ABS 2007 6401.0 Index Numbers  Housing  Australia A2325981V;
* %let HCPI03to06=(132.9/119.6); *(=1.111204013); *abbreviated
  as HCPI for our purpose - it is housing CPI;
* HPI - House Price Index is;
```

```
/*get changes in index numbers to calculate uprating factor
  for housing cost */;

PROC IMPORT OUT= WORK.HCPI_change
      DATAFILE= "C:\HCPI data file path"; /* for the latest
        HCPI.xls */
      DBMS=EXCEL REPLACE;
      SHEET="HCPIchange";
RUN;

proc transpose data =HCPI_change
                               out=HCPI_change;
      ID UpFactor;
run;

data HCPI_change (drop= _label_);
      set HCPI_change;
      label _name_ = "State";
      rename _name_ = state;
run;

data sih35;
      set sih35;
            state2=put (state, state.);
run;

proc sort data=sih35 ; by state2; run;

data hcpi_change;
 length state2 $ 5;
 set hcpi_change;
 drop state ;
  state2 = upcase(state);
run;

proc sort data=HCPI_change; by state2; run;

data SIH35;
      merge SIH35 (in=state_exist) HCPI_change ;
      by state2;
      if state_exist;
run;

data SIH35;
      set SIH35;

        if sih_yr=5 then do;
                      TotIncUP    =   TotInc*&WAGE05to06.;
                      DispIncUP   =       DispInc*&WAGE05to06.;
                      RentUP      =       Rent*HCPI05to06;
                      MortRepUP   =       MortRep*HCPI05to06;
```

```
                              RateGWup     =           RateGW*HCPI05to06;
                              HCostUP      =           HCost*HCPI05to06;
            if sih_yr=3 then do;
                              TotIncUP     =    TotInc*&WAGE03to06.;
                              DispIncUP    =           DispInc*&WAGE03to06.;
                              RentUP       =           Rent*HCPI03to06;
                              MortRepUP    =           MortRep*HCPI03to06;
                              RateGWup     =           RateGW*HCPI03to06;
                              HCostUP      =           HCost*HCPI03to06;
            if sih_yr=5 and tenure in (3,5) then do;
                              RentUP       =           Rent*&WAGE05to06.;
                              HCostUP      =           HCost*&WAGE05to06.;
                    end;
            if sih_yr=3 and tenure in (3,5) then do;
                              RentUP       =           Rent*&WAGE03to06.;
                              HCostUP      =           HCost*&WAGE03to06.;
                    end;

MortRentUp=sum(RentUP,MortRepUP);
run;

data SIH35;
        set SIH35;
        SumofHcost = sum(Rent, MortRep, RateGW);
        SumofHcostUP = sum(RentUP, MortRepUP, RateGWUP);
        SumForBuyer = sum(MortRep, RateGW);
        SumForBuyerUP = sum(MortRepUP, RateGWUP);
run;

proc sort data=SIH35; by hh_id; run;

data SIH35;
                set SIH35;
                        persWT=0;
                        persWT = HHWT*sum(adults, kids);
                        childWT=0;
                        childWT = HHWT*kids;
                        oldWT=0;
                        oldWT = HHWT*old;
                        count_HH=1;
                        count_pers= sum (adults, kids);
                        count_child= kids;
                        count_old= old;
run;

/* Bring on nonclassifiable HH IDs to be able to merge with
   area linkage file */

data SIH35x;
        set SIH35;
```

```
          output;
              sih_yr=substr(hh_id,length(hh_id),1);
              hh_id=(hh_id * 100000)+sih_yr;
              output;
run;

proc sort data=SIH35x; by hh_id; run;

*****microdata creation ************************************;

/*******base weights files*******/;

data wts_&state.;
  set wts&state..areawts_conv_&datewt.;
run;

%let indi=Hstress; *indi=hstress to calculate housing stress;
%let hcostVar = hcostUp;
%let Year=2006;
%let wt=&&lev.wt;
%put &wt.;

/*******small area weights (wts)*********/;
data wts_&state.;
set wts&state..areawts_090727; /* the new weights file name */
run;

%macro Income_type (incVar= , hcostVar=);
/* housing stress based on gross income or disposable income */;
/* Generate housing stress flags from deciles */;

%equivalised_decile_macro(indsn=sih35x,
                                    outdsn=hstress_flags,
                                     adults=adults,
                                      kids=kids,
                                      income=dispInc,wt=persWt);

%housing_stress_flags_from_decile (indsn=hstress_flags,
                outdsn=hstress_flags, income=&incVar.,
                  HCOST=&hcostVar.,
                   incDecile=equivincdec,
                      rule1=r30only, rule2=r30_40, rule3=r30_10_40,
                          GTvsGE=gt, benchVal=0.3);

%macro states( year=&year.);

/**rename and sort weight files;*/
proc sort data = wts_&state._base;
       by HH_ID pers_id;
run;
```

```
/**====== inflate weights======;*/

%inflate_wts (
      indsn=wts_&state._base,
      outdsn=wts_&state.&year.,
      base_yr=2001,
      tgt_yr= &YEAR.);

/* *merge weight files into  micro data files ;*/

proc sort data = wts_&state. ; by hh_id pers_id; run;
proc sort data = hstress_flags; by hh_id;
run;

*calculate the number of hhs and hh level housing stress flag;

data interim_&state.;
      merge hstress_flags (in =a) wts_&state. ;
      by hh_id;
      IF a and first.hh_id then output;
run;

/* *======== modify flags======; */
%macro rules (rule=);
data interim_&state.;
      set interim_&state.;
       rename hStressFlag_&rule. = hStressFlag_&rule._HH;
run;

data interim_&state.;
      set interim_&state.;
      count_HH=1;
      count_pers= sum (adults, kids);
      hStressFlag_&rule._pers = count_pers*hStressFlag_
        &rule._HH;
      hStressFlag_&rule._child = count_child*hStressFlag_
        &rule._HH;
run;

%macro level (lev=);                        *no default;
%macro indic (indi=hstress);    *default is hstress;

%mend indic;
%indic (indi=hStress);

%mend level;
*%level (lev=pers);
%level (lev=HH);
*%level (lev=child);
```

```
%mend rules;
*%rules (rule=R30Only);
%rules (rule=R30_40);  /* choose appropriate rule of HS
  measure; */
*%rules (rule=R30_10_40);

%mend states;
%states ( year=2006);

%mend Income_type;
*%Income_type (incVar=dispIncUp, hcostVar=hcostUp);
  *disposable income  ;
%Income_type (incVar=TotIncUP, hcostVar=MortRentUp);*gross
  income ; /*hcost05*/

*===========================================================;
*estimation of housing stress by tenure type ;
*===========================================================;

/* calculate total population by SLA by tenure type by rule ; */

proc means data=interim_&state. noprint;
      class tenure;
      weight count_&lev.;
      output out=tenure sum=;
run;

proc transpose data=tenure
                         out=tenure1;
      var _all_;
run;

data tenure2 (drop= _label_ );
      set tenure1;
            if (substr (_name_,1,2))='wt' and (substr
              (_name_,3,1))>0  then output;
run;

data tenure3 (drop = _name_);
            attrib SLA_ID label="SLA_ID";
            set tenure2(rename=(col1=Total&lev.
                                        col2=Owner
                                        col3=Buyer
                                        col4=RenterPub
                                        col5=RenterPriv
                                        col6=OtherTenure));
            SLA_ID=input(substr(_name_,3),12.);
            Total&lev.=round(total&lev.);
```

```
                owner=round(owner);
                Buyer=round(Buyer);
                RenterPub=round(renterPub);
                RenterPriv=round(renterPriv);
                OtherTenure=round(otherTenure);
run;

/*----------------------------------------------------------------
calculates number of households in housing stress by tenure type
----------------------------------------------------------;*/

proc means data=interim_&state. noprint;
        class tenure;
        weight &indi.Flag_&rule._&lev.;
        output out=&indi._Tenure sum=;
run;

proc transpose data=&indi._tenure
                             out=&indi._tenure1;
        var _all_;
run;

data &indi._tenure2 (drop= _label_ );
        set &indi._tenure1;
                if (substr (_name_,1,2))='wt' and (substr
                (_name_,3,1))>0  then output;
run;

data &indi._tenure3(drop = _name_);
                attrib SLA_ID label="SLA_ID";
                set &indi._tenure2(rename=(col1=Total&indi.
                                           col2=Owner&indi.
                                           col3=Buyer&indi.
                                           col4=RenterPub&indi.
                                           col5=RenterPriv&indi.
                                           col6=OtherTenure&indi.
        ));
                SLA_ID=input(substr(_name_,3),12.);
                Total&indi. = round(total&indi. );
                Owner&indi. =round(owner&indi.);
                Buyer&indi. =round(Buyer&indi. );
                RenterPub&indi. =round(renterPub&indi. );
                RenterPriv&indi. =round(renterPriv&indi. );
                OtherTenure&indi. =round(otherTenure&indi. );
run;

/* merge files containing numerators and denominators and
   calculate percentages; */

proc sort data=&indi._tenure3; by sla_id; run;
proc sort data=tenure3; by sla_id; run;
```

```
data pc&lev.&indi.Tenure&state.;

        merge tenure3 &indi._tenure3;
               by sla_id;
                       pc_Total&indi.=round(Total&indi. /
                         Total&lev.*100,.1);
                       pc_Owner&indi. =round(owner&indi. /
                         owner*100,.1);
                       pc_Buyer&indi. =round(Buyer&indi. /
                         Buyer*100,.1);
                       pc_RenterPub&indi. =round(renterPub&indi./
                         renterPub*100,.1);
                       pc_RenterPriv&indi. =round(renterPriv&indi./
                         renterPriv*100,.1);
                       pc_OtherTenure&indi.
                         =round(otherTenure&indi. /
                         otherTenure*100,.1);
run;

/* add SLA names; */

proc sort data=slaname;
        by sla_id;
run;
data libsas.&rule.&incVar._Tenure&state. ;
             %let var=Tenure;
        length State $ 5   Rule $ 12 Income $ 8 Var $ 8;
        State ="&state.";
        Rule ="&rule.";
        Income = "&IncVar.";
        Var = "&var.";
set pc&lev.&indi.Tenure&state.;
        merge slaname pc&lev.&indi.Tenure&state. (in=a) ;
        by sla_id;
        if a;
run;
PROC EXPORT DATA= libsas.&rule.&incVar._Tenure&state.
             OUTFILE= "C:\file path …\&rule.&lev.&state..xls"
             DBMS=EXCEL REPLACE;
        SHEET="&lev.&state.";
RUN;

-----------------------------------------------------------------------
© Authors
  *-----------------------------------------------------------------*
```

10.6 Concluding Remarks

This book has fulfilled the primary purposes by reviewing all existing small area estimation methodologies and then by developing a range of novel methodological capacities in small area estimation, including an alternative spital microdata simulation technique, MMT and new methods for validations and statistical reliability. All of the methodologies have been successfully applied on various real-world problems using the Australian data. The alternative methodologies developed for small area estimation and tested in this book are robust and generalizable to other types of microsimulation modeling. Ultimately, this book has contributed to the existing body of knowledge in small area estimation and microsimulation modeling through the following ways:

- It reviews exclusively the small area estimation methodologies including the indirect geographic approach of small area estimation.

- It develops an alternative methodology for small area microdata simulation in the spatial microsimulation modeling approach of small area estimation.

- It develops an effective MMT-based spatial model for small area estimation and successfully generates small area housing stress estimates at a range of geographic levels.

- It composes and discusses various results of the housing stress estimates in SLAs, major capital cities, different states, and throughout Australia.

- It develops new validation tools to test the statistical significance of small area estimates produced by the MMT-based spatial model and establishes the measures of statistical reliability such as CIs of the synthetic small area estimates.

References

ABS. 1999. Demographic estimates and projections: Concepts, sources and methods, Catalogue no. 3228.0, Australian Bureau of Statistics, Canberra, Australian Capital Territory, Australia.

ABS. 2002a. The 1998–1999 household expenditure survey, Australia: Confidentialised unit record files (CURF), Technical manual (2nd edn.), ABS Catalogue no. 6544.0, Australian Bureau of Statistics, Canberra, Australian Capital Territory, Australia.

ABS. 2002b. Measuring Australia's progress, ABS Catalogue no. 1370.0, Australian Bureau of Statistics, Canberra, Australian Capital Territory, Australia.

ABS. 2004. Statistical matching of the HES and NHS: An exploration of issues in the use of unconstrained and constrained approaches in creating a base-file for a microsimulation model of the pharmaceutical benefits scheme, ABS Methodology Advisory Committee Paper, Australian Bureau of Statistics, Canberra, Australian Capital Territory, Australia.

ABS. 2005. Housing occupancy and costs of Australia—2002–03, ABS Catalogue no. 4130.0.55.001, Australian Bureau of Statistics, Canberra, Australian Capital Territory, Australia.

ABS. 2006. Census dictionary Australia, ABS Catalogue no. 2901.0, Australian Bureau of Statistics, Canberra, Australian Capital Territory, Australia.

ABS. 2007a. *Information Paper: 2006 Census of Population and Housing—Census Data Products*, Australian Bureau of Statistics, Canberra, Australia. http://www.abs.gov.au/ausstats/abs@.nsf/lookup/2011.0.55.001Main%20Features1042011. Accessed on November 15, 2009.

ABS. 2007b. Information paper: Survey of income and housing, Australia: User guide 2005–06, ABS Catalogue no. 6553.0, Australian Bureau of Statistics, Canberra, Australian Capital Territory, Australia.

ABS. 2008a. Year Book Australia, ABS Catalogue no. 1301.0, Australian Bureau of Statistics, Canberra, Australian Capital Territory, Australia.

ABS. 2008b. Survey of income and housing—Confidentialised unit record files: Technical manual—Australia 2005–06, ABS Catalogue no. 6541.0, Australian Bureau of Statistics, Canberra, Australian Capital Territory, Australia.

ABS. 2009. Consumer price index: Concepts, sources and methods—Information paper, ABS Catalogue no. 6461.0, Australian Bureau of Statistics, Canberra, Australian Capital Territory, Australia.

AIHW. 2008. Housing assistance in Australia, Catalogue no. HOU 173, Australian Institute of Health and Welfare, Canberra, Australian Capital Territory, Australia.

AIHW. 2009. Public rental housing 2007–08: Commonwealth state housing agreement national data report, Housing assistance data development series, Catalogue no. HOU 187, Australian Institute of Health and Welfare, Canberra, Australian Capital Territory, Australia.

Aitkin, M. 2008. Applications of the Bayesian bootstrap in finite population inference. *Journal of Official Statistics*, 24(1): 21–51.

Alderman, H., Babita, M., Demombynes, G., Makhatha, N., and Ozler, B. 2002. How low can you go?: Combining census and survey data for mapping poverty in South Africa. *Journal of African Economics*, 11(2): 169–200.

Alegre, J., Arcarons, J., Calonge, S., and Manresa, A. 2000. Statistical matching between different datasets: An application to the Spanish household survey (EPF90) and the income tax file (IRPF90), http://selene.uab.es/mmercader/workshop/cuerpo.html. Accessed on April 15, 2008.

Anderson, B. 2007a. Creating small-area income estimates: Spatial microsimulation modelling, http://www.communities.gov.uk/publications/communities/creatingsmallareaincome. Accessed on April 3, 2008.

Anderson, B. 2007b. Estimating time spent on-line at small area levels: A spatial microsimulation approach (no. CWP-2007-01). Chimera, University of Essex, Ipswich, U.K.

Anderson, B. 2011. Estimating small area income deprivation: An iterative proportional fitting approach. Centre for Research in Economic Sociology and Innovation (CRESI) Working Paper 2011-02, University of Essex, Colchester, U.K.

Anderson, B. 2013. Estimating small-area income deprivation: An iterative proportional fitting approach. In R. Tanton and K. Edwards (eds.), *Spatial Microsimulation: A Reference Guide for Users*. Understanding Population Trends and Processes, vol. 6. Springer, London, U.K., pp. 49–67.

Anderson, B., Agostini, P.D., Laidoudi, S., Weston, A., and Zong, P. 2007a. Time and money in space. In paper presented at the *First General Conference of the International Microsimulation Association*, Vienna, Austria, August 20–22, 2007.

Anderson, B., Paola De Agostini, P., Laidoudi, S., Weston, A., and Zong, P. 2007b. Time and money in space. University of Essex Chimera Working Paper Number: 2007-09, University of Essex, Colchester, U.K.

Bakker, B. 2011. Micro integration. Statistical Methods paper 201108, Statistics Netherlands, The Hague/Heerlen, the Netherlands. https://www.cbs.nl/NR/rdonlyres/DE0239B4-39C6-4D88-A2BF-21DB3038B97C/0/2011x3708.pdf.

Bakker, B.F.M., van Rooijen, J., and van Toor, L. 2014. The System of social statistical data sets of Statistics Netherlands: An integral approach to the production of register-based social statistics. *Statistical Journal of the IAOS*, 30: 411–424.

Ballas, D. 2001. A spatial microsimulation approach to local labour market policy analysis. Unpublished PhD thesis, School of Geography, University of Leeds, Leeds, U.K.

Ballas, D. and Clarke, G.P. 2000. GIS and microsimulation for local labour market policy analysis. *Computers, Environment and Urban Systems*, 24(2): 305–330.

Ballas, D. and Clarke, G.P. 2001. Modelling the local impacts of national social policies: A spatial microsimulation approach. *Environment and Planning C: Government and Policy*, 19(4): 587–606.

Ballas, D., Clarke, G., and Commins, P. 2001. Building a dynamic spatial microsimulation model for the Irish rural economy. In paper presented at the *12th European Colloquium on Theoretical and Quantitative Geography*, Rouen, France, September 7–11, 2001.

Ballas, D., Clarke, G., and Dewhurst, J. 2006. Modelling the socio-economic impacts of major job loss or gain at the local level: A spatial microsimulation framework. *Spatial Economic Analysis*, 1(1): 127–146.

Ballas, D., Clarke, G.P., Dorling, D., Eyre, H., Thomas, B., and Rossiter, D. 2005a. SimBritain: A spatial microsimulation approach to population dynamics. *Population, Space and Place*, 11(1): 13–34.

Ballas, D., Clarke, G., Dorling, D., and Rossiter, D. 2007. Using SimBritain to model the geographical impact of National Government Policies. *Geographical Analysis*, 39(1): 44–77.

Ballas, D., Clarke, G., and Turton, I. 1999. Exploring microsimulation methodologies for the estimation of household attributes. In paper presented at the *Fourth International Conference on Geo-Computation*, Fredericksburg, VA, July 25–28.

Ballas, D., Clarke, G.P., and Turton, I. 2003. A spatial microsimulation model for social policy evaluation. In B. Boots and R. Thomas (eds.), *Modelling Geographical Systems*, vol. 70. Kluwer, Dordrecht, the Netherlands, pp. 143–168.

Ballas, D., Clarke, G.P., and Wiemers, E. 2005b. Building a dynamic spatial micro-simulation model for Ireland. *Population, Space and Place*, 11(2): 157–172.

Ballas, D., Rossiter, D., Thomas, B., Clarke, G.P., and Dorling, D. 2005c. Geography matters: Simulating the local impacts of national social policies. Joseph Rewntree Foundation, York, U.K.

Baschieri, A., Jane, F., Duncan, H., and Craig, H. 2005. Creating a poverty map for Azerbaijan. World Bank Policy Research Working Paper 3793. The World Bank Group, Washington, DC.

Battese, G.E., Harter, R.M., and Fuller, W.A. 1988. An error component model for prediction of county crop areas using survey and satellite data. *Journal of the American Statistical Association*, 83(1): 28–36.

Bayes, T. 1763. An essay towards solving a problem in the doctrine of chances. *Philosophical Transactions of the Royal Society of London*, 53(2): 370–418.

Bell, P. 2000a. GREGWT and TABLE macros—User guide, Australian Bureau of Statistics, Canberra, , Australian Capital Territory, Australia, Unpublished.

Bell, P. 2000b. Weighting and standard error estimation for ABS household surveys. Australian Bureau of Statistics, Canberra, Australian Capital Territory, Australia.

Bell, W., Basel, W., Cruse, C., Dalzell, L., Maples, J., O'Hara, B., and Powers, D. 2007. Use of ACS data to produce SAIPE model-based estimates of poverty for counties. U.S. Census Bureau, Washington, DC.

Bell, W.R. and Huang, E.T. 2006. Using the t-distribution to deal with outliers in small area estimation. In *Proceedings of Statistics Canada Symposium on Methodological Issues in Measuring Population Health*, Statistics Canada, Ottawa, Ontario, Canada.

Bennett, V.J., Beard, M., Zollner, P.A., Fernandez-Juricic, E., Westphal, L., and LeBlanc, C.L. 2009. Understanding wildlife responses to human disturbance through simulation modelling: A management tool. *Ecological Complexity*, 6(1): 113–134.

Benson, T. 2005. An investigation of the spatial determinants of the local prevalence of poverty in rural Malawi. *Food Policy*, 30(4): 532–550.

Benson, T. 2006. Insights from poverty maps for development and food relief program targeting. Discussion Paper 205. IFPRI Food Consumption and Nutrition Division, Washington, DC.

Berg, E. and Fuller, W. 2014. Small area prediction of proportions with applications to the Canadian Labour Force Survey. *Journal of Survey Statistics and Methodology*, 2: 227–256.

Berg, E.J. and Fuller, W.A. 2012. Estimation of error covariance matrices for small area prediction. *Computational Statistics and Data Analysis*, 56: 2949–2962.

Bernardo, J. and Rueda, R. 2002. Bayesian hypothesis testing: A reference approach. *International Statistical Review*, 70(3): 351–372.

Birkin, M. and Clarke, M. 1985. Comprehensive dynamic urban models: Integrating macro- and microapproaches. In D.A. Griffith and R.P. Haining (eds.), *Transformations through Space and Time: An Analysis of Nonlinear Structures, Bifurcation Points and Autoregressive Dependencies*. Martinus Nijhoff, Dordrecht, the Netherlands, pp. 165–292.

Birkin, M. and Clarke, M. 1988. SYNTHESIS—A synthetic spatial information system for urban and regional analysis: Methods and examples. *Environment and Planning Analysis*, 20(12): 1645–1671.

Birkin, M., Wu, B., and Rees, P. 2009. Moses: Dynamic spatial microsimulation with demographic interactions. In P. Williamson, A. Zaidi, and A. Harding (eds.), *New Frontiers in Microsimulation Modelling*. Ashgate, Oxford, U.K.

Boonstra, H.J. 2004. A simulation study of repeated weighting estimation. Discussion paper 04003. Statistics Netherlands, Voorburg/Heerlen, the Netherlands.

Brackstone, G.J. 1987. Small area data: Policy issues and technical challenges. In R. Platek, J.N.K. Rao, C.E. Sarndal, and M.P. Singh (eds.), *Small Area Statistics*, vol. 3. Wiley, New York.

Bradley, M. and Bowman, J. 2006. A summary of design features of activity-based microsimulation models for U.S. MPOs. In white paper for the *Conference on Innovations in Travel Demand Modeling*, Austin, TX, May 2006.

Breidt, F.J. 2004. Small area estimation for natural resource surveys. In *Monitoring Science and Technology Symposium*, Denver, CO.

Brouwers, L. 2005. MicroPox: A large-scale and spatially explicit microsimulation model for smallpox planning. In V. Ingalls (ed.), *The Proceedings of the 15th International Conference on Health Sciences Simulation*, SCS International, San Diego, CA, pp. 70–77.

Brouwers, L., Ekenberg, L., and Hansson, K. 2004. Multi-criteria decision-making of policy strategies with public-private re-insurance systems. *Risk Decision and Policy*, 9(1): 23–45.

Brown, L. and Harding, A. 2002. Social modelling and public policy: Application of microsimulation modelling in Australia. *Journal of Artificial Societies and Social Simulation*, 5(1): 1–14.

Brown, L. and Harding, A. 2005. The new frontier of health and aged care: Using microsimulation to assess policy options. In *Proceedings of the Conference on Tools for Microeconomic Policy Analysis*, Productivity Commission, Canberra, Australian Capital Territory, Australia.

Brown, L., Yap, M., Lymer, S., Chin, S.F., Leicester, S., Blake, M., and Harding, A. 2004a. Spatial microsimulation modelling of care needs, costs and capacity for self-provision: Detailed regional projection for older Australians to 2020. In paper presented at the *Australian Population Association Conference*, Canberra, Australian Capital Territory, Australia, September 28, 2004.

Brown, T., Beyeler, W., and Barton, D. 2004b. Assessing infrastructure interdependencies: The challenge of risk analysis for complex adaptive systems. *International Journal of Critical Infrastructures*, 1(1): 108–117.

Burdett, C.L., Kraus, B.R., Garza, S.J., Miller, R.S., and Bjork, K.E. 2015. Simulating the distribution of individual livestock farms and their populations in the United States: An example using domestic swine (*Sus scrofa domesticus*) Farms. *PLoS One*, 10(11): e0140338.

Cai, L., Creedy, J., and Kalb, G. 2004. Reweighting the survey of income and housing costs for tax microsimulation modelling. The University of Melbourne, Melbourne, Victoria, Australia.

Cassells, R., Harding, A., Miranti, R., Tanton, R., and McNamara, J. 2010. Spatial microsimulation: Preparation of sample survey and census Data for SpatialMSM/08 and SpatialMSM/09. Online Technical Paper—TP36. NATSEM, University of Canberra, Canberra, Australian Capital Territory, Australia.

Chambers, R., Chandra, H., Salvati, N., and Tzavidis, N. 2014. Outlier robust small area estimation. *Journal of the Royal Statistical Society B*, 76(1): 47–69.

Chambers, R. and Tzavidis, N. 2006. M-quantile models for small area estimation. *Biometrika*, 93(2): 255–268.

Chambers, R.L. and Clark, R.G. 2012. *An Introduction to Model-Based Survey Sampling with Applications*. Oxford University Press, New York.

Chan, W.T. 1963. *A Source Book in Chinese Philosophy*. Princeton University Press, Princeton, NJ.

Chandra, H. and Chambers, R. 2009. Multipurpose weighting for small area estimation. *Journal of Official Statistics*, 25(3): 379–395.

Chen, X., Meaker, J.W., and Zhan, F.B. 2006. Agent-based modelling and analysis of hurricane evacuation procedures for the Florida keys. *Natural Hazards*, 38(2): 321–338.

Chin, S.F. and Harding, A. 2006. Regional dimensions: Creating synthetic small-area microdata and spatial microsimulation models. Online Technical Paper—TP33. NATSEM, University of Canberra, Canberra, Australian Capital Territory, Australia.

Chin, S.F. and Harding, A. 2007. SpatialMSM. In A. Gupta and A. Harding (eds.), *Modelling Our Future: Population Ageing, Health and Aged Care*. North-Holland, Amsterdam, the Netherlands.

Chin, S.F., Harding, A., and Bill, A. 2006. Regional dimensions: Preparation of 1998–99 HES for reweighting to small-area benchmarks. Online Technical Paper—TP34. NATSEM, University of Canberra, Canberra, Australian Capital Territory, Australia.

Chin, S.F., Harding, A., Lloyd, R., McNamara, J., Phillips, B., and Vu, Q.N. 2005. Spatial microsimulation using synthetic small area estimates of income, tax and social security benefits. *Australasian Journal of Regional Studies*, 11(3): 303–335.

Christen, P. 2012. A survey of indexing techniques for scalable record linkage and deduplication. *IEEE Transactions on Knowledge and Data Engineering*, 24(9): 1537–1555.

Clancy, D., Breen, J., Butler, A.M., Morrissey, K., O'Donoghue, C., and Thorne, F. 2012. Modelling the location economics of biomass production for electricity generation. In C. O'Donoghue, S. Hynes, K. Morrissey, D. Ballas, and G. Clarke (eds.), *Modelling the Local Economy: A Simulation Approach*. Springer Verlag, Berlin, Germany.

Clarke, G., Kashti, A., McDonald, A., and Williamson, P. 1997. Estimating small area demand for water: A new methodology. *Journal of the Chartered Institution of Water and Environmental Management*, 11: 186–192.

Clarke, M. and Holm, E. 1987. Microsimulation methods in spatial analysis and planning. *Geografiska Annaler: Series B, Human Geography*, 69(2): 145–164.

Cleveland, S.W. and McGill, R. 1987. Graphical perception: The visual decoding of quantitative information on graphical displays of data. *Journal of the Royal Statistical Society. Series A*, 150(3): 192–229.

Cochran, W.G. 1977. *Sampling Techniques*, 3rd edn. John Wiley & Sons, Inc., New York.

Cochran, W. G. 1993. *Sampling Techniques*, John Wiley and Sons Inc., New York.

Commonwealth Department of Employment and Workplace Relations. 2007. Small Area Labour Markets. Commonwealth of Australia, Canberra, Australian Capital Territory, Australia.

Cormen, T.H., Leiserson, C.E., Rivest, R.L., and Stein, C. 2001. *Introduction to Algorithms: Inverting Matrices*. MIT Press, Cambridge, MA.

Coutinho, W., de Waal, T., and Shlomo, N. 2013. Calibrated hot-deck donor imputation subject to edit restrictions. *Journal of Official Statistics*, 29(2): 299–321.

Cox, D.R. and Hinkley, D.V. 2000. *Theoretical Statistics*. CRC Press, Boca Raton, FL.

Creedy, J., Duncan, A., Harris, M., and Scutella, R. 2002. *Microsimulation Modelling of Taxation and the Labour Market*. Edward Elgar, Cheltenham, U.K.

Creedy, J. and Kalb, G. 2006. *Labour Supply and Microsimulation: The Evaluation of Tax Policy Reforms*. Edward Elgar Publishing Limited, Cheltenham, U.K.

Cullinan, J., Hynes, S., and O'Donoghue, C. 2006. The use of spatial microsimulation and geographic information systems (GIS) in benefit function transfer—An application to modelling the demand for recreational activities in Ireland. In paper presented at the *Eighth Nordic Seminar on Microsimulation Models*, Oslo, Norway, June 7–9, 2006.

Cullinan, J., Hynes, S., and O'Donoghue, C. 2008. Estimating catchment area population indicators using network analysis: an application to two small-scale forests in County Galway. *Irish Geography*, 41(3): 279–294.

Dagum, C. 1980. The generation and distribution of income, the Lorenz curve and the Gini ratio. *Economie Appliquée*, 33(2): 327–367.

Daalmans, J. 2015. Estimating detailed frequency tables from registers and sample surveys. Discussion paper 201503. Statistics Netherlands, Voorburg/Heerlen, the Netherlands.

D'Alò, M., Consiglio, D.L., Falorsi, S., and Solari, F. 2005. Small area estimation of the Italian poverty rate, http://www.stat.jyu.fi/sae2005/abstracts/istat1.pdf. Accessed on April 1, 2008.

Datta, D.S. and Lahiri, P. 2000. A unified measure of uncertainty of estimated best linear unbiased predictors in small area estimation problems. *Statistica Sinica*, 10(2): 613–627.

Datta, G.S. 2009. Model-based approach to small area estimation. In D. Pfeffermann and C.R. Rao (eds.), *Handbook of Statistics: Sample Surveys: Inference and Analysis*. vol. 29B. North-Holland, Amsterdam, the Netherlands, pp. 251–288.

Datta, G.S., Ghosh, M., Steorts, R., and Maples, J. 2011. Bayesian benchmarking with applications to small area estimation. *Test*, 20(4): 574–588.

Davis, B. 2003. Choosing a method for poverty mapping. Research Reports, Agriculture and Economic Development Analysis Division, Food and Agriculture Organization of the United Nations, Rome, Italy, http://www.fao.org/3/a-y4597e.pdf. Accessed on November 7, 2009.

De Waal, T. 2014. General approaches to combining administrative data and surveys. Discussion paper 201402. Statistics Netherlands, Voorburg/Heerlen, the Netherlands.

Deming, W.E. and Stephan, F.F. 1940. On a least squares adjustment of a sampled frequency table when the expected marginal totals are known. *The Annals of Mathematical Statistics*, 11(4): 427–444.

DeVeaux, R., Vellerman, P., and Block, D. 2004. *Intro Stats*. Pearson, Boston, MA.

Deville, J.C. and Sarndal, C.E. 1992. Calibaration estimators in survey sampling. *Journal of the American Statistical Association*, 87(3): 376–382.

Ditto, W.L., Spano, M.L., and Lindner, J.L. 1995. Techniques for the control of chaos. *Physica D: Nonlinear Phenomena*, 86: 198–211.

Doherty, S.T. and Miller, E.J. 2000. A computerized household activity scheduling survey. *Transportation*, 27(1): 75–97.

Duley, C.J. 1989. A model for updating census-based population and household information for inter-censal years. School of Geography, University of Leeds, Leeds, U.K.

Edwards, K.L. and Clarke, G.P. 2009. The design and validation of a spatial micro-simulation model of obesogenic environments for children in Leeds, UK: SimObesity. *Social Science and Medicine*, 69(7): 1127–1134.

Efron, B. and Morris, C. 1975. Data analysis using Stein's estimator and its generalizations. *Journal of the American Statistical Association*, 70(3): 311–313.

Elazar, D. and Conn, L. 2005. *Small Area Estimates of Disability in Australia.* ABS Publication no. 1351.0.55.006. Canberra, Australian Capital Territory, Australia.

Elazar, D.N. 2004. Small area estimation of disability in Australia. *Statistics in Transition*, 6(5): 667–684.

Elbers, C., Lanjouw, P., and Leite, P.G. 2008. Testing the poverty map methodology in Minas Gerais. World Bank Policy Research Working Paper 4513. The World Bank Group, Washington, DC.

Elbers, C., Olsen Lanjouw, J., and Lanjouw, P. 2003. Micro-level estimation of poverty and inequality. *Econometrica*, 71(1): 355–364.

Eliasson, J. and Mattsson, L.G. 2006. Equity effects of congestion pricing quantitative methodology and a case study for Stockholm. *Transportation Research Part A: Policy and Practice*, 40(7): 602–620.

Epstein, J.M. 1999. Agent-based computational models and generative social science. *Complexity*, 4(5): 41–60.

Erciulescu, A.L. and Fuller, W.A. 2013. Small area prediction of the mean of a binomial random variable. In *JSM Proceedings: Survey Research Methods Section*, pp. 855–863. Montreal, CA, August 3-8, 2013.

Ericson, W.A. 1969. Subjective Bayesian models in sampling finite populations. *Journal of the Royal Statistical Society: Series B*, 31(2): 195–233.

Estevao, V. and Sarndal, C.E. 2004. Borrowing strength is not the best technique within a wide class of design-consistent domain estimators. *Journal of Official Statistics*, 20: 645–660.

Estevao, V.M. and Sarndal, C.E. 2006. Survey estimates by calibration on complex auxiliary information. *International Statistical Review* 74(1): 127–147.

EURAREA. 2005. *Enhancing small area estimation techniques to meet European needs.* http://www.ine.es/en/docutrab/eurarea/eurarea_05_en.doc. Accessed on July 7, 2014.

EURAREA Consortium. 2004a. Enhancing small area estimation techniques to meet European needs, http://www.statistics.gov.uk/eurarea/download.asp. Accessed on December 11, 2007.

EURAREA Consortium. 2004b. EURAREA project reference volume. Office for National Statistics, Newport, U.K.

Evans, S.P. and Kirby, H.R. 1974. A three dimensional farness procedure for calibrating gravity models. *Transportation Research*, 8(1): 105–122.

Fabrizi, E., Ferrante, M.R., and Pacei, S. 2008. Estimation of poverty rates for the Italian population classified by household type and administrative region. In paper presented at the *30th General Conference of the International Association for Research in Income and Wealth*, Portoroz, Slovenia, August 24, 2008.

FaHCSIA. 2008. Making housing affordable again, http://www.facs.gov.au/internet/facsinternet.nsf/vIA/housing_affordable/ $File/making_housing_affordable_again.pdf. Accessed on July 26, 2009.

Farrell, N., Morrissey, K., and O'Donoghue, C. 2013. Creating a spatial microsimulation model of the Irish Local Economy. In R. Tanton and K. Edwards (eds.), *Spatial Microsimulation: A Reference Guide for Users. Understanding Population Trends and Processes*, vol. 6. Springer, London, U.K., pp. 105–125.

Fay, R.E. and Herriot, R.A. 1979. Estimation of income from small places: An application of James-Stein procedures to census data. *Journal of the American Statistical Association*, 74(2): 269–277.

Federal Committee on Statistical Methodology. 1993. Indirect estimators in federal programs. Statistical Policy Working Paper 21. U.S. Office of Management and Budget, Washington, DC.

Felsenstein, D., Ashbel, E., and Ben-Nun, A. 2007. Microsimulation of metropolitan employment deconcentration: Application of the UrbanSim model in the Tel Aviv region. In E. Koomen, J. Stillwell, A. Bakema, and H.J. Scholten (eds.), *Modelling Land-Use Change Progress and Applications*. Springer (The GeoJournal Library), Dordrecht, the Netherlands.

Ferreira, F.H.G., Lanjouw, P., and Neri, M. 2003. A robust poverty profile for Brazil using multiple data sources. *Revista Brasileira de Economia*, 57(1): 27–41.

Field, A. 2000. *Discovering Statistics Using SPSS for Windows*. SAGE Publication Ltd., London, U.K.

Fienberg, S.E. 1970. An iterative procedure for estimation in contingency tables. *The Annals of Mathematical Statistics*, 41(7): 907–917.

Fines, P. 2006. Modelling life expectancy at birth in small cities in Canada. In *Proceedings of Statistics Canada's Symposium 2006: Methodological Issues in Measuring Population Health*, Statistics Canada, Ottawa, Ontario, Canada.

Flood, L. 2007. Can we afford the future? An evaluation of the new Swedish pension system. In A. Harding and A. Gupta (eds.), *Modelling Our Future: Population Ageing Social Security and Taxation*. Elsevier, Amsterdam, the Netherlands.

Foard, G., Karmel, R., Collett, S., Bosworth, E., and Hulmes, D. 1994. *Public Housing in Australia*. Australian Institute for Health and Welfare, Canberra, Australian Capital Territory, Australia.

Franklin, J.P. 2005. Nonparametric distributional analysis of a transportation policy: Stockholm's congestion pricing trial. In paper presented at the *Transportation Research Board 84th Annual Meeting*, Washington, DC.

Fuller, W.A. 2009. *Sampling Statistics*. Wiley, New York.

Ghosh, M., Kim, D., Sinha, K., Maiti, T., Katzoff, M., and Parsons, V.L. 2009. Hierarchical and Empirical Bayes small domain estimation of the proportion of persons without health insurance or minority subpopulations. *Survey Methodology*, 35: 53–66.

Ghosh, M. and Maiti, T. 2004. Small area estimation based on natural exponential family quadratic variance function models and survey weights. *Biometrika*, 91: 95–112.

Ghosh, M., Natarajan, K., Stroud, T.W.F., and Carlin, B.P. 1998. Generalized linear models for small area estimation. *Journal of the American Statistical Association*, 93: 273–282.

Ghosh, M. and Rao, J.N.K. 1994. Small area estimation: An appraisal. *Statistical Science*, 9(1): 55–93.

Gilbert, N. and Troitzsch, K.G. 2005. *Simulation for the Social Scientist*. Open University Press, Milton Keynes, U.K.

Goerndt, M.E., Monleon, V.J., and Temesgen, H. 2011. A comparison of small-area estimation techniques to estimate selected stand attributes using LiDAR-derived auxiliary variables. *Canadian Journal of Forest Research*, (6): 1189–1201.

Gomez-Rubio, V., Best, N., and Richardson, S. 2008. A comparison of different methods for small area estimation. NCRM Working Paper. ESRC National Centre for Research Methods, Southampton, U.K., pp. 1–26.

Gonzalez, J.F., Placek, P.J., and Scott, C. 1966. Synthetic estimation of follow back survey at the national centre for health statistics. In W.L. Schaible (ed.), *Indirect Estimators in U.S. Federal Programs*, vol. 16. Springer-Verlag, New York.

Gonzalez, M.E. 1973. Use and evaluation of synthetic estimates. In *Proceedings of the Social Statistics Section*, American Statistical Association, Alexandria, VA.

Gouweleeuw, J. and Hartgers, M. 2004. The method of repeated weighting in the 2001 Census. In E. Schulte Nordholt, M. Hartgers, and R. Gircour (eds.), *The Dutch Virtual Census of 2001. Analysis and Methodology*. Statistics Netherlands, Voorburg/Heerlen, the Netherlands, pp. 261–276.

Grab, J. and Grimm, M. 2008. Spatial inequalities explained—Evidence from Burkina Faso. Ibero America Institute for Econ, Research (IAI) Discussion Papers 173. Ibero-America Institute for Economic Research, Berlin, Germany.

Grattan, M. 2008. Rudd determined to act on housing affordability. *The Age*, http://www.theage.com.au/news/national/pm-acts-on-affordability/2008/03/02/1204402272856.html. Accessed on March 3, 2008.

Gupta, A. and Harding, A. 2007. Introduction and overview. In A. Harding and A. Gupta, (eds.), *Modelling Our Future: Population Ageing, Health and Aged Care. International Symposia in Economic Theory and Econometrics*. Emerald Publishing, Bingley, U.K., pp. 1–40.

Gupta, S.C. and Kapoor, V.K. 2008. *Fundamentals of Mathematical Statistics*. Sultan Chand & Sons, New Delhi, India.

Guttman, I. and Hougaard, P. 1985. Studentization and prediction problems in multivariate multiple regression. *Communications in Statistics: Theory and Methods*, 14(8): 1251–1258.

Halpin, B. 1999. Simulations in sociology. *The American Behavioral Scientist*, 42(10): 1488–1508.

Hanaoka, K. and Clarke, G.P. 2007. Spatial microsimulation modelling for retail market analysis at the small-area level. *Computers, Environment and Urban Systems*, 31(1): 162–187.

Harding, A. 1993. *Lifetime Income Distribution and Redistribution: Applications of a Microsimulation Model*. North-Holland, Amsterdam, the Netherlands.

Harding, A. (ed.). 1996. *Microsimulation and Public Policy*. Contributions to Economic Analysis. North-Holland, Amsterdam, the Netherlands.

Harding, A. 2000. Dynamic microsimulation: Recent trends and future prospects. In A. Gupta and V. Kapur (eds.), *Microsimulation in Government Policy and Forecasting*. North-Holland, Amsterdam, the Netherlands.

Harding, A. 2007. *APPSIM: The Australian Dynamic Population and Policy Microsimulation Model.* In paper presented at *First General Conference of the International Microsimulation Association,* Vienna, Austria August 20–22, 2007.

Harding, A. and Gupta, A. (eds.). 2007. Modelling our future: Population aging, social security and taxation. In *International Symposia in Economic Theory and Econometrics* (ISETE) Series. Elsevier, Amsterdam, the Netherlands.

Harding, A., Keegan, M., and Kelly, S. 2010. Validating a dynamic population microsimulation model: Recent experience in Australia. *International Journal of Microsimulation (Special edition on Methodology),* 3(2): 46–64.

Harding, A., Kelly, S., Percival, R., and Keegan, M. 2009a. Population ageing and government age pension outlays: Using microsimulation models to inform policy making. Kansai Institute for Social and Economic Research, Osaka, Japan.

Harding, A., Lloyd, R., Bill, A., and King, A. 2003. Assessing poverty and inequality at a detailed regional level: New advances in spatial microsimulation. In M. McGillivray and M. Clarke (eds.), *Understanding Human Well-Being,* vol. 1. United Nation University Press, Helsinki, Finland, pp. 239–261.

Harding, A., Lloyd, R., Bill, A., and King, A. 2006. Assessing poverty and inequality at a detailed regional level: New advances in spatial microsimulation. In M. McGillivray and M. Clarke (eds.), *Understanding Human Well-Being.* UN University Press, Tokyo, Japan.

Harding, A., Phillips, B., and Kelly, S. 2004. Trends in housing stress. In paper presented at the *National Summit on Housing Affordability,* Canberra, Australian Capital Territory, Australia, June 28, 2004.

Harding, A., Vu, Q.N., Tanton, R., and Vidyattama, Y. 2009b. Improving work incentives and incomes for parents: The national and geographic impact of liberalising the family tax benefit income test. *The Economic Record,* 85(1): 48–58.

Harland, K., Heppenstall, A., Smith, D., and Birkin, M. 2012. Creating realistic synthetic populations at varying spatial scales: A comparative critique of population synthesis techniques. *Journal of Artificial Societies and Social Simulation* 15(1): 1, http://jasss.soc.surrey.ac.uk/15/1/1.html. Accessed on 08/02/2016.

Hassan, S., Pavon, J., and Gilbert, N. 2008. Injecting data into simulation: Can agent-based modelling learn from microsimulation? Mimeo. The Centre for Research in Social Simulation Mimeo, New York.

Heady, P., Clarke, P., Brown, G., Ellis, K., Heasman, D., Hennell, S., Longhurst, J., and Mitchell, B. 2003. Model-based small area estimation series no. 2: Small area estimation project report. Office for National Statistics, Newport, U.K.

Heady, P. and Ralphs, M. 2005. EURAREA: An overview of the project and its findings. In *SAE2005 Conference* Finland, August 28–31. http://www.stat.jyu.fi/sae2005/abstracts/heady.pdf. Accessed on March 11, 2008.

Henderson, C.R. 1950. Estimation of genetic parameters (abstract). *Annals of Mathematical Statistics,* 21(3): 309–310.

Hennessy, T., Shrestha, S., and Hynes, S.P. 2007. The effect of decoupling on farming in Ireland: A regional analysis. *Irish Journal of Agricultural & Food Research,* 46(1): 76–87.

Henninger, N. 1998. Mapping and geographic analysis of poverty and human welfare—Review and assessment. Report prepared for the UNEP/CGIAR Initiative on GIS. World Resources Institute, Washington, DC.

Hentschel, J., Lanjouw, J.O., Lanjouw, P., and Poggi, J. 2000. Combining census and survey data to trace spatial dimensions of poverty: A case study of Ecuador. *World Bank Economic Review*, 14(1): 147–165.

Hentschel, J., Olson, L.J., and Lanjouw, P. 1998. Combining census and survey data to study spatial dimensions of poverty. World Bank Policy Research Working Paper Series no. 1928. World Bank, Washington, DC.

Hermes, K. and Poulsen, M. 2012. A review of current methods to generate synthetic spatial microdata using reweighting and future directions. *Computers, Environment and Urban Systems*, 36(4): 281–290.

Holm, E., Holme, K., Lindgren, U., and Makila, K. 2002. The SVERIGE spatial microsimulation model. Model Report. Department of Social and Economic Geography, Umea University, Umea, Sweden.

Holt, D. and Smith, T.M.F. 1979. Post stratification. *Journal of the Royal Statistical Society: Series A*, 142(1): 33–46.

Hooimeijer, P. 1996. A life course approach to urban dynamics: State of the art in and research. In G.P. Clarke (ed.), *Microsimulation for Urban and regional Policy Analysis*. Pion Limited, London, U.K., pp. 28–64.

Hooimeijer, P. and Oskamp, A. 2000. Locsim: Microsimulation of households and housing market. In paper presented at the *10th Biennial Conference of the Australian Population Association*, Melbourne, Victoria, Australia, November 28–December 1, 2000.

Hoshen, M.B., Burton, A.H., and Bowcock, T.J.V. 2007. Simulating disease transmission dynamics at a multi-scale level. *International Journal of Microsimulation*, 1(1): 26–34.

Houbiers, M. 2004. Towards a social statistical database and unified estimates at Statistics Netherlands. *Journal of Official Statistics*, 20: 55–75.

Houbiers, M., Knottnerus, P., Kroese, A.H., Renssen, R.H., and Snijders, V. 2003. Estimating consistent table sets: Position paper on repeated weighting. Discussion paper 03005. Statistics Netherlands, Voorburg/Heerlen, the Netherlands.

Huang, Z. and Williamson, P. 2001. A comparison of synthetic reconstruction and combinatorial optimisation approaches to the creation of small-area microdata. Working Paper 2001/2. Population Microdata Unit, Department of Geography, University of Liverpool, Liverpool, U.K.

Hughes, A., Sathe, N., and Spagnola, K. 2008. State estimates of substance use from the 2005–2006 national surveys on drug use and health. Substance Abuse and Mental Health Services Administration, Office of Applied Studies, Rockville, MD.

Hynes, S., Farrelly, N., Murphy, E. and O'Donoghue, C. 2008. Modelling habitat conservation and participation in agri-environmental schemes: A spatial microsimulation approach. *Ecological Economics*, 66(3): 258–269.

Hynes, S., Morrissey, K., and O'Donoghue, C. 2006. Building a static farm level spatial microsimulation model: Statistically matching the Irish national farm survey to the Irish census of agriculture. In paper presented at the *46th Congress of European Regional Science Association*, Volos, Greece, August 30–September 3, 2006.

Hynes, S., Morrissey, K., O'Donoghue, C., and Clarke, G. 2009. Building a static farm level spatial microsimulation model for rural development and agricultural policy analysis in Ireland. *International Journal of Agricultural Resources, Governance and Ecology*, 8(3): 282–299.

IMA (International Microsimulation Association). 2007. *New frontiers in microsimulation modelling: Introduction.* http://www.euro.centre.org/ima2007/papers.html. Accessed on April 9, 2008.

James, W. and Stein, C. 1961. Estimation with quadratic loss. In *Proceedings of the Fourth Berkeley Symposium on Mathematical Statistics and Probability,* University of California Press, Berkeley, CA.

Jeffreys, H. 1961. *Theory of Probability.* Oxford University Press, Oxford, U.K.

Jiang, J. 2007. *Linear and Generalized Linear Mixed Models and Their Applications.* Springer, New York.

Jiang, J. and Lahiri, P. 2006. Mixed model prediction and small area estimation. *Test,* 15: 1–96.

Joshi, H., Guhathakurta, S., Konjevod, G., Crittenden, J. and Li, K. (2006). Simulating the effect of light rail on urban growth in phoenix: An application of the urbansim modeling environment, *Journal of Urban Technology,* 13(2): 91–111.

Kavroudakis, D., Ballas, D., and Birkin, M. 2009. Use of spatial microsimulation for public policy: The case of education. In paper presented to the *Second General Conference of the International Microsimulation Association,* Ottawa, Ontario, Canada.

Keegan, M. 2011. Mandatory superannuation and self-sufficiency in retirement: An application of the APPSIM dynamic microsimulation model. *Social Science Computer Review,* 29(1): 67–84.

Kelly, D. 2004. SMILE static simulator software user manual teagasc. Rural Economy Research Centre (RERC), Teagasc Athenry Publication, Galway, Republic of Ireland.

Kelly, S., Phillips, B., and Taylor, E. 2006. Baseline small area projections of the demand for housing assistance. AHURI Final Report no. 92. Australian Housing and Urban Research Institute, Melbourne, Victoria, Australia.

Kerani, R.P., Handcock, M.S., Handsfield, H.H., and Holmes K, K. 2005. Comparative geographic concentrations of four sexually transmitted infections. *American Journal of Public Health,* 95(2): 324–330.

Kim, J.K. and Park, M. 2010. Calibration estimation in survey sampling. *International Statistical Review,* 78(1): 21–39.

King, A. 1994. *Towards Indicators of Housing Stress.* Monograph Series no. 2. Department of Housing and Regional Development, Canberra, Australian Capital Territory, Australia.

King, A. 2007. Providing income support services to a changing aged population in Australia: Centrelink's regional microsimulation model. In A. Gupta and A. Harding (eds.), *Modelling Our Future: Population Ageing, Health and Aged Care.* North-Holland, Amsterdam, the Netherlands.

King, A., Mclellan, J., and Lloyd, R. 2002. Regional microsimulation for improved service delivery in Australia: Centrelink's CUSP model. In paper prepared for the *27th General Conference of The International Association for Research in Income and Wealth,* Stockholm, Sweden, August 18–24, 2002.

Kirkpatrick, S., Gelatt Jr., C.D., and Vecchi, M.P. 1983. Optimisation by simulated annealing. *Science,* 220(4): 671–680.

Kitamura, R., Pendyala, R.M., Pas, E.I., and Reddy, P. 1995. Application of AMOS, an activity-based TCM evaluation tool, to the Washington, DC, Metropolitan area. In *23rd European Transport Forum: Proceedings of Seminar E Transportation Planning Methods,* PTRC Education and Research Services, Ltd., London, U.K., pp. 177–190.

Knottnerus, P. and Van Duin, C. 2006. Variances in repeated weighting with an application to the Dutch Labour Force Survey. *Journal of Official Statistics*, 22(3): 565–584.

Kongmuang, C., Clarke, G.P., Evans, A.J., and Jin, J. 2006. SimCrime: A spatial micro-simulation model for the analysing of crime in Leeds. Working Paper 06/1. The School of Geography, University of Leeds, Leeds, U.K.

Kordos, J. 2005. Impact of the EURAREA project on research in small area estimation in Poland, http://www.stat.jyu.fi/sae2005/abstracts/kordos.pdf. Accessed on March 21, 2008.

Kott, P.S. 2009. Calibration weighting: Combining probability samples and linear prediction models. In D. Pfeffermann and C.R. Rao (eds.), *Sample Surveys: Inference and Analysis. Handbook of Statistics* 29B. North-Holland, Amsterdam, the Netherlands, pp. 55–82.

Kroese, A.H. and Renssen, R.H. 1999. Weighting and imputation at Statistics Netherlands. In *Proceedings of the IASS Satellite Conference on Small Area Estimation*, Riga, Latvia, pp. 109–120.

Kroese, A.H. and Renssen, R.H. 2000. New applications of old weighting techniques; constructing a consistent set of estimates based on data from different surveys. In *Proceedings of the ICES II*, American Statistical Association, Buffalo, NY, pp. 831–840.

Kruseman, G., Blokland, P.W., Bouma, F., Luesink, H., Mokveld, L., and Vrolijk, H. 2008. Micro-simulation as a tool to assess policy concerning non-point source pollution: The case of ammonia in Dutch agriculture. In paper prepared for presentation at the *107th EAAE Seminar "Modelling of Agricultural and Rural Development Policies"*, Sevilla, Spain, January 29–February 1, 2008.

Laird, J., Druitt, S., and Fraser, D. 1999. Edinburgh city centre: A microsimulation case-study. *Traffic Engineering and Control*, 40(2): 72–76.

Landis, J. and Zhang, M.M. 1998. The second generation of the California urban futures model. Part 2: Specification and calibration results of the land use change module. *Environment and Planning B: Planning and Design*, 25(3): 657–666.

Landis, J.D. 1994. The California urban futures model: A new generation of metro-politan simulation models. *Environment and Planning B*, 21(2): 399–420.

Landt, J and Bray, R. 1997. Alternative approaches to measuring rental housing afford-ability in Australia. Online Discussion Paper—DP16. NATSEM, University of Canberra, Canberra, Australian Capital Territory, Australia.

Lanjouw, J.O., Hentschel, J., Lanjouw, P., and Poggi, J. 1999. Combining census and survey data to study spatial dimensions of poverty. Working Paper Series—Policy Research Working Papers. The World Bank Group, Washington, DC, pp. 1–35.

Lehtonen, R., Sarndal, C.E., and Veijanen, A. 2003. The effect of model choice in estimation for domains, including small domains. *Survey Methodology*, 29(1): 33–44.

Lehtonen, R., Särndal, C.E., and Veijanen, A. 2005. Does the model matter? Comparing model-assisted and model-dependent estimators of class frequencies for domains. *Statistics in Transition*, 7: 649–673.

Lehtonen, R. and Veijanen, A. (1999) Domain estimation with logistic generalized regression and related estimators. In *Proceedings of the IASS Satellite Conference on Small Area Estimation*, Riga, Latvia, pp. 121–128.

Lehtonen, R. and Veijanen, A. 2009. Design-based methods of estimation for domains and small areas. In D. Pfeffermann and C.R. Rao (eds.), *Sample Surveys: Inference and Analysis. Handbook of Statistics*, vol. 29B. North-Holland, Amsterdam, the Netherlands, pp. 219–249.

Lehtonen, R. and Veijanen, A. 2012. Small area poverty estimation by model calibration. *Journal of the Indian Society of Agricultural Statistics (special issue on Small Area Estimation)*, 66: 125–133.

Lehtonen, R. and Veijanen, A. 2014. Small area estimation of poverty rate by model calibration and "hybrid" calibration. In *NORDSTAT 2014 Conference*, Turku, Finland, June 2014.

Lehtonen, R. and Veijanen, A. 2015a. Design-based methods to small area estimation and calibration approach. In M. Pratesi (ed.), *Analysis of Poverty Data by Small Area Estimation*. Wiley, Chichester, England.

Lehtonen, R. and Veijanen, A. 2015b. Estimation of poverty rate and quintile share ratio for domains and small areas. In G. Alleva and A. Giommi (eds.), *Topics in Theoretical and Applied Statistics*. Springer, New York.

Lehtonen, R., Veijanen, A., Myrskylä, M., and Valaste, M. 2011. Small area estimation of indicators on poverty and social exclusion. Advanced Methodology for European Laeken Indicators (AMELI) Research Project Report WP2 (D2.2, FP7-SSH-2007-217322 AMELI). University of Trier, Trier, Germany, https://www.uni-trier.de/index.php?id=24676&L=2. Accessed on 06/02/2016.

Levy, P.S. 1979. Small area estimation—Synthetic and other procedures, 1968–1978. In J. Steinberg (ed.), *Synthetic Estimates for Small Areas: Statistical Workshop Papers and Discussion*. National Institute on Drug Abuse, Washington, DC.

Lindgren, U. 1999. Simulating the long-term labour market effects of an industrial investment: A microsimulation approach. *Erdkunde*, 53(2): 150–162.

Lindgren, U. and Elmquist, H. 2005. Environmental and economic impacts of decision-making at an arable farm: An integrative modeling approach. *Ambio: A Journal of the Human Environment*, 34(4): 35–46.

Lindgren, U., Strömgren, M., Holm, E., and Häggström, L.E. 2007. Analyzing socioeconomic impacts of large investments by spatial microsimulation. In paper presented at *First General Conference of the International Microsimulation Association: Celebrating 50 Years of Microsimulation* Vienna, Austria, August 20–22, 2007.

Little, R. 2007. An objective Bayesian view of survey weights. In *O'Bayes 07*, http://3w.eco.uniroma1.it/OB07/papers/little.ppt. Accessed on June 27, 2008.

Liu, B., Lahiri, P., and Kalton, G. 2007. Bayes modelling of survey-weighted small area proportions. In *Proceedings of the Section on Survey Research Methods American Statistical Association*, Salt Lake City, Utah, July 29–August 2, pp. 3181–3186.

Liu, T.P. and Kovacevic, M.S. 1997. An empirical study on categorically constrained matching. In *Proceedings of the Survey Methods Section*, Statistical Society of Canada, Ottawa, Ontario, Canada.

Lloyd, R. 2007. STINMOD: Use of a static microsimulation model in the policy process in Australia. In A. Harding and A. Gupta (eds.), *Modelling Our Future: Population Ageing Social Security and Taxation*. Elsevier, Amsterdam, the Netherlands.

Lloyd, R., Harding, A., and Hellwig, O. 2000. Regional divide a study of incomes in regional Australia. *Australasian Journal of Regional Studies*, 6(3): 271–283.

Lo, A.Y. 1986. Bayesian statistical inference for sampling a finite population. *The Annals of Statistics*, 14(3): 1226–1233.

Lohr, S.L. and Prasad, N.G.N. 2003. Small area estimation with auxiliary survey data. *Canadian Journal of Statistics*, 31(4): 383–396.

Longford, N.T. 2004. Missing data and small area estimation in the UK labour force survey. *Journal of the Royal Statistical Society: Series A*, 167(2): 341–373.

Longford, N.T. 2005. *Missing Data and Small-Area Estimation: Modern Analytical Equipment for the Survey Statistician*. Springer, Amsterdam, the Netherlands, 360pp.

Lorenz, M.O. 1905. Methods for measuring the concentration of wealth. *American Statistical Association*, 9(2): 209–219.

Lovelace, R. 2014. Introducing spatial microsimulation with R: A practical. National Centre for Research Methods NCRM Working Paper 08. University of Leeds, Leeds, U.K.

Lovelace, R. and Ballas, D. 2013. "Truncate, replicate, sample": A method for creating integer weights for spatial microsimulation. *Computers, Environment and Urban Systems*, 41: 1–11.

Lovelace, R., Birkin, M., Ballas, B., and van Leeuwen, E. 2015. Evaluating the performance of iterative proportional fitting for spatial microsimulation: New tests for an established technique. *Journal of Artificial Societies and Social Simulation*, 18(2): 21, http://jasss.soc.surrey.ac.uk/18/2/21.html. Accessed on 10/02/2016.

Lu, L. and Larsen, M. 2007. Small area estimation in a survey of high school students in Iowa. In *Proceedings of the American Statistical Association Section on Survey Research Methods*, Salt Lake City, Utah, July 29–August 2, pp. 2627–2634.

Lundevaller, E.H., Holm, E., Stromgren, M., and Lindgren, U. 2007. Spatial dynamic micro-simulation of demographic development. In paper presented at *First General Conference of the International Microsimulation Association: Celebrating 50 Years of Microsimulation*, Vienna, Austria, August 20–22.

Lundgren, A. 2004. Micro-simulation modelling of domestic tourism travel patterns in Sweden. In paper presented to *Seventh International Forum on Tourism Statistics*, Stockholm, Sweden, June 9–11, 2004.

Lymer, S. and Brown, L. 2012. Developing a dynamic microsimulation model of the Australian health systems: A means to explore impacts of obesity over the next 50 years. *Epidemiology Research International*, 2012(1): 1–13.

Lymer, S., Brown, L., Harding, A., and Yap, M. 2009. Predicting the need for aged care services at the small area level: The CAREMOD spatial microsimulation model. *International Journal of Microsimulation*, 2(2): 27–42.

Lymer, S., Brown, L., Yap, M., and Harding, A. 2008. 2001 regional disability estimates for New South Wales, Australia, using spatial microsimulation. *Applied Spatial Analysis*, 1(1): 99–116.

Magnus, J. and Neudecker, H. 1988. *Matrix Differential Calculus with Applications in Statistics and Econometrics*. John Wiley & Sons, Inc., Chichester, U.K.

Maiti, T. 2001. Robust generalized linear mixed models for small area estimation. *Journal of Statistical Planning and Inference*, 98: 225–238.

Malec, D. 2005. Small area estimation from the American Community Survey using a hierarchical logistic model of persons and housing units. *Journal of Official Statistics*, 21(3): 411–432.

Malec, D., Davis, W.W., and Cao, X. 1999. Model-based small area estimates of overweight prevalence using sample selection adjustment. *Statistics in Medicine*, 18(2): 189–200.

Mason, R.D. and Lind, D.A. 1996. *Statistical Techniques in Business and Economics.* Times Mirror Higher Education Group, Inc., Chicago, IL.

McNamara, J., Tanton, R., and Phillips, B. 2006. The regional impact of housing costs and assistance on financial disadvantage: Positioning paper. Australian Housing and Urban Research Institute, Melbourne, Victoria, Australia.

McNelis, S. 2006. Rental systems in Australia and overseas. AHURI Final Report no. 95. Australian Housing and Urban Research Institute, Melbourne, Victoria, Australia.

McRoberts, R.E. 2011. Estimating forest attribute parameters for small areas using nearest neighbors techniques. *Forest Ecology and Management*, 272: 3–12.

Meeden, G. 2003. A noninformative Bayesian approach to small area estimation. *Survey Methodology*, 29(1): 19–24.

Melhuish, T., King, A., and Taylor, E. 2004. The regional impact of Commonwealth Rent Assistance. Report no. 71. Australian Housing and Urban Research Institute, Melbourne, Victoria, Australia.

Merz, J. 1991. Microsimulation—A survey of principles, developments and applications. *International Journal of Forecasting*, 7(1): 77–104.

Metropolis, N., Rosenbluth, A.W., Rosenbluth, M.N., Teller, A.H., and Teller, E. 1953. Equation of state calculations by fast computing machines. *Journal of Chemical Physics*, 21(8): 1087–1092.

Miller, E.J., Hunt, J.D., Abraham, J.E., and Salvini, P.A. 2004. Microsimulating urban systems. *Computers, Environment and Urban Systems*, 28(1): 9–44.

Miller, E.J., and Salvini, P.A. (2001). The Integrated Land Use, Transportation, Environment (ILUTE) microsimulation modelling system: Description and current status. In D. Hensher (Ed.), The leading edge in travel behavior research, selected papers from the *9th International Association for Travel Behaviour Research Conference*, Gold Coast, Queensland, Australia, July 2–5, 2000.

Mitchell, R., Shaw, M., and Dorling, D. 2000. *Inequalities in Life and Death: What If Britain Were More Equal?* Policy Press, London, U.K.

Moeckel, R., Schwarze, B., Spiekermann, K., and Wegener, M. 2007. Simulating interactions between land use, transport and environment. In paper presented at the *11th World Conference on Transport Research*, University of California, Berkeley, CA, June 24–28, 2007.

Mohadjer, L., Rao, J.N.K., and Liu, B. 2007. Hierarchical Bayes small area estimates of adult literacy using unmatched sampling and linking models. In *Proceedings of the Section on Survey Research Methods*, American Statistical Association, Alexandria, VA, pp. 3203–3210.

Molina, I. and Rao, J.N.K. 2010. Small area estimation of poverty indicators. *Canadian Journal of Statistics*, 38(3): 369–385.

Molina, I., Saei, A., and Lombardía, J.M. 2007. Small area estimates of labour force participation under a multinomial logit mixed model. *Journal of the Royal Statistical Society: Series A*, 170(4): 975–1000.

Montanari, G.E. and Ranalli, M.G. 2005. Nonparametric model calibration estimation in survey sampling. *Journal of the American Statistical Association*, 100(9): 1429–1442.

Montanari, G.E., Ranalli, G.M., and Vicarelli, C. 2010. A comparison of small area estimators of counts aligned with direct higher level estimates. In *Scientific Meeting of the Italian Statistical Society* Padua, Italy, June 29–July 1, http://homes.stat.unipd.it/mgri/SIS2010/Program/contributedpaper/678 1393-1-DR.pdf. Accessed on 08/02/2016.

Moriarity, C. and Scheuren, F. 2001. Statistical matching: A paradigm for assessing the uncertainty in the procedure. *Journal of Official Statistics*, 17(3): 407–422.

Moriarity, C. and Scheuren, F. 2003. A note on Rubin's statistical matching using file concatenation with adjusted weights and multiple imputations. *Journal of Business and Educational Studies*, 21(1): 65–73.

Morrissey, K., Clarke, G., Hynes, S., and O'Donoghue, C. 2010. Examining the factors associated with depression at the small area level in Ireland using spatial microsimulation techniques. *Irish Geography*, 43(1): 1–22.

Morrissey, K. and O'Donoghue, C. 2009. SMILE—A spatial microsimulation model for the Irish economy. NATSEM at the University of Canberra, Canberra, Australian Capital Territory, Australia, http://www.canberra.edu.au/centres/natsem/publications. Accessed on 06/08/2010.

Morrissey, K. and O'Donoghue, C. 2011. The spatial distribution of labour force participation and market earnings at the sub-national level in Ireland. *Review of Economic Analysis*, 3(1): 80–101.

Nakaya, T., Fotheringham, A.S., Hanaoka, K., Clarke, G.P., Ballas, D., and Yano, K. 2007. Combining microsimulation and spatial interaction models for retail location analysis. *Journal of Geographical Systems*, 4(2): 345–369.

Nandram, B. and Sayit, H. 2011. A Bayesian analysis of small area probabilities under a constraint. *Survey Methodology*, 37(2): 137–152.

National Center for Health Statistics. 1968. *Synthetic State Estimates of Disability*. P.H.S. Publications, Government Printing Office, Washington, DC.

National Housing Strategy. 1991. *The Affordability of Australian Housing*. Australian Government Publishing Service, Canberra, Australian Capital Territory, Australia.

National Housing Strategy. 1992. *National Housing Strategy: Summary of Papers*. Australian Government Publishing Service, Canberra, Australian Capital Territory, Australia.

Nepal, B., Tanton, R., and Harding, A. 2010. Measuring housing stress: How much do definitions matter? *Urban Policy and Research*, 28(2): 211–224.

Nepal, B., Tanton, R., Harding, A., and McNamara, J. 2008. Measuring housing stress at small area levels: How much do definitions matter? In paper presented at the *Third Australasian Housing Researchers Conference*, Melbourne, Victoria, Australia, June 18–20, 2008.

Nguyen, L.D. and Moran, M.S. 2008. An integrated land use-transport model for the Paris region (SIMAURIF): Ten lessons learned after four years of development. In *Transportation Research Board Annual Meeting 2009*, Paper no. 09-0024 Washington, DC, January 11–15.

Nielsen, O.A. 2002. A stochastic route choice model for car travellers in the Copenhagen region. *Networks and Spatial Economics*, 2(2): 327–346.

Nishuri, H., Lawpoolsri, S., Kittitrakul, C., Leman, M.M., Maha, M.S., and Maungnoicharoen, S. 2004. Health inequalities in Thailand: Geographic distribution of medical supplies in the provinces. *Southeast Asian Journal Tropical Medicine and Public Health*, 35(4): 735–740.

Noble, A., Haslett, S., and Arnold, G. 2002. Small area estimation via generalized linear models. *Journal of Official Statistics*, 18: 45–60.

Norman, P. 1999. Putting iterative proportional fitting on the researcher's desk. Working Paper 99/03. School of Geography, University of Leeds, Leeds, U.K.

O'Donoghue, C., Dillon, E., Green, S., Hennessy, T., Hynes, S., and Morrissey, K. 2012a. Assessing the sustainability of Irish farming across space. In C. O'Donoghue, S. Hynes, K. Morrissey, D. Ballas, and G. Clarke (eds.), *Modelling the Local Economy: A Simulation Approach*. Springer Verlag, London, U.K.

O'Donoghue, C., Hynes, S., Morrissey, K., Ballas, D., and Clarke, G. 2012b. *Spatial Microsimulation for Rural Policy Analysis*. Springer-Verlag, London, U.K.

O'Donoghue, C., Lennon, J., Loughrey, J., and Meredith, D. 2012c. Short and medium-term projections of household income in Ireland using a spatial microsimulation model. Working Paper Series, Rural Economy and Development Programme. Teagasc, Carlow, Ireland.

Opsomer, J.D., Breidt, F.J., Claeskens, G., Kauermann, G., and Ranalli, M.G. 2008a. Nonparametric small area estimation using penalized spline regression. *Journal of the Royal Statistical Society B*, 70(1): 265–286.

Opsomer, J.D., Claeskens, G., Ranalli, M.G., Kauermann, G., and Breidt, F.J. 2008b. Non-parametric small area estimation using penalized spline regression. *Journal of the Royal Statistical Society: Series B*, 70(1): 265–286.

Orcutt, G.H. 1957. A new type of socio-economic system. *Review of Economics and Statistics*, 39(2): 116–123.

Orcutt, G.H. 2007. A new type of socio-economic system. *International Journal of Microsimulation* (Reprinted), 1(1): 3–9.

Orcutt, G.H., Caldwell, S., and Wertheimer, R. 1976. *Policy Exploration through Micro-Analytic Simulation*. The Urban Institute, Washington, DC.

Orcutt, G.H., Greenberg, J.K., and Rivlin, A. 1961. *Microanalysis of Socioeconomic Systems: A Simulation Study*. Harper & Row, New York.

Oskamp, A. 1997. *Local Housing Market Simulation: A Micro Approach*. Thesis Publishers, Amsterdam, the Netherlands, 204pp.

Percival, R., Phillips, B., and King, A. Unaffordable housing in the ACT in 2001: A report for the ACT affordable housing taskforce. NATSEM, University of Canberra, Canberra, Australian Capital Territory, Australia.

Petersa, I., Brassel, K.H., and Sporria, C. 2002. A microsimulation model for assessing urine flows in urban wastewater management. In *iEMSs 2002 Sessions (Part Two) Techniques and Methodologies*, The International Environmental Modelling and Software Society (iEMSs), Switzerland. http://www.iemss.org/iemss2002/proceedings/pdf/volume%20due/25.pdf. Accessed on June 25, 2008.

Pfeffermann, D. 2002. Small area estimation—New developments and directions. *International Statistical Review*, 70(1): 125–143.

Pfeffermann, D. 2013. New important developments in small area estimations. *Statistical Science*, 28(1): 40–68.

Pfeffermann, D. and Burck, L. 1990. Robust small area estimation combining time series and cross-sectional data. *Survey Methodology*, 16: 217–237.

Pfeffermann, D. and Correa, S. 2012. Empirical bootstrap bias correction and estimation of prediction mean square error in small area estimation. *Biometrika*, 99(4): 457–472.

Pfeffermann, D. and Tiller, R. 2006. Small-area estimation with state-space models subject to benchmark constraints. *Journal of American Statistical Association*, 101(8): 1387–1397.

Pfeffermann, D., Terryn, B., and Moura, F.A.S. 2008. Small area estimation under a two-part random effects model with application to estimation of literacy in developing countries. *Survey Methodology*, 34(2): 235–249.

Pham, D.T. and Karaboga, D. 2000. *Intelligent Optimisation Techniques: Genetic Algorithms, Tabu Search, Simulated Annealing and Neural Networks.* Springer, London, U.K.

Phillips, B. and Kelly, S. 2006. Housemod: A regional microsimulation projections model of housing in Australia. In paper presented to *Australian Housing Research Conference,* Adelaide, Australia, June 19–21.

Pickett, W., Koushik, A., Faelker, T., and Brown, K.S. 2000. Estimation of youth smoking behaviours in Canada. *Chronic Diseases in Canada,* 21(3): 119–127.

Potter, S.R., Jay, D., Atwood, J.L., and Robert, L.K. 2009. A national assessment of soil carbon sequestration on cropland: A microsimulation modeling approach. In *Soil Carbon Sequestration and the Greenhouse Effect,* 2nd edn. SSSA Special Publication 57, Madison, WI.

Press, J. 1986. *Applied Multivariate Analysis.* Holt, Rinehart & Winston, Inc., New York.

Pritchard, D.R. and Miller, E.J. 2012. Advances in population synthesis: Fitting many attributes per agent and fitting to household and person margins simultaneously. *Transportation,* 39(3): 685–704.

Pudney, S. and Sutherland, H. 1994. How reliable are microsimulation results? An analysis of the role of sampling error in a U.K. tax-benefit model. *Journal of Public Economics,* 53(3): 327–365.

Rahman, A. 2007. Prediction distribution for linear regression model with multivariate Student-t errors under the Bayesian approach. In *Proceedings of the Third International Conference on Research and Education in Mathematics (ICREM3),* Kuala Lumpur, Malaysia, vol. 1, pp. 188–193, April 10–12.

Rahman, A. 2008a. A review of small area estimation problems and methodological developments. Online Discussion Paper—DP66. NATSEM, University of Canberra, Canberra, Australian Capital Territory, Australia.

Rahman, A. 2008b. The possibility of using Bayesian prediction theory in small area estimation. In presentation to at the *ANZRSAI 32nd Annual Conference/ ARCRNSISS National Conference,* Adelaide, South Australia, Australia, November 30–December 3, 2008.

Rahman, A. 2008c. *Bayesian Predictive Inference for Some Linear Models under Student-t Errors.* VDM Verlag, Saarbrucken, Germany.

Rahman, A. 2009a. Small area estimation through spatial microsimulation models: Some methodological issues. In paper presented at the *Second General Conference of the International Microsimulation Association,* Ottawa, Ontario, Canada, October 6, 2009.

Rahman, A. 2009b. Objective Bayesian prediction for the matrix-T error regression model. In paper presented at the *2009 International Workshop on Objective Bayes Methodology (O-Bayes09),* Philadelphia, PA, May 6, 2009.

Rahman, A. (2010). Small area housing stress estimation in Australia: Spatial microsimulation modelling and statistical reliability, In paper presented at the *Australian Housing and Urban Research Institute Postgraduate Scholar's Symposium,* University of Auckland, New Zealand, November 14–17.

Rahman, A. (2011) *Small area housing stress estimation in Australia: Microsimulation modelling and statistical reliability,* PhD thesis, University of Canberra, Canberra.

Rahman, A. (2014). Small area estimation and micro simulation modelling for social policy analysis, In paper presented at the *Applied Statistics and Public Policy Analysis Conference 2014 (ASPPAC2014),* Wagga Wagga, Australia, December 11–12.

Rahman, A. (2015). Calculating confidence intervals for a spatial microsimulation model using a Z-statistic error, In paper presented at the 5th *World Congress of the International Microsimulation Association (IMA)*, Luxembourg, September 2–4.

Rahman, A. (n.d.), Small area housing stress estimation in Australia: Calculating confidence intervals for a spatial microsimulation model. *Communications in Statistics—Simulation and Computation*, (forthcoming).

Rahman, A. and Harding, A. 2012. A new analysis of the characteristics of households in housing stress: Results and tools for validation. In paper presented at the *Sixth Australasian Housing Researchers' Conference 2012 (AHRC12)*, The University of Adelaide, Adelaide, South Australia, Australia, February 8–10, 2012, pp. 1–23.

Rahman, A. and Harding, A. 2014. Spatial analysis of housing stress estimation in Australia with statistical validation. *Australasian Journal of Regional Studies*, 20(3): 452–486.

Rahman, A. and Upadhyay, S. 2015. A Bayesian reweighting technique for small area estimation. In S.K. Upadhyay, U. Singh, D.K. Dey, and A. Loganathan (eds.), *Current Trends in Bayesian Methodology with Applications*. Statistics Series in Statistical Theory and Methods. Chapman & Hall/CRC, London, U.K., pp. 503–519.

Rahman, A., Harding, A., Tanton, R., and Liu, S. 2010a. Methodological issues in spatial microsimulation modelling for small area estimation. *The International Journal of Microsimulation*, 3(2): 3–22.

Rahman, A., Harding, A., Tanton, R., and Liu, S. 2010b. Simulating the characteristics of populations at the small-area level: New validation techniques for a spatial microsimulation model in Australia. In *JSM Proceedings, Social Statistics Section*, American Statistical Association, Alexandria, VA, pp. 2022–2036, July 31–August 5, Vancouver.

Rahman, A., Harding, A., Tanton, R. and Liu, S. (2013). Simulating the characteristics of populations at the small area level: New validation techniques for a spatial microsimulation model in Australia, *Computational Statistics and Data Analysis*, 57(1): 149–165.

Rahman, A., Gao, J., D'Este, C. and Ahmed, S.E. (2016). An assessment of the effects of prior distributions on the Bayesian predictive inference. *The International Journal of Statistics and Probability*, 5(5), (in press).

Raney, B., Cetin, N., Voellmy, A., Vrtic, M. Axhausen, K., and Nagel, K. 2003. An agent-based microsimulation model of Swiss travel: First results. *Networks and Spatial Economics*, 3: 23–41.

Rao, J.N.K. 1999. Some current trends in sample survey theory and methods (with discussion). *Sankhya: The Indian Journal of Statistics: Series B*, 61(1): 1–57.

Rao, J.N.K. 2002. Small area estimation: Update with appraisal. In N. Balakrishnan (ed.), *Advances on Methodological and Applied Aspects of Probability and Statistics*, vol. 1. Taylor & Francis, New York, pp. 113–139.

Rao, J.N.K. 2003a. *Small Area Estimation*. John Wiley & Sons, Inc., Hoboken, NJ.

Rao, J.N.K. 2003b. Some new developments in small area estimation. *JIRSS*, 2(2): 145–169.

Rao, J.N.K. 2005. Inferential issues in small area estimation: Some new developments. *Statistics in Transition*, 7(4): 513–526.

Rao, J.N.K. 2008. Some methods for small area estimation. *Revista Internazionale di Siencze Sociali*, 4(3): 387–406.

Rao, J.N.K. and Molina, I. 2015. *Small Area Estimation*, 2nd edn. Wiley, New York.

Rao, J.N.K., Sinha, S.K., and Roknossadati, M. 2009. Robust small area estimation using penalized spline mixed models. In *Proceedings of the Survey Research Methods Section*, American Statistical Association, Alexandria, VA, pp. 145–153.

Rao, J.N.K. and Yu, M. 1994 Small area estimation by combining time series and cross-sectional data. *Canadian Journal of Statistics*, 22(4): 511–528.

Rassler, S. 2002. *Statistical Matching: A Frequentist Theory, Practical Applications, and Alternative Bayesian Approaches*. Springer Verlag, New York.

Rassler, S. 2004. Data fusion: Identification problems, validity, and multiple imputation. *Austrian Journal of Statistics*, 33(2): 153–171.

Rees, P., Martin, D., and Williamson, P. (eds.) 2002. *The Census Data System*. Wiley, Chichester, England.

Renssen, R.H., Kroese, A.H., and Willeboordse, A.J. 2001. Aligning estimates by repeated weighting. Research paper. Statistics Netherlands, Hague/Heerlen, the Netherlands.

Rephann, T.J. 2001. Economic-demographic effects of immigration: Results from a dynamic, spatial microsimulation model. In paper presented at the *2001 Annual Meeting of the Mid-Atlantic Division of the Association of American Geographers*, Frostburg, MD, October 5, 2001.

Rephann, T.J. 2002. The importance of geographical attributes in the decision to attend college. *Socio-Economic Planning Sciences*, 36(2): 291–307.

Rephann, T.J., Makila, K., and Holm, E. 2005. Microsimulation for local impact analysis: An application to plant shutdown. *Journal of Regional Science*, 45(1): 183–222.

Rephann, T.J. and Ohman, M. 1999. Building a microsimulation model for crime in Sweden: Issues and applications. In paper presented at the *eminarium om Ekobrottsforskning*, Stockholm, Sweden, February 22, 1999.

Robinson, G.K. 1991. That BLUP is a good thing: The estimation of random effects. *Statistical Science*, 6(1): 15–51.

Rodgers, W.L. 1984. An evaluation of statistical matching. *Journal of Business and Economic Statistics*, 2(1): 91–102.

Rubin, D.B. 1987. *Multiple Imputation for Nonresponse in Surveys*. John Wiley & Sons, New York.

Saarloos, D.J.M. 2006. A framework for a multi-agent planning support system. Unpublished PhD thesis, Eindhoven University Press, Eindhoven, the Netherlands.

Saei, A. and Chambers, R. 2003. Small area estimation: A review of methods-based on the application of mixed models. S3RI Methodology Working Paper M03/15. Southampton Statistical Sciences Research Institute, Southampton, U.K.

Salvati, N., Chandra, H., Ranalli, M.G., and Chambers, R. 2010. Small area estimation using a nonparametric model-based direct estimator. *Computational Statistics and Data Analysis*, 54: 2159–2171.

Salvati, N., Tzavidis, N., Pratesi, M., and Chambers, R. 2012. Small area estimation via M-quantile geographically weighted regression. *Test*, 21(1): 1–28.

Sandel, M. and Wright, R.J. 2006. When home is where the stress is: Expanding the dimensions of housing that influence asthma morbidity. *Archives of Disease in Childhood*, 91(11): 942–948.

Santos, G. and Rojey, L. 2004. Distributional impacts of road pricing: The truth behind the myth. *Transportation*, 31(1): 21–42.

Sarndal, C.E. 2007. The calibration approach in survey theory and practice. *Survey Methodology*, 33(2): 99–119.

Sarndal, C.E., Swensson, B., and Wretman, J. 1992. *Model Assisted Survey Sampling*. Springer-Verlag Inc., New York.

Scarborough, P., Allender, S., Rayner, M., and Goldacre, M. 2009. Validation of model-based estimates (synthetic estimates) of the prevalence of risk factors for coronary heart disease for wards in England. *Health & Place*, 15(2): 596–605.

SCRGSP. 2007. Report on government services 2006. Productivity Commission (Steering Committee for the Review of Government Service Provision), Melbourne, Victoria, Australia.

Siegel, C.L. (1935). Ueber die analyticche theorie der quadratischen formen, *Annals of Mathematics*, 36(4): 527–606.

Simpson, L. and Tranmer, M. 2005. Combining sample and census data in small area estimates: Iterative proportional fitting with standard software. *The Professional Geographer*, 57(2): 222–234.

Singh, A.C. and Mohl, C.A. 1996. Understanding calibration estimators in survey sampling. *Survey Methodology*, 22(2): 107–115.

Sinha, S. 2004. Robust analysis of generalized linear mixed models. *Journal of the American Statistical Association*, 99: 451–460.

Sinha, S.K. and Rao, J.N.K. 2009. Robust small area estimation. *Canadian Journal of Statistics*, 37(3): 381–399.

Smith, D.M., Harland, K., and Clarke, G.P. 2007. SimHealth: Estimating small area populations using deterministic spatial microsimulation in Leeds and Bradford. Working Paper 07/06. University of Leeds School of Geography, Leeds, U.K.

Smith, K.S., Nogle, J., and Cody, S. 2002. A regression approach to estimating the average number of persons per household. *Demography*, 39(4): 697–712.

Smith, T.M.F. 1991. Post-stratification. *Journal of the Royal Statistical Society: Series D*, 40(3): 315–323.

Statistics Netherlands. 2000. Special issue on *Integrating Administrative Registers and Household Surveys*. Netherlands Official Statistics 15. Statistics Netherlands, Voorburg/Heerlen, the Netherlands.

Statistics Netherlands. 2014. Dutch 2011 Census: Analysis and methodology. Statistics Netherlands, Hague/Heerlen, the Netherlands, http://www.cbs.nl/NR/rdonlyres/5FDCE1B4-0654-45DA-8D7E-807A0213DE66/0/2014b57pub.pdf. Accessed on 10/02/2016.

Steel, G.D., Tranmer, M., and Holt, M. 2003. Analysis combining survey and geographically aggregated data. In R.L. Chambers and C.J. Skinner (eds.), *Analysis of Survey Data*, vol. 1. John Wiley & Sons Ltd., England, U.K., pp. 323–343.

Stimson, R. and McCrea, R. 2004. A push-pull framework for modelling the relocation of retirees to a retirement village: The Australian experience. *Environment and Planning Analysis*, 36(8): 1451–1470.

Stimson, R., Robson, A., and Shyy, T.K. 2008. Modelling the determinants of spatial differentials in endogenous regional employment growth and decline across regional Australia, 1996–2006. In paper presented at the *ANZRSAI 32nd Annual Conference/ARCRNSISS National Conference*, Adelaide, South Australia, Australia, November 30–December 3, 2008.

Stimson, R., Stough, R., and Roberts, B. 2006. *Regional Economic Analysis and Planning Strategy*. Springer, Heidelberg, Germany.

Strauch, D., Moeckel, R., Wegener, M., Gräfe, J., Muhlhans, H., Rindsfüser, G., and Beckmann, K. 2005. Linking transport and land use planning: The microscopic dynamic simulation model ILUMASS. In P.M. Atkinson, G.M. Foody, S.E. Darby, and F. Wu (eds.), *Geodynamics*. CRC Press, Boca Raton, FL, pp. 295–311.

Svoray, T. and Benenson, I. 2009. Scale and adequacy of environmental microsimulation. *Ecological Complexity*, 6(1): 77–79.

Tanton, R. 2007. SPATIALMSM: The Australian spatial microsimulation model. In *First General Conference of the International Microsimulation Association*, Vienna, Austria, August 20–21, 2007.

Tanton, R. and Edwards, K. 2013. *Spatial Microsimulation: A Reference Guide for Users*. Springer, London, U.K.

Tanton, R., Jones, R., and Lubulwa, G. 2001. Analyses of the 1998 Australian national crime and safety survey. In paper presented at the *2001 Conference of the Character, Impact and Prevention of Crime in Regional Australia*, Townsville, Queensland, Australia, August 2–3, 2001.

Tanton, R., Nepal, B., and Harding, A. 2008. Trends in housing affordability and housing stress, 1995–96 to 2005–06. AMP.NATSEM Income and Wealth Report Issue 19. AMP Financial Services, Sydney, New South Wales, Australia.

Tanton, R. and Vidyattama, Y. 2010. Pushing it to the edge: Extending generalised regression as a spatial microsimulation method. *The International Journal of Microsimulation*, 3(2): 23–33.

Tanton, R., Vidyattama, Y., McNamara, J., Vu, Q., and Harding, A. 2009. Old, single and poor: Using microsimulation and microdata to analyse poverty and the impact of policy change among older Australians. *Economic Papers*, 28(2): 102–120.

Tanton, R., Vidyattama, Y., Nepal, B., and McNamara, J. 2011. Small area estimation using a reweighting algorithm. *Journal of the Royal Statistical Society: Series A*, 174(4): 931–951.

Tanton, R., Williamson, P., and Harding, A. 2007. Comparing two methods of reweighting a survey file to small area data—Generalised regression and combinatorial optimisation. In *The First General Conference of the International Microsimulation Association*, Vienna, Austria, August 20–22, 2007.

Taylor, E., Harding, A., Lloyd, R., and Blake, M. 2004. Housing unaffordability at the statistical local area level: New estimates using spatial microsimulation. *Australian Journal of Regional Studies*, 10(3): 279–300.

Teh, Y.W. and Welling, M. 2003. On improving the efficiency of the iterative proportional fitting procedure. In *Proceedings of the International Workshop on Artificial Intelligence and Statistics*, FL, vol. 9, January 3–6.

Tiao, D.C. and Zellner, A. 1964. On the Bayesian estimation of multivariate regression. *Journal of Royal Statistical Society: Part B*, 26(2): 277–285.

Tiglao, N.C. 2002. Small area estimation and spatial microsimulation of household characteristics in developing countries with a focus on informal settlements in metro manila. Unpublished PhD thesis, University of Tokyo, Tokyo, Japan.

Torabi, M. and Rao, J.N.K. 2008. Small area estimation under a two-level model. *Survey Methodology*, 34(1): 11–17.

Torabi, M. and Shokoohi, F. 2015. Non-parametric generalized linear mixed models in small area estimation. *The Canadian Journal of Statistics*, 43: 82–96.

Torelli, N. and Trevisani, M. 2008. Labour force estimates for small geographical domains in Italy: Problems data and models. Working Paper no. 118, Dipartimento di Scienze Economiche e Statistiche, Universita' di Trieste. http://citeseerx.ist.psu.edu/viewdoc/download;jsessionid=6119E69634C6A2E58471C 3837C0790A0?doi=10.1.1.489.8349&rep=rep1&type=pdf. Accessed on September 13, 2010.

Tranmer, M., Pickles, A., Fieldhouse, E., Elliot, M., Dale, A., Brown, M., Martin, D., Steel, D., and Gardiner, C. 2001. Microdata for small areas. The Cathie Marsh Centre for Census and Survey Research (CCSR). University of Manchester, Manchester, U.K.

Tranmer, M., Pickles, A., Fieldhouse, E., Elliot, M., Dale, A., Brown, M., Martin, D., Steel, D., and Gardiner, C. 2005. The case for small area microdata. *Journal of the Royal Statistical Society: Series A*, 168(1): 29–49.

Tsutsumi, J. and O'Connor, K. 2005. International students and the changing character of the inner area of a city: A case study of Melbourne. In paper presented at the *State of Australian Cities: National Conference*, Brisbane, Queensland, Australia, November 30–December 2, 2005.

Tzavidis, N., Marchetti, S., and Chambers, R. 2010. Robust estimation of small-area means and quantiles. *Australian and New Zealand Journal of Statistics*, 52(1): 167–186.

Tzavidis, N., Ranalli, M.G., Salvati, N., Dreassi, E., and Chambers, R. 2015. Robust small area prediction for counts. *Statistical Methods in Medical Research*, 24(3): 373–395.

Ugarte, M.D., Militino, A.F., and Goicoa, T. 2009. Benchmarked estimates in small areas using linear mixed models with restrictions. *Test*, 18(2): 342–364.

USDA. 2007. Statistical highlights of U.S. agriculture 2006 and 2007. National Agricultural Statistics Service (NASS), Washington, DC.

Van, I.E. and Post, W. 1998. Microsimulation methods for population projection. *Population*, 10(1): 97–136.

van Laarhoven, P.J. and Aarts, E.H. 1987. *Simulated Annealing: Theory and Applications*. Springer, New York.

van Leeuwen, E. and Nijkamp, P. 2009. *A Micro-simulation Model for E-Services in Cultural Heritage Tourism*. VU University Mimeo, Amsterdam, the Netherlands.

van Leeuwen, E., Clarke, G., and Rietveld, P. 2009. Microsimulation as a tool in spatial decision making: Simulation of retail developments in a Dutch town. In A. Zaidi, A. Harding, and P. Williamson (eds.), *New Frontiers in Microsimulation Modelling*, vol. 1. Ashgate, Vienna, Austria, pp. 97–122.

Van Leeuwen, E., Dekkers, J. and Rietveld, P. 2008. The development of a static farm-level spatial microsimulation model to analyse on and off farm activities of Dutch farmers presenting the research framework. Paper presented at the 3rd Israeli—Dutch Regional Science Workshop, Jerusalem, November 4–6.

van Leeuwen, E.S., Hagens, J.E., and Nijkamp, P. 2007. Multi-agent systems: A tool in spatial planning: An example of microsimulation use in retail development. *disP*, 170(3): 19–32.

Van Soest, A. 1995. Structural models of family labor supply: A discrete choice approach. *Journal of Human Resources*, 30(1): 63–88.

Van Wissen, L. 2000. A micro-simulation model of firms: Applications of concepts of the demography of the firm. *Papers in Regional Science*, 79(1): 111–124.

Vega, A., Miller, A.C., and O'Donoghue, C. (2014). The seafood sector in Ireland: Economic impacts of seafood production growth targets. Working Papers No. 163051, Socio-Economic Marine Research Unit, National University of Ireland, Galway.

Veldhuisen, J., Timmermans, H., and Kapoen, L. 2000. RAMBLAS: A regional planning model based on the microsimulation of daily activity travel patterns. *Environment and Planning A*, 32(3): 427–443.

Vizcaino, E.L., Lombardia, M.J., and Morales, D. 2013. Multinomial-based small area estimation of labour force indicators. *Statistical Modelling*, 13(2): 153–178.

Voas, D. and Williamson, P. 2000. An evaluation of the combinatorial optimisation approach to the creation of synthetic microdata. *International Journal of Population Geography*, 6(3): 349–366.

Waddell, P. (2002). UrbanSim: Modeling urban development for land use, transportation and environmental planning. *Journal of the American Planning Association*, 68(3): 297–314.

Waddell, P. and Borning, A. 2004. A case study in digital government: Developing and applying UrbanSim, a system for simulating urban land use, transportation, and environmental impacts. *Social Science Computer Review*, 22(1): 37–51.

Wang, J., Fuller, W.A., and Qu, Y. 2008. Small area estimation under a restriction. *Survey Methodology*, 34(1): 29–36.

Wegener, M. 2004. Overview of land-use transport models. In D.A. Hensher, K.J. Button, K.E. Haynes, and P. Stopher (eds.), *Handbook of Transport Geography and Spatial Systems*, vol. 5. Elsevier Science, Kidlington, U.K.

Williamson, P. 1992. Community health care policies for the elderly: A microsimulation approach. PhD thesis, School of Geography, University of Leeds, Leeds, U.K.

Williamson, P. 1999. Microsimulation: An idea whose time has come? In paper presented at the *39th European Congress of the European Regional Science Association*, Dublin, Ireland, August 23–27, 1999.

Williamson, P. 2007. CO Instruction manual. Working Paper 2007/1, Population Microdata Unit, Department of Geography, University of Liverpool, Liverpool, U.K.

Williamson, P. 2013. An evaluation of two synthetic small area microdata simulation methodologies: Synthetic reconstruction and combinatorial optimisation. In R. Tanton and K. Edwards (eds.), *Spatial Microsimulation: A Reference Guide for Users SE-3. Understanding Population Trends and Processes*, vol. 6. Springer, Dordrecht, the Netherlands, pp. 19–47.

Williamson, P., Birkin, M., and Rees, P. 1998. The estimation of population microdata by using data from small area statistics and sample of anonymised records. *Environment and Planning Analysis*, 30(6): 785–816.

Williamson, P., Clarke, G.P., and McDonald, A.T. 1996. Estimating small area demands for water with the use of microsimulation. In G.P. Clarke (ed.), *Microsimulation for Urban and Regional Policy Analysis*. Pion, London, U.K.

Wong, D.W.S. 1992. The reliability of using the iterative proportional fitting procedure. *The Professional Geographer*, 44(3): 340–357.

World Bank. 2004. Mapping poverty, http://siteresources.worldbank.org/INTPGI/Resources/342674-1092157888460/poverty_mapping.pdf. Access on July 25, 2015.

Wu, B. and Birkin, M. 2013. Moses: A dynamic spatial microsimulation model for demographic planning. In R. Tanton and K. Edwards (eds.), *Spatial Microsimulation: A Reference Guide for Users. Understanding Population Trends and Processes*, vol. 6. Springer, London, U.K., pp. 171–193.

Wu, B., Birkin, M., and Rees, P. 2008. A spatial microsimulation model with student agents. *Computers Environment and Urban Systems*, 32(2): 440–453.

Wu, B.M. and Birkin, M. 2012. Moses: A dynamic spatial microsimulation model for demographic planning. In R. Tanton and K. Edwards (eds.), *Spatial Microsimulation: A Reference Guide for Users*. Springer, Dordrecht, the Netherlands.

Wu, C. 2003. Optimal calibration estimators in survey sampling. *Biometrika*, 90: 937–951.

Wu, C. and Sitter, R.R. 2001. A model-calibration to using complete auxiliary information from survey data. *Journal of the American Statistical Association* 96: 185–193.

Yan, G. and Sedransk, J. 2010. A note on Bayesian residuals as a hierarchical model diagnostic technique. *Statistical Papers*, 51(1): 1–10.

Yates, J. and Gabriel, M. 2006. National research venture 3: Housing affordability for lower income Australians. Australian Housing and Urban Research Institute, Melbourne, Victoria, Australia.

Ybarra, L.M.R. and Lohr, S.L. 2008. Small area estimation when auxiliary information is measured with error. *Biometrika*, 95(7): 919–931.

You, Y. and Rao, J.N.K. 2002a. A pseudo-empirical best linear unbiased prediction approach to small area estimation using survey weights. *Canadian Journal of Statistics*, 30(3): 431–439.

You, Y. and Rao, J.N.K. 2002b. Small area estimation using unmatched sampling and linking models. *Canadian Journal of Statistics*, 30(1): 3–15.

You, Y., Rao, J.N.K., and Gambino, J. 2003. Model-based unemployment rate estimation for the Canadian labour force survey: A hierarchical Bayes approach. *Survey Methodology*, 29(1): 25–32.

Zaidi, A., Harding, A., and Williamson, P. 2009. (eds.), *New Frontiers in Microsimulation Modelling*. Ashgate, London, U.K.

Zellner, A. 1971. *An Introduction to Bayesian Inference in Econometrics*. John Wiley & Sons, New York.

Zhang, L.C. 2009. Estimates for small area compositions subjected to informative missing data. *Survey Methodology*, 35(2): 191–201.

Zhang, L.C. and Chambers, R.L. 2004. Small area estimates for cross-classifications. *Journal of the Royal Statistical Society: Series B*, 66(3): 479–496.

Zhao, F. and Chung, S. 2006. A study of alternative land use forecasting models—Final report. Technical Report BD015-10, Florida Department of Transportation, Tallahassee, FL.

Zhou, B. and Kockelman, K.M. 2010. Land use change through microsimulation of market dynamics: An agent-based model of land development and locator bidding in Austin, Texas. In paper presented at the *89th Annual Meeting of the Transportation Research Board*, Washington, DC, January 26, 2010.

Appendix A: The Newton–Raphson Iteration Method

The Newton–Raphson iteration method is a root-finding algorithm for a nonlinear equation. The method is based on the first few terms of the Taylor series of a function. Although it is a well-known iteration method, the basic theory is provided here.

Let for a single variable nonlinear equation $f(z) = 0$, the Taylor series of $f(z)$ about the point $z = z_0 + \varepsilon$ is expressed as

$$f(z_0 + \varepsilon) = f(z_0) + f'(z_0)\varepsilon + f''(z_0)\varepsilon^2 + \cdots \qquad (A.1)$$

where
z_0 is an initial assumed root of $f(z)$
f' represents the first-order derivative
ε is a very small arbitrary positive quantity

Keeping terms only to the first-order derivative, we have

$$f(z_0 + \varepsilon) \approx f(z_0) + f'(z_0)\varepsilon. \qquad (A.2)$$

This is the equation of the tangent line to the curve of $f(z)$ at the point $\{z_0, f(z_0)\}$, and hence $(0, z_0)$ is the interval where that tangent line intersects the *horizontal* axis at z_1 (see, e.g., Figure A.1).

The expression in (A.2) that can be used to estimate the amount of adjustment for ε should require to converge to the accepted root starting from an initial assumed root value, z_0. From the relation in (A.2), after setting $f(z_0 + \varepsilon) = 0$ and considering an arbitrary quantity $\varepsilon = \varepsilon_0$, we get

$$\varepsilon_0 = -\frac{f(z_0)}{f'(z_0)},$$

which is the first-order adjustment to the original root.

By considering $z_i = z_{i-1} + \varepsilon_{i-1}$ for $i = 1, 2, \ldots, r, \ldots$, we can subsequently obtain a new ε_i, for which

$$\varepsilon_i = -\frac{f(z_i)}{f'(z_i)}; \quad \forall i. \qquad (A.3)$$

Let the process should be repeated until $(r + 1)$ times when the value of the arbitrary quantity, ε_r, is reached to the accuracy level. In other words, the

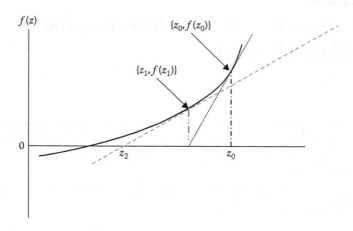

FIGURE A.1
Graphical representation of the Newton–Raphson iteration process.

process should be repeated until $(r + 1)$ times when an estimated root of the function—say z_{r+1}—will converge to a precisely stable number or to an accepted root value. Hence, the following algorithm can be applied iteratively to obtain an accepted root:

$$z_{r+1} = z_r - \left\{f'(z_r)\right\}^{-1} f(z_r); \quad \text{for all } r = 1, 2, 3, \ldots \tag{A.4}$$

It is worth noting that the Newton–Raphson method uses this iterative process to approach one root of a function, and a well-chosen initial root value can lead the convergence quickly (see Figure A.1). However, the procedure can be unstable near a horizontal asymptote or a local extremum. Besides, the Newton–Raphson iteration method is easily adapted to deal with a set of equations for a function with vector variables when its second-order derivative also exists.

Now, Equation 4.11 can be written as a function of the vector λ, which is as follows:

$$l_j(\lambda) = C_j - \sum_{k \in s} d_k \left\{f^{-1}(x'_k \lambda) - 1\right\} x_{k,j} = 0 \tag{A.5}$$

for $j = 1, 2, \ldots, p$, where $C = T_x - \hat{t}_{x,s}$ is a known vector, $d_k\{f^{-1}(x'_k\lambda) - 1\}$ is a scalar, and the equation is nonlinear in the Lagrange multipliers vector, λ. Equation A.5 can be solved by the aforementioned Newton–Raphson iterative procedure. Hence, the iteration algorithm can be expressed as

$$\lambda_{[r+1]} = \lambda_{[r]} - [l'(\lambda)]^{-1}_{\lambda_{[r]}} [l(\lambda)]_{\lambda_{[r]}}; \quad \text{for all } r = 1, 2, 3, \ldots \tag{A.6}$$

where

$\lambda_{[r]}$ is the value of the vector λ in the rth iteration

$l'(\lambda) = [\partial l_j(\lambda)/\partial \lambda_n]$ represents the Hessian matrix

$[l'(\lambda)]_{\lambda_{[r]}}$ defines the values of vector $l'(\lambda)$, which are determined by the rth iteration values of vector $\lambda_{[r]}$

Note that GREGWT stops the iteration process when $|\lambda_{[r+1]} - \lambda_{[r]}| < \varepsilon_r = 0.0001$ is satisfied or predefined maximum iteration has reached. However, the ε_r can take any suitable positive arbitrary value, and the choice fully depends on our desired accuracy.

Appendix B: Topics Index of the 2005–2006 Survey of Income and Housing: CURFs

TABLE B.1

Topics in the 2006 Census of Population and Housing

Topic index
- Name of all residents *including visitors* on the census night at the dwelling
- Sex, date of birth, or age at last birthday
- Relationship to other persons in the household (HH)
- Marital status of respondent
- **Number of children ever given birth by female respondent**
- Aboriginal or Torres Strait Islander status
- Usual address of respondent
- Country of birth
- Year of first arrival in Australia for the overseas born respondents
- Country of birth of parents
- Main language other than English spoken at home/proficiency in spoken English
- Ancestry of respondent
- Religion
- **Core activity need for assistance**
- Educational, vocational, and training qualifications
- Labor force status
- **Employment type**
- Status in employment
- Occupation
- Industry or business of a respondent's employment
- Name of employer
- Hours worked last week in all jobs
- Addresses of workplace (including building/property name) and usual residence
- Mode of travel to work
- Number of registered motor vehicles owned or used by members of the HH, garaged at or near their dwelling
- Income of the respondent from all sources
- **Information about unpaid work (including hours worked, type of work, etc.)**
- Residential status of respondent at nonprivate dwelling

Topic for housing
- The address of dwelling, including street name, suburb, postcode, etc.
- Number of bedrooms in the dwelling
- Tenure type
- Housing loan repayments
- Type of landlord for rented dwelling and rental payments
- **Status of Internet connection at dwelling**
- Structure and location of private dwellings
- Nonprivate dwellings

Source: ABS, Census dictionary Australia, ABS Catalogue no. 2901.0, Australian Bureau of Statistics, Canberra, Australian Capital Territory, Australia, 2006.
Note: Bold indicates a new topic compared with the previous census.

TABLE B.2

BCP Tables for the 2006 Census Available at *Census Data Online*

Basic Community Profile Data Set Tables for the 2006 *CPH*, Australia	
B01	Selected person characteristics by sex
B02	Selected medians and averages
B03	Place of usual residence on census night by age
B04	Age by sex
B05	Registered marital status by age by sex
B06	Social marital status by age by sex
B07	Indigenous status by age by sex
B08	Ancestry by country of birth of parents
B09	Country of birth of person by sex
B10	Country of birth of person by year of arrival in Australia
B11	Proficiency in spoken English/language by year of arrival in Australia by sex
B12	Language spoken at home by sex
B13	Religious affiliation by sex
B14	Type of educational institution attending (F/P-time student status by age) by sex
B15	Highest year of school completed by age by sex
B16	Gross individual income by age by sex
B17	Core activity need for assistance by age by sex
B18	Voluntary work for an organization or group by age by sex
B19	Unpaid domestic work: number of hours by age by sex
B20	Unpaid assistance to a person with a disability by age by sex
B21	Unpaid child care by age by sex
B22	Relationship in the HH by age by sex
B23	Number of children ever born by age of female
B24	Family composition
B25	Family composition by sex of person in family
B26	Gross family income by family composition
B27	Family blending
B28	Gross HH income by HH composition
B29	Number of motor vehicles by dwellings
B30	HH composition by number of persons, usually residents
B31	Dwelling structure
B32	Tenure type and landlord type by dwelling structure
B33	Housing loan repayment by dwelling structure
B34	Rent by landlord type
B35	Type of internet connection by dwelling structure
B36	Selected person characteristics by sex (second release processing)
B37	Place of usual residence 1 year ago by sex
B38	Place of usual residence 5 years ago by sex
B39	Nonschool qualification: level of education by age by sex
B40	Nonschool qualification: field of study by age by sex

(Continued)

TABLE B.2 (*Continued*)

BCP Tables for the 2006 Census Available at *Census Data Online*

Basic Community Profile Data Set Tables for the 2006 *CPH*, Australia	
B41	Labor force status by age by sex
B42	Industry of employment by age by sex
B43	Industry of employment by occupation
B44	Occupation by age by sex
B45	Method of travel to work by sex

Source: ABS. 2007a. *Information Paper: 2006 Census of Population and Housing—Census Data Products*, Australian Bureau of Statistics, Canberra, Australia. http://www.abs.gov.au/ ausstats/abs@.nsf/lookup/2011.0.55.001Main%20Features1042011.   Accessed   on November 15, 2009; and Rahman, A. (2011) *Small area housing stress estimation in Australia: Microsimulation modelling and statistical reliability*, PhD thesis, University of Canberra, Canberra.

TABLE B.3

Selected Topics in the SIH—CURFs

Topics in the 2005–2006 Survey of Income and Housing—CURFs Data

Data item labels

Identifiers
Unique HH number—unique number allocated to all members in the HH
Family number—in the HH
Income unit (IU) number—within each family in the HH
Person number within each IU
HH level identifier
IU level identifier
Person level identifier

Person, IU, and HH characteristics
Area characteristics/geography
    Area of usual residence
    Index of relative socioeconomic disadvantage—decile—Australia
    Remoteness area
    Section of state
    State or territory
Demographics
    Age
    Age of the HH reference person
    Age of the oldest dependent child in the HH
    Age of the oldest dependent child in the IU
    Age of the youngest dependent child in the HH
    Age of the youngest dependent child in the IU
    Country of birth
    Country of birth by main language
    Country of birth by main language of the HH reference person
    Country of birth of the HH reference person
    Family composition of the HH (alternative)
    Family composition of the HH
    Family type
    IU type—IU
    IU type—person
    Life cycle group—HH
    Number of dependents aged less than 25 years in the HH
    Number of dependent children aged 0–2 years in the IU
    Number of dependent children aged 10–12 years in the IU
    Number of dependent children aged 10–14 years in the IU
    Number of dependent children aged 13–14 years in the IU
    Number of dependent children aged 15–24 years in the HH

(Continued)

TABLE B.3 (*Continued*)

Selected Topics in the SIH—CURFs

Topics in the 2005–2006 Survey of Income and Housing—CURFs Data

Data item labels

    Number of dependent children aged 15–24 years in the IU

    Number of dependent children aged less than 15 years in the HH

    Number of dependent children aged 5–9 years in the IU

    Number of dependent children aged 3–4 years in the IU

    Number of dependent children in the IU

    Number of families in the HH

    Number of females in the HH

    Number of IUs in the HH

    Number of males in the HH

    Number of nondependent children in the HH

    Number of nonfamily members in the HH

    Number of other relatives in the HH

    Number of persons aged 15–64 years in the IU

    Number of persons aged 15 years and over in the HH

    Number of persons aged 65 years or more in the IU

    Number of persons in the HH

    Number of persons in the IU

    Position in the IU (relationship to the IU reference person)

    Position in the HH (publication definition)

    Relationship in the HH

    Sex of the respondent

    Sex of the HH reference person

    Social marital status

    Year of arrival in Australia

    Year of arrival in Australia of the HH reference person

Education

    Education status

    Educational institution attending

    Highest year of school completed

    Level of highest nonschool qualification

    Main field of highest nonschool qualification

    Number of full-time and part-time students in the HH aged 15–24 years

    Number of full-time and part-time students in the HH aged 25 years and over

    Type of study in the current year

Labor force status

    Duration of unemployment

    Full-time or part-time status

    Industry of main job

(*Continued*)

TABLE B.3 (*Continued*)

Selected Topics in the SIH—CURFs

Topics in the 2005–2006 Survey of Income and Housing—CURFs Data
Data item labels

Labor force status
Labor force status of the HH reference person
Looked for work
Not in the labor force status
Number of earners in the HH
Number of employed persons in the HH
Number of hours usually worked per week in main and second jobs
Number of hours usually worked per week in main job
Number of hours usually worked per week in second job
Number of jobs currently held
Number of unemployed persons in the HH
Occupation in main job
Status in employment (main job)
Status in employment (second job)

Housing information
Amount owing on mortgages for alterations/additions—HH
Amount owing on mortgages for alterations/additions—HH (prorata)
Amount owing on mortgages for other purposes (excl. business and investment loans)—HH
Amount owing on mortgages for other purposes (excl. business and investment loans)—HH (prorata)
Amount owing on mortgages to purchase/build—HH
Amount owing on mortgages to purchase/build—HH (prorata)
Amount owing on unsecured loans for housing purposes—HH
Amount owing on unsecured loans for housing purposes—HH (prorata)
Canadian National Occupancy Standard
Dwelling structure—HH
Landlord type—HH
Number of bedrooms—HH
Tenure type—HH
Weekly body corporate payments with refunds deducted—HH
Weekly combined rate payments—HH
Weekly combined rate payments with refunds deducted—HH
Weekly general and water rate payments—HH
Weekly general and water rate payments with refunds deducted—HH
Weekly general rate payments—HH
Weekly general rate payments with refunds deducted—HH
Weekly housing costs (after refunds, interest only, incl. body corp, loans prorated by purpose, no tenure adjustment)

(Continued)

TABLE B.3 *(Continued)*

Selected Topics in the SIH—CURFs

Topics in the 2005–2006 Survey of Income and Housing—CURFs Data

Data item labels

Weekly housing costs (SIH basis)—HH

Weekly mortgage repayments for alterations/additions—HH

Weekly mortgage repayments for alterations/additions with refunds deducted—HH (prorata)

Weekly mortgage repayments for alterations/additions (interest component with refunds deducted)—HH (prorata)

Weekly mortgage repayments for other purposes (excl. business and investment loans)—HH

Weekly mortgage repayments for other purposes (excl. business and investment loans) (interest component with refunds deducted)—HH (prorata)

Weekly mortgage repayments for other purposes with refunds deducted (excl. bus and investment loans)—HH (prorata)

Weekly mortgage repayments to purchase/build—HH

Weekly mortgage repayments to purchase/build with refunds deducted—HH (prorata)

Weekly mortgage repayments to purchase/build (interest component with refunds deducted)—HH (prorata)

Weekly rent payments—HH

Weekly rent payments with refunds deducted—HH

Weekly repayments on unsecured loans for housing purposes—HH

Weekly repayments on unsecured loans for housing purposes (interest component with refunds deducted)—HH (prorata)

Weekly repayments on unsecured loans for housing purposes with refunds deducted—HH (prorata)

Whether dwelling purchased/built in last 3 years is first home owned—HH

Whether dwelling purchased/built in last 3 years was new or established—HH

Weekly water rate payments—HH

Weekly water rate payments with refunds deducted—HH

Year of purchased dwelling—HH

Income

Employee income

    Current weekly benefit from the employer-provided car park (nonsalary sacrifice)

    Current weekly benefit from the employer-provided child care (nonsalary sacrifice)

    Current weekly benefit from the employer-provided computer (nonsalary sacrifice)

    Current weekly benefit from the employer-provided housing (nonsalary sacrifice)

    Current weekly benefit from the employer-provided low-interest loans (nonsalary sacrifice)

    Current weekly benefit from the employer-provided shares (nonsalary sacrifice)

    Current weekly benefit from the employer-provided superannuation (above minimum to nonsalary sacrifice)

    Current weekly benefit from the employer-provided telephone (nonsalary sacrifice)

    Current weekly benefit from the employer-provided vehicle (nonsalary sacrifice)

    Current weekly employee cash income from regular bonuses

(Continued)

TABLE B.3 (*Continued*)

Selected Topics in the SIH—CURFs

Topics in the 2005–2006 Survey of Income and Housing—CURFs Data

Data item labels

    Current weekly employee income salary sacrificed for superannuation

    Current weekly employee income salary sacrificed for child care

    Current weekly employee income salary sacrificed for computer

    Current weekly employee income salary sacrificed for housing

    Current weekly employee income salary sacrificed for other benefits

    Current weekly employee income salary sacrificed for telephone charges

    Current weekly HH employee income (2003–2004 basis)

    Current weekly HH employee income (2005–2006 basis)

    Current weekly income salary sacrificed for vehicle

    Current weekly IU employee income (2003–2004 basis)

    Current weekly IU employee income (2005–2006 basis)

    Current weekly other noncash benefit from employer (nonsalary sacrifice)

    Previous financial year employee income from all jobs

    Previous financial year HH employee income

    Previous financial year IU employee income

    Total current weekly employee income (as reported)

    Total current weekly employee income (inclusive salary sacrifice)

    Whether reported employee income included the amount salary sacrificed

Own unincorporated business income

    Current weekly cash income from own unincorporated business flag

    Current weekly cash income from own unincorporated business (previous SIH basis using PFY data)

    Current weekly cash income from own unincorporated business (reported)

    Current weekly HH income from own unincorporated business

    Current weekly HH income from own unincorporated business (previous SIH basis using PFY data)

    Current weekly HH income from own unincorporated business flag

    Current weekly IU income from own unincorporated business

    Current weekly IU income from own unincorporated business (previous SIH basis using PFY data)

    Current weekly IU income from own unincorporated business flag

    Previous financial year HH income from own unincorporated business

    Previous financial year HH income from own unincorporated business flag

    Previous financial year income from own unincorporated business

    Previous financial year income from own unincorporated business flag

    Previous financial year IU income from own unincorporated business

    Previous financial year IU income from own unincorporated business flag

(Continued)

TABLE B.3 (*Continued*)

Selected Topics in the SIH—CURFs

Topics in the 2005–2006 Survey of Income and Housing—CURFs Data
Data item labels

Government pensions and allowances
   Current weekly HH income from government pensions and allowances
   Current weekly income from age pension
   Current weekly income from Austudy/Abstudy
   Current weekly income from carer allowance
   Current weekly income from carer payment
   Current weekly income from disability pension (DVA)
   Current weekly income from disability support pension
   Current weekly income from family tax benefits (modeled)
   Current weekly income from maternity payment
   Current weekly income from mature age allowance
   Current weekly income from newstart allowance
   Current weekly income from one-off carer bonus
   Current weekly income from one-off payment to older Australians
   Current weekly income from other government pensions and allowances
   Current weekly income from overseas pensions and benefits
   Current weekly income from parenting payment
   Current weekly income from partner allowance
   Current weekly income from seniors' concession allowance
   Current weekly income from service pension (DVA)
   Current weekly income from sickness allowance
   Current weekly income from special benefit
   Current weekly income from utilities allowance
   Current weekly income from war widows pension (DVA)
   Current weekly income from widow allowance
   Current weekly income from wife pension
   Current weekly income from youth allowance
   Current weekly IU income from government pensions and allowances
   Previous financial year HH income from government pensions and allowances
   Previous financial year income from age pension
   Previous financial year income from Austudy/Abstudy
   Previous financial year income from carer allowance
   Previous financial year income from carer payment
   Previous financial year income from disability pension (DVA)
   Previous financial year income from disability support pension
   Previous financial year income from family tax benefits
   Previous financial year income from maternity payment
   Previous financial year income from mature age allowance

(Continued)

TABLE B.3 (*Continued*)

Selected Topics in the SIH—CURFs

Topics in the 2005–2006 Survey of Income and Housing—CURFs Data

Data item labels

    Previous financial year income from newstart allowance

    Previous financial year income from other government pensions and allowances

    Previous financial year income from overseas pensions and benefits

    Previous financial year income from parenting payment

    Previous financial year income from partner allowance

    Previous financial year income from service pension (DVA)

    Previous financial year income from sickness allowance

    Previous financial year income from special benefit

    Previous financial year income from war widows pension (DVA)

    Previous financial year income from widow allowance

    Previous financial year income from wife pension

    Previous financial year income from youth allowance

    Previous financial year IU income from government pensions and allowances

    Total current weekly income from government pensions and allowances

    Total previous financial year income from government pensions and allowances

Investment income

    Current weekly HH income from investments

    Current weekly HH income from investments (previous SIH basis using data)

    Current weekly income from dividends (based on PFY)

    Current weekly income from dividends (reported)

    Current weekly income from financial institution account interest (based on PFY)

    Current weekly income from financial institution account interest (reported)

    Current weekly income from interest on debentures and bonds (based on PFY)

    Current weekly income from interest on debentures and bonds (reported)

    Current weekly income from interest on loans to persons not in this HH (based on PFY)

    Current weekly income from interest on loans to persons not in this HH (reported)

    Current weekly income from nonresidential property flag (based on PFY)

    Current weekly income from nonresidential property flag (reported)

    Current weekly income from nonresidential property (based on PFY)

    Current weekly income from nonresidential property (reported)

    Current weekly income from other financial investments (reported)

    Current weekly income from other financial investments (based on PFY)

    Current weekly income from residential property flag (reported)

    Current weekly income from residential property flag (based on PFY)

    Current weekly income from residential property (based on PFY)

    Current weekly income from residential property (reported)

    Current weekly income from royalties (based on PFY)

    Current weekly income from royalties (reported)

(Continued)

TABLE B.3 (*Continued*)

Selected Topics in the SIH—CURFs

Topics in the 2005–2006 Survey of Income and Housing—CURFs Data
Data item labels

Current weekly income from trusts (based on PFY)

Current weekly income from trusts (reported)

Current weekly interest paid on money borrowed to purchase shares or units in trusts

Current weekly IU income from investments (previous SIH basis using data)

Current weekly IU income from investments (reported)

Previous financial year HH income from investments

Previous financial year income from dividends

Previous financial year income from financial institution account interest

Previous financial year income from interest on debentures and bonds

Previous financial year income from interest on loans to persons not in this HH

Previous financial year income from nonresidential property

Previous financial year income from nonresidential property flag

Previous financial year income from other financial investments

Previous financial year income from residential property

Previous financial year income from residential property flag

Previous financial year income from royalties

Previous financial year income from trusts

Previous financial year interest paid on money borrowed to purchase shares or units in trusts

Previous financial year IU income from investments

Total current weekly income from investments (based on PFY)

Total current weekly income from investments (reported)

Total current weekly income from property flag (based on PFY)

Total current weekly income from property flag (reported)

Total current weekly income from property (based on PFY)

Total current weekly income from property (reported)

Total previous financial year income from investments

Total previous financial year income from property

Total previous financial year income from property flag

Other income

Current weekly HH income from other regular sources (excluding superannuation)

Current weekly HH income from superannuation/annuity/allocated pension

Current weekly income from accident compensation and sickness insurance

Current weekly income from child support/maintenance

Current weekly income from persons not living in the HH

Current weekly income from regular sources n.e.c.

Current weekly income from regular workers' compensation

Current weekly income from scholarships

Current weekly income from superannuation/annuity/allocated pension

(*Continued*)

TABLE B.3 (*Continued*)

Selected Topics in the SIH—CURFs

Topics in the 2005–2006 Survey of Income and Housing—CURFs Data

---

Data item labels

Current weekly IU income from other regular sources (excluding superannuation)

Current weekly IU income from superannuation/annuity/allocated pension

Previous financial year HH income from other regular sources (excluding superannuation)

Previous financial year HH income from superannuation/annuity/allocated pension

Previous financial year income from accident compensation and sickness insurance

Previous financial year income from child support/maintenance

Previous financial year income from persons not living in the HH

Previous financial year income from regular sources n.e.c.

Previous financial year income from regular workers' compensation

Previous financial year income from scholarships

Previous financial year income from superannuation/annuity/allocated pension

Previous financial year IU income from other regular sources (excluding superannuation)

Previous financial year IU income from superannuation/annuity/allocated pension

Total current weekly income from other regular sources

Total previous financial year income from other regular sources

Other income items

Current weekly HH disposable income (2003–2004 basis)

Current weekly HH disposable income (2005–2006 basis)

Current weekly HH equivalized disposable income (2003–2004 basis)

Current weekly HH equivalized disposable income (2005–2006 basis)

Equivalizing factor (modified OECD)

HH previous financial year exclusion flag

Imputed current weekly tax payable

Imputed tax payable for previous financial year

IU previous financial year exclusion flag

Number of pension/benefit recipients in the HH

Person previous financial year exclusion flag

Previous financial year HH disposable income

Principal source of current HH income (2003–2004 basis)

Principal source of current HH income (2005–2006 basis)

Principal source of current HH income (pre 2003–2004 SIH basis)

Principal source of current income (2003–2004 basis)

Principal source of current income (2005–2006 basis)

Principal source of current income (pre 2003–2004 SIH basis)

Principal source of current IU income (2003–2004 basis)

Principal source of current IU income (2005–2006 basis)

Principal source of current IU income (pre 2003–2004 SIH basis)

Principal source of previous financial year HH income

Principal source of previous financial year income

(Continued)

TABLE B.3 (*Continued*)

Selected Topics in the SIH—CURFs

Topics in the 2005–2006 Survey of Income and Housing—CURFs Data
Data item labels

Principal source of previous financial year IU income
Total current weekly HH income from all sources (2003–2004 basis)
Total current weekly HH income from all sources (2005–2006 basis)
Total current weekly HH income from all sources (pre 2003–2004 SIH basis)
Total current weekly income from all sources (2003–2004 basis)
Total current weekly income from all sources (2005–2006 basis)
Total current weekly income from all sources (pre 2003–2004 SIH basis)
Total current weekly IU income from all sources (2003–2004 basis)
Total current weekly IU income from all sources (2005–2006 basis)
Total current weekly IU income from all sources (pre 2003–2004 SIH basis)
Total previous financial year HH income from all sources
Total previous financial year income from all sources
Total previous financial year IU income from all sources

Other receipts and payments
Current weekly payments for child support/maintenance
Current weekly payments for maintenance/alimony to former spouse
Current weekly payments to family members not in the HH
Previous financial year payments for child support/maintenance
Previous financial year payments for maintenance/alimony to former spouse
Previous financial year payments to family members not in the HH

Wealth
Amount of credit card debt—HH level
Amount of HECS liability
Amount of Student Financial Supplement liability
Balance of accounts with government superannuation funds
Balance of accounts with government superannuation funds—HH level
Balance of accounts with nongovernment superannuation funds
Balance of accounts with nongovernment superannuation funds—HH level
Estimated sale price of dwelling—HH
Net wealth of HH
Principal outstanding on investment loans (excl. business and rental property loans)
Principal outstanding on loans for other properties (excl. business and investment loans)
Principal outstanding on loans for other purposes (excl. business and investment loans)
Principal outstanding on loans for selected dwelling
Principal outstanding on loans for vehicle purchases (excl. business and investment loans)
Principal outstanding on rental property loans
Value of accounts held with financial institutions
Value of accounts held with financial institutions—HH level
Value of assets n.e.c.

(Continued)

TABLE B.3 (*Continued*)

Selected Topics in the SIH—CURFs

Topics in the 2005–2006 Survey of Income and Housing—CURFs Data

Data item labels

Value of children's assets

Value of contents of selected dwelling

Value of debentures and bonds

Value of debentures and bonds—HH level

Value of loans to persons not in the same HH

Value of loans to persons not in the same HH—HH level

Value of nonresidential property

Value of other financial investments

Value of other financial investments—HH level

Value of own incorporated business (net of liabilities)

Value of own incorporated business (net of liabilities)—HH level

Value of own unincorporated business (net of liabilities)

Value of own unincorporated business (net of liabilities)—HH level

Value of residential property excl. selected dwelling

Value of shares

Value of shares—HH level

Value of trusts

Value of trusts—HH level

Value of vehicles

Imputation flags

Flag to indicate HH level imputation

Flag to indicate HH contains person level imputation

Flag to indicate person level imputation

Imputation flag for different selected modules (see ABS 2008b for a more details)

Weights

SIH replicate weight—HH 01-60

SIH replicate weight—IU 01-60

SIH replicate weight—Person 01-60

Weight—HH (SIH)

Weight—IU (SIH)

Weight—Person (SIH)

Other items

Number of credit cards in the HH

Quarter of interview

Imputed rent

Weekly gross imputed rent

Weekly net imputed rent

Source: ABS, Survey of income and housing—Confidentialised unit record files: Technical manual—Australia 2005–2006, ABS Catalogue no. 6541.0, Australian Bureau of Statistics, Canberra, Australian Capital Territory, Australia, 2008b.

Appendix C: Tables of the Housing Stress for 50 SLAs with the Highest Numbers and Percentages Estimates

TABLE C.1

Fifty SLAs with the Highest Number of Households in Housing Stress

| SLA Name | SLA ID | Estimated No. of Households | | |
		Total	HS	%
Canterbury (C)	105201550	44,257	7852	17.7
Fairfield (C)—East	105252851	34,376	7219	21.0
Stirling (C)—Central	505157914	40,936	5294	12.9
Randwick (C)	105106550	44,848	5081	11.3
Liverpool (C)—East	105254901	31,165	5056	16.2
Wollongong (C)—Inner	115058451	36,646	4777	13.0
Gosford (C)—West	105703104	34,808	4736	13.6
Blacktown (C)—South-West	105530753	27,928	4595	16.5
Holroyd (C)	105403950	30,783	4345	14.1
Rockdale (C)	105156650	32,873	4223	12.9
Blacktown (C)—South-East	105530752	30,559	3962	13.0
Auburn (A)	105400200	19,054	3928	20.6
Brimbank (C)—Sunshine	205101182	27,060	3869	14.3
Warringah (A)	105658000	48,437	3857	8.0
Wyong (A)—North-East	105708551	27,156	3839	14.1
Penrith (C)—East	105456351	30,185	3821	12.7
Ryde (C)	105556700	35,666	3760	10.5
Darebin (C)—Preston	205301892	30,037	3754	12.5
Frankston (C)—West	205852174	29,077	3718	12.8
Wyong (A)—South and West	105708554	25,160	3565	14.2
Bankstown (C)—North-East	105200351	16,935	3484	20.6
Swan (C)	505108050	31,103	3475	11.2
Gr. Dandenong (C) Bal	205752674	23,953	3471	14.5
Marrickville (A)	105055200	28,087	3443	12.3
Rockingham (C)	505207490	29,250	3415	11.7
Gosnells (C)	505253780	31,779	3385	10.7
Glen Eira (C)—Caulfield	205652311	29,931	3335	11.1
Brimbank (C)—Keilor	205101181	27,171	3265	12.0

(Continued)

TABLE C.1 (*Continued*)

Fifty SLAs with the Highest Number of Households in Housing Stress

SLA Name	SLA ID	Estimated No. of Households		
		Total	HS	%
Wollongong (C) Bal	115058454	31,686	3259	10.3
Maribyrnong (C)	205104330	24,195	3255	13.5
Hume (C)—Broadmeadows	205353271	19,585	3230	16.5
Campbelltown (C)—North	105301501	23,481	3180	13.5
Kingston (C)—North	205653431	33,411	3165	9.5
Casey (C)—Cranbourne	205801613	21,152	3114	14.7
Hurstville (C)	105154150	26,100	3098	11.9
Coffs Harbour (C)—Pt A	125011801	18,289	3055	16.7
Melbourne (C)—Remainder	205054608	18,968	3034	16.0
Tweed (A)—Tweed-Heads	120057554	20,526	3030	14.8
Sutherland Shire (A)—East	105157151	36,628	3021	8.3
Campbelltown (C)—South	105301504	22,264	2934	13.2
Ipswich (C)—Central	305253962	25,031	2933	11.7
Canning (C)	505251330	27,998	2930	10.5
Penrith (C)—West	105456354	26,748	2929	11.0
Casey (C)—Berwick	205801612	27,770	2929	10.6
Bankstown (C)—North-West	105200353	19,757	2917	14.8
Gr. Dandenong (C)—Dandenong	205752671	18,772	2913	15.5
Wyndham (C)—North	205207261	24,525	2912	11.9
Newcastle (C)—Inner City	110055903	19,820	2826	14.3
Shoalhaven (C)—Pt B	115106952	22,564	2811	12.5
Launceston (C)—Pt B	615054012	23,359	2798	12.0

TABLE C.2

Fifty SLAs with the Highest Number of Buyer Households in Housing Stress

SLA Name	SLA ID	Estimated No. of Households		
		Buyer	HS	%
Fairfield (C)—East	105252851	10,118	2455	24.3
Canterbury (C)	105201550	11,933	2175	18.2
Casey (C)—Cranbourne	205801613	12,178	1797	14.8
Brimbank (C)—Sunshine	205101182	9,316	1774	19.0
Casey (C)—Berwick	205801612	16,309	1756	10.8
Brimbank (C)—Keilor	205101181	11,129	1718	15.4
Swan (C)	505108050	16,091	1710	10.6
Blacktown (C)—South-West	105530753	11,385	1683	14.8
Liverpool (C)—East	105254901	10,593	1645	15.5
Gosnells (C)	505253780	15,793	1613	10.2
Hume (C)—Craigieburn	205353274	9,743	1582	16.2
Liverpool (C)—West	105254904	10,398	1572	15.1
Hume (C)—Broadmeadows	205353271	7,067	1563	22.1
Penrith (C)—East	105456351	13,210	1514	11.5
Wyndham (C)—North	205207261	12,148	1437	11.8
Gr. Dandenong (C) Bal	205752674	7,578	1395	18.4
Fairfield (C)—West	105252854	7,723	1361	17.6
Blacktown (C)—South-East	105530752	11,080	1316	11.9
Bankstown (C)—North-East	105200351	5,149	1305	25.3
Wyong (A)—North-East	105708551	9,162	1294	14.1
Rockdale (C)	105156650	9,043	1290	14.3
Campbelltown (C)—North	105301501	10,095	1289	12.8
Holroyd (C)	105403950	9,147	1286	14.1
Stirling (C)—Central	505157914	12,836	1279	10.0
Blacktown (C)—North	105530751	14,008	1271	9.1
Gosford (C)—West	105703104	11,614	1267	10.9
Warringah (A)	105658000	17,206	1239	7.2
Auburn (A)	105400200	5,614	1229	21.9
Frankston (C)—West	205852174	10,346	1210	11.7
Rockingham (C)	505207490	12,779	1149	9.0
Manningham (C)—West	205504214	10,630	1146	10.8
Whittlesea (C)—South-West	205407076	6,093	1137	18.7
Kingston (C)—North	205653431	11,661	1130	9.7
Joondalup (C)—South	505154174	16,151	1126	7.0
Cockburn (C)	505201820	12,240	1126	9.2
Bankstown (C)—North-West	105200353	5,841	1123	19.2

(Continued)

TABLE C.2 (*Continued*)

Fifty SLAs with the Highest Number of Buyer Households in Housing Stress

SLA Name	SLA ID	Estimated No. of Households		
		Buyer	HS	%
Casey (C)—Hallam	205801616	7,553	**1115**	14.8
Yarra Ranges (S)—Lilydale	205607453	11,207	**1108**	9.9
Melton (S)—East	205204651	8,399	**1098**	13.1
Wollongong (C) Bal	115058454	10,944	**1089**	10.0
Campbelltown (C)—South	105301504	9,511	**1085**	11.4
Sutherland Shire (A)—West	105157152	16,102	**1081**	6.7
Knox (C)—North-East	205553672	10,104	**1051**	10.4
Penrith (C)—West	105456354	11,883	**1042**	8.8
Hurstville (C)	105154150	7,960	**1021**	12.8
Wyong (A)—South and West	105708554	8,348	**1012**	12.1
Shoalhaven (C)—Pt B	115106952	5,901	**1008**	17.1
Darebin (C)—Preston	205301892	7,817	**1005**	12.9
Frankston (C)—East	205852171	8,483	**1002**	11.8
Gr. Dandenong (C)—Dandenong	205752671	5,700	**975**	17.1

TABLE C.3

Fifty SLAs with the Highest Number of Public Renter Households in Housing Stress

SLA Name	SLA ID	Estimated No. of Households		
		Public Renter	HS	%
Blacktown (C)—South-West	105530753	4785	909	19.0
Sydney (C)—South	105057205	3350	750	22.4
Liverpool (C)—East	105254901	3992	677	17.0
Randwick (C)	105106550	3288	661	20.1
Campbelltown (C)—North	105301501	3371	653	19.4
Blacktown (C)—South-East	105530752	3176	602	19.0
Wollongong (C)—Inner	115058451	3363	580	17.3
Canterbury (C)	105201550	2833	560	19.8
Moonee Valley (C)—Essendon	205105063	2804	551	19.7
Fairfield (C)—East	105252851	2878	503	17.5
Bankstown (C)—North-West	105200353	2552	488	19.1
Campbelltown (C)—South	105301504	2675	465	17.4
Stirling (C)—Central	505157914	2451	423	17.3
Wollongong (C) Bal	115058454	2725	420	15.4
Parramatta (C)—North-East	105406252	1887	416	22.1
Holroyd (C)	105403950	2167	402	18.6
Yarra (C)—North	205057351	2127	393	18.5
Playford (C)—Elizabeth	405055683	2587	379	14.7
Darebin (C)—Preston	205301892	1933	328	17.0
Fairfield (C)—West	105252854	1533	323	21.1
Gosford (C)—West	105703104	1731	308	17.8
Botany Bay (C)	105051100	1491	308	20.7
Melbourne (C)—Remainder	205054608	1403	301	21.5
Newcastle (C)—Inner City	110055903	1460	301	20.6
Whyalla (C)	435058540	2109	298	14.1
Bankstown (C)—North-East	105200351	1607	296	18.4
Bankstown (C)—South	105200355	1562	295	18.9
Lake Macquarie (C)—East	110054651	2015	291	14.4
Maribyrnong (C)	205104330	1651	288	17.4
Ryde (C)	105556700	1443	287	19.9
Parramatta (C)—South	105406254	1283	287	22.4
Sydney (C)—West	105057206	1694	284	16.8
Newcastle (C)—Throsby	110055905	1431	274	19.2
Launceston (C)—Pt B	615054012	1959	273	13.9
Marion (C)—Central	405204061	1811	272	15.0

(Continued)

TABLE C.3 (*Continued*)

Fifty SLAs with the Highest Number of Public Renter Households in Housing Stress

SLA Name	SLA ID	Estimated No. of Households		
		Public Renter	HS	%
Marion (C)—North	405204064	1700	267	15.7
Yarra (C)—Richmond	205057352	1289	262	20.3
Shellharbour (C)	115056900	1715	252	14.7
Penrith (C)—West	105456354	1363	244	17.9
Port Adel. Enfield (C)—Park	405105896	1648	243	14.8
Glenorchy (C)	605052610	1599	240	15.0
Stonnington (C)—Prahran	205056351	1128	236	20.9
Corio—Inner	210052752	1565	236	15.1
Port Phillip (C)—West	205055902	1397	235	16.8
Hume (C)—Broadmeadows	205353271	1386	234	16.9
Banyule (C)—Heidelberg	205300661	1425	234	16.4
Inala	305071288	1590	233	14.7
Charles Sturt (C)—North-East	405101068	1481	231	15.6
Wyong (A)—South and West	105708554	1256	230	18.3
Swan (C)	505108050	1439	221	15.4

TABLE C.4

Fifty SLAs with the Highest Number of Private Renter Households in Housing Stress

SLA Name	SLA ID	Estimated No. of Households		
		Private Renter	HS	%
Canterbury (C)	105201550	12,954	5105	43.7
Fairfield (C)—East	105252851	8,987	4256	53.0
Randwick (C)	105106550	16,568	3670	26.1
Stirling (C)—Central	505157914	11,170	3588	35.9
Wollongong (C)—Inner	115058451	9,136	3322	42.7
Gosford (C)—West	105703104	8,249	3144	41.9
Marrickville (A)	105055200	11,194	2791	26.4
Rockdale (C)	105156650	9,362	2773	31.2
Liverpool (C)—East	105254901	7,615	2731	44.8
Holroyd (C)	105403950	8,583	2653	35.6
Glen Eira (C)—Caulfield	205652311	9,676	2618	27.7
Ryde (C)	105556700	9,939	2591	29.0
Auburn (A)	105400200	6,572	2538	41.0
Melbourne (C)—Remainder	205054608	10,304	2484	27.0
Warringah (A)	105658000	11,374	2463	23.0
Wyong (A)—North-East	105708551	6,023	2418	41.8
Darebin (C)—Preston	205301892	7,016	2412	39.1
Port Phillip (C)—St Kilda	205055901	12,073	2341	20.5
Wyong (A)—South and West	105708554	5,726	2312	44.4
Frankston (C)—West	205852174	6,792	2310	36.8
Maribyrnong (C)	205104330	7,178	2181	34.4
Tweed (A)—Tweed-Heads	120057554	4,655	2154	49.1
Rockingham (C)	505207490	6,763	2131	33.3
Coffs Harbour (C)—Pt A	125011801	4,789	2123	48.5
Penrith (C)—East	105456351	6,369	2084	36.2
Newcastle (C)—Inner City	110055903	6,459	2064	36.6
Blacktown (C)—South-East	105530752	6,347	2039	41.6
Blacktown (C)—South-West	105530753	5,432	2000	53.6
Parramatta (C)—Inner	105406251	7,916	1968	27.0
Brimbank (C)—Sunshine	205101182	5,079	1964	41.1
Gr. Dandenong (C) Bal	205752674	5,389	1945	38.3
Ipswich (C)—Central	305253962	6,411	1939	33.1
Kingston (C)—North	205653431	6,791	1911	29.8
Sutherland Shire (A)—East	105157151	8,345	1910	25.3
Bankstown (C)—North-East	105200351	4,191	1882	52.0

(Continued)

TABLE C.4 (*Continued*)

Fifty SLAs with the Highest Number of Private Renter Households in Housing Stress

SLA Name	SLA ID	Estimated No. of Households		
		Private Renter	HS	%
Hurstville (C)	105154150	6,001	1864	34.5
Sydney (C)—East	105057204	10,240	1809	19.8
Canning (C)	505251330	6,177	1809	32.7
Hornsby (A)—South	105604004	7,145	1792	26.5
Waverley (A)	105108050	9,479	1788	19.7
Moreland (C)—Brunswick	205255251	6,940	1787	27.3
Launceston (C)—Pt B	615054012	5,343	1780	38.4
Bayswater (C)	505100420	5,920	1774	32.4
Port Stephens (A)	110056400	4,884	1774	39.1
Hervey Bay (C)—Pt A	315073751	4,580	1769	40.5
Gr. Dandenong (C)—Dandenong	205752671	4,854	1768	39.8
Shoalhaven (C)—Pt B	115106952	4,042	1739	44.1
Wollongong (C) Bal	115058454	4,254	1731	50.6
Stonnington (C)—Prahran	205056351	9,060	1684	21.2
Hastings (A)—Pt A	125033751	3,927	1669	45.8

TABLE C.5

Fifty SLAs with the Highest Number of Total Renter Households in Housing Stress

SLA Name	SLA ID	Estimated No. of Households		
		Total Renter	HS	%
Canterbury (C)	105201550	15,787	5665	35.9
Fairfield (C)—East	105252851	11,865	4759	40.1
Randwick (C)	105106550	19,856	4331	21.8
Stirling (C)—Central	505157914	13,621	4011	29.5
Wollongong (C)—Inner	115058451	12,499	3902	31.2
Gosford (C)—West	105703104	9,980	3452	34.6
Liverpool (C)—East	105254901	11,607	3408	29.4
Holroyd (C)	105403950	10,750	3055	28.4
Marrickville (A)	105055200	12,046	2951	24.5
Rockdale (C)	105156650	10,173	2922	28.7
Blacktown (C)—South-West	105530753	10,217	2909	28.5
Ryde (C)	105556700	11,382	2878	25.3
Melbourne (C)—Remainder	205054608	11,707	2785	23.8
Darebin (C)—Preston	205301892	8,949	2740	30.6
Auburn (A)	105400200	7,391	2694	36.5
Glen Eira (C)—Caulfield	205652311	9,972	2678	26.9
Blacktown (C)—South-East	105530752	9,523	2641	27.7
Warringah (A)	105658000	12,301	2611	21.2
Wyong (A)—South and West	105708554	6,982	2542	36.4
Wyong (A)—North-East	105708551	6,509	2520	38.7
Frankston (C)—West	205852174	7,955	2497	31.4
Port Phillip (C)—St Kilda	205055901	12,724	2474	19.4
Maribyrnong (C)	205104330	8,829	2469	28.0
Newcastle (C)—Inner City	110055903	7,919	2365	29.9
Coffs Harbour (C)—Pt A	125011801	5,939	2321	39.1
Penrith (C)—East	105456351	7,654	2303	30.1
Tweed (A)—Tweed-Heads	120057554	5,323	2285	42.9
Sydney (C)—South	105057205	11,736	2267	19.3
Rockingham (C)	505207490	7,554	2255	29.9
Bankstown (C)—North-East	105200351	5,798	2178	37.6
Wollongong (C) Bal	115058454	6,979	2151	30.8
Moonee Valley (C)—Essendon	205105063	9,303	2145	23.1
Parramatta (C)—Inner	105406251	8,760	2133	24.4
Ipswich (C)—Central	305253962	7,728	2123	27.5
Sutherland Shire (A)—East	105157151	9,339	2109	22.6

(Continued)

TABLE C.5 *(Continued)*

Fifty SLAs with the Highest Number of Total Renter Households in Housing Stress

SLA Name	SLA ID	Estimated No. of Households		
		Total Renter	HS	%
Brimbank (C)—Sunshine	205101182	5,852	2087	35.7
Hurstville (C)	105154150	7,129	2072	29.1
Gr. Dandenong (C) Bal	205752674	6,129	2064	33.7
Launceston (C)—Pt B	615054012	7,302	2053	28.1
Kingston (C)—North	205653431	7,422	2026	27.3
Sydney (C)—East	105057204	11,478	2024	17.6
Canning (C)	505251330	7,338	2017	27.5
Gr. Dandenong (C)—Dandenong	205752671	5,926	1932	32.6
Stonnington (C)—Prahran	205056351	10,188	1920	18.9
Bayswater (C)	505100420	6,807	1918	28.2
Port Stephens (A)	110056400	5,746	1907	33.2
Hornsby (A)—South	105604004	7,723	1896	24.6
Moreland (C)—Brunswick	205255251	7,498	1896	25.3
Campbelltown (C)—North	105301501	7,341	1889	25.7
Penrith (C)—West	105456354	7,171	1879	26.2

TABLE C.6

Fifty SLAs with the Highest Percentage of Overall Households in
Housing Stress

| SLA Name | SLA ID | Estimated No. of Households | | |
		Overall	HS	%
Melbourne (C)—Inner	205054601	4,998	1347	27.0
City (in Canberra)	805051449	328	76	23.2
Fairfield (C)—East	105252851	34,376	7219	21.0
Sydney (C)—Inner	105057201	6,015	1240	20.6
Bankstown (C)—North-East	105200351	16,935	3484	20.6
Auburn (A)	105400200	19,054	3928	20.6
Woodridge	305304656	6,531	1345	20.6
Biggera Waters-Labrador	307103508	9,000	1798	20.0
Parramatta (C)—South	105406254	9,503	1884	19.8
City—Inner (in Brisbane)	305011143	1,181	231	19.6
Playford (C)—West Central	405055688	4,653	914	19.6
St Lucia	305031506	3,251	633	19.5
Byron (A)	120101350	10,713	2070	19.3
City—Remainder	305011146	1,716	332	19.3
Southport	307103585	9,911	1913	19.3
Playford (C)—Elizabeth	405055683	10,110	1949	19.3
Beenleigh	307053461	2,967	544	18.3
Eagleby	307053466	3,285	589	17.9
Canterbury (C)	105201550	44,257	7852	17.7
Surfers Paradise	307103587	7,157	1268	17.7
Tweed (A)—Tweed Coast	120057556	3,336	581	17.4
Carrara-Merrimac	307153525	5,916	1025	17.3
Waterford West	305304654	2,060	354	17.2
Kingsholme-Upper Coomera	307153551	4,477	768	17.2
Kingston	305304612	4,038	691	17.1
Palm Beach	307103573	5,579	952	17.1
Marsden	305304623	5,763	981	17.0
Parramatta (C)—Inner	105406251	14,974	2538	16.9
Caboolture (S)—Central	305202008	6,514	1101	16.9
Redland (S) Bal	305506283	2,994	507	16.9
Varsity Lakes	307153592	4,165	705	16.9
Margate-Woody Point	305456204	4,588	773	16.8
Coffs Harbour (C)—Pt A	125011801	18,289	3055	16.7
Morayfield	305202018	6,780	1123	16.6
Coombabah	307153531	4,098	679	16.6
Maroochy (S)—Coastal North	309054905	8,561	1422	16.6

(Continued)

TABLE C.6 (*Continued*)

Fifty SLAs with the Highest Percentage of Overall Households in Housing Stress

SLA Name	SLA ID	Estimated No. of Households		
		Overall	HS	%
Maroochy (S)—Maroochydore	309054907	7,138	1182	**16.6**
Blacktown (C)—South-West	105530753	27,928	4595	**16.5**
Hume (C)—Broadmeadows	205353271	19,585	3230	**16.5**
Bilinga-Tugun	307103511	2,515	415	**16.5**
Coolangatta	307103527	2,188	360	**16.5**
Molendinar	307153564	1,855	306	**16.5**
Cairns (C)—Central Suburbs	350052065	8,194	1352	**16.5**
Port Adel. Enfield (C)—Park	405105896	5,739	943	**16.4**
Noosa (S)—Sunshine-Peregian	309055755	3,477	568	**16.3**
Liverpool (C)—East	105254901	31,165	5056	**16.2**
Strathfield (A)	105357100	10,363	1675	**16.2**
Magnetic Island	345057031	845	137	**16.2**
South Brisbane	305011525	1,618	261	**16.1**
Miami	307103563	2,562	412	**16.1**

TABLE C.7

Fifty SLAs with the Highest Percentage of Buyer Households in
Housing Stress

SLA Name	SLA ID	Estimated No. of Households		
		Buyers	HS	%
Symonston	805357929	15	9	60.0
Tableland	710353409	4	2	50.0
Petermann-Simpson	710403009	5	2	40.0
Isisford (S)	335054050	16	6	37.5
Willawong	305111615	20	7	35.0
Nhulunbuy	710252409	26	8	30.8
Nungarin (S)	525156860	23	7	30.4
Conargo (A)	155151860	180	52	28.9
Westonia (S)	525159030	11	3	27.3
Bankstown (C)—North-East	105200351	5,149	1305	25.3
Parramatta (C)—South	105406254	2,762	678	24.5
Fairfield (C)—East	105252851	10,118	2455	24.3
Dumbleyung (S)	520053010	50	12	24.0
Perry (S)	315105900	46	11	23.9
Biggenden (S)	315100700	150	35	23.3
Tiaro (S)	315106850	790	180	22.8
Central Darling (A)	160101700	76	17	22.4
Wickepin (S)	520059100	49	11	22.4
Hume (C)—Broadmeadows	205353271	7,067	1563	22.1
Auburn (A)	105400200	5,614	1229	21.9
Kilkivan (S)	315104300	402	86	21.4
Kolan (S)	315104400	574	122	21.3
Mukinbudin (S)	525155950	33	7	21.2
Clarence Valley (A) Bal	125051738	705	149	21.1
Mount Magnet (S)	535105810	19	4	21.1
Karoonda East Murray (DC)	420103080	115	24	20.9
Tambellup (S)	515058120	29	6	20.7
Rochedale	305111495	59	12	20.3
Colac-Otway (S)—South	210151755	410	83	20.2
Cook (S)	350102500	252	51	20.2
Tasman (M)	610055210	257	52	20.2
Pingelly (S)	520057140	104	21	20.2
Isis (S)	315104000	697	140	20.1
Pine Creek (CGC)	710203030	15	3	20.0
Remainder of ACT	810059009	25	5	20.0

(*Continued*)

TABLE C.7 (*Continued*)

Fifty SLAs with the Highest Percentage of Buyer Households in
Housing Stress

SLA Name	SLA ID	Estimated No. of Households		
		Buyers	HS	%
Mingenew (S)	535155530	35	7	20.0
Kyogle (A)	120104550	1,004	197	19.6
Wondai (S)	315107450	493	96	19.5
Clifton (S)	320052400	273	53	19.4
Warroo (S)	325057200	93	18	19.4
Bankstown (C)—North-West	105200353	5,841	1123	19.2
Nambucca (A)	125055700	1,777	341	19.2
Monto (S)	315105150	224	43	19.2
Tammin (S)	525108190	26	5	19.2
Walgett (A)	135107900	371	71	19.1
Barunga West (DC)	415050430	194	37	19.1
Greenough (S)—Pt B	535153854	141	27	19.1
Byron (A)	120101350	3,094	587	19.0
Inala	305071288	1,078	205	19.0
Brimbank (C)—Sunshine	205101182	9,316	1774	19.0

TABLE C.8

Fifty SLAs with the Highest Percentage of Public Renter Households in Housing Stress

| SLA Name | SLA ID | Estimated No. of Households | | |
		Public Renter	HS	%
E. Gippsland (S)—South-West	250052115	3	3	100
Riverhills	305071487	4	4	100
Pallara-Heathwood-Larapinta	305111456	3	3	100
Pittwater (A)	105656370	3	3	100
Chapman	805201089	3	3	100
Dalrymple (S)	345152700	5	4	80.0
Bridgeman Downs	305071075	5	4	80.0
Deagon	305071173	9	7	77.8
Cambooya (S)—Pt A	320012151	4	3	75.0
Fairfield	305091214	4	3	75.0
Clarence Valley (A) Bal	125051738	6	4	66.7
Boonah (S)	312100800	9	6	66.7
Pullenvale	305071473	9	6	66.7
Gunn-Palmerston City	705102811	13	8	61.5
Nakara	705051078	34	20	58.8
Palerang (A)—Pt B	145106184	14	8	57.1
Currumbin	307103533	8	4	50.0
City (in Canberra)	805051449	8	4	50.0
Wolffdene-Bahrs Scrub	307053493	4	2	50.0
Chapel Hill	305071127	4	2	50.0
Isaacs	805154419	26	13	50.0
Holder	805204059	39	17	43.6
Kingsholme-Upper Coomera	307153551	14	6	42.9
Marrara	705051064	7	3	42.9
Waramanga	805208469	134	56	41.8
Macedon Ranges (S)—Romsey	235204134	10	4	40.0
Rocklea	305111498	10	4	40.0
Melbourne (C)—Inner	205054601	46	18	39.1
Murrindindi (S)—West	240205622	13	5	38.5
Ashmore-Benowa	307153502	26	10	38.5
Calamvale	305111094	13	5	38.5
Quairading (S)	525107350	21	8	38.1
Fannie Bay	705051028	111	42	37.8
Gilmore	805253159	106	40	37.7
Geebung	305071236	19	7	36.8

(Continued)

Appendix C: Tables of the Housing Stress for 50 SLAs

TABLE C.8 (*Continued*)

Fifty SLAs with the Highest Percentage of Public Renter Households in Housing Stress

| SLA Name | SLA ID | Estimated No. of Households | | |
		Public Renter	HS	%
South Brisbane	305011525	19	7	36.8
Calwell	805250819	52	19	36.5
Melton (S)—East	205204651	11	4	36.4
Robina	307153582	22	8	36.4
Fremantle (C)—Inner	505203431	22	8	36.4
Wongan-Ballidu (S)	525109310	25	9	36
Donnybrook-Balingup (S)	510102870	37	13	35.1
Symonston	805357929	3	1	33.3
Etheridge (S)	350103100	3	1	33.3
Kingborough (M)—Pt B	610053612	3	1	33.3
Colac-Otway (S)—North	210151754	3	1	33.3
Hope Island	307153547	3	1	33.3
Melbourne (C)—S'bank-D'lands	205054605	9	3	33.3
Whittlesea (C)—North	205407071	51	17	33.3
Thuringowa (C)—Pt B	345156831	3	1	33.3

TABLE C.9

Fifty SLAs with the Highest Percentage of Private Renter Households in Housing Stress

SLA Name	SLA ID	Estimated No. of Households		
		Private Renter	HS	%
Stuart-Roseneath	345057068	34	20	58.8
Hall	805403689	12	7	58.3
Croydon (S)	350102600	9	5	55.6
Pinjarra Hills	305071465	9	5	55.6
Playford (C)—Elizabeth	405055683	1978	1098	55.5
Playford (C)—West Central	405055688	918	481	52.4
Port Adel. Enfield (C)—Park	405105896	867	449	51.8
Nungarin (S)	525156860	10	5	50.0
Hervey Bay (C)—Pt B	315103754	250	122	48.8
Fairfield (C)—East	105252851	8987	4256	47.4
Byron (A)	120101350	2997	1419	47.3
Tweed (A)—Tweed-Heads	120057554	4655	2154	46.3
Nambucca (A)	125055700	1442	666	46.2
Bellingen (A)	125050600	1017	468	46.0
Great Lakes (A)	110103400	2806	1287	45.9
Cooloola (S) (excl. Gympie)	315102532	1210	552	45.6
Redland (S) Bal	305506283	791	358	45.3
West Tamar (M)—Pt B	615105812	60	27	45.0
Bankstown (C)—North-East	105200351	4191	1882	44.9
Clarence Valley (A)—Coast	125051736	1675	749	44.7
Parramatta (C)—South	105406254	2060	917	44.5
Bankstown (C)—North-West	105200353	2932	1304	44.5
Eurobodalla (A)	145152750	2881	1278	44.4
Hepburn (S)—East	220102911	460	204	44.3
Coffs Harbour (C)—Pt A	125011801	4789	2123	44.3
Lismore (C)—Pt B	120104854	754	330	43.8
Kempsey (A)	125104350	2138	937	43.8
Hastings (A)—Pt B	125103754	1884	819	43.5
Hume (C)—Broadmeadows	205353271	3293	1427	43.3
Alexandrina (DC)—Coastal	410200221	854	369	43.2
Shoalhaven (C)—Pt B	115106952	4042	1739	43.0
Tweed (A)—Tweed Coast	120057556	1034	441	42.6
Maroochy (S) Bal	309104918	1286	547	42.5
Hastings (A)—Pt A	125033751	3927	1669	42.5
Tweed (A)—Pt B	120107558	1387	590	42.5

(Continued)

TABLE C.9 (*Continued*)

Fifty SLAs with the Highest Percentage of Private Renter Households in Housing Stress

SLA Name	SLA ID	Estimated No. of Households		
		Private Renter	HS	%
Richmond Valley (A) Bal	120106612	807	341	42.3
Greater Taree (C)	125103350	3499	1473	42.1
Bribie Island	305202002	1815	763	42.0
Playford (C)—West	405055686	319	134	42.0
Coffs Harbour (C)—Pt B	125051804	1299	543	41.8
Caboolture (S)—Hinterland	305202014	288	120	41.7
Port Adel. Enfield (C)—Inner	405055894	1667	692	41.5
Yarra Ranges (S)—Central	205607451	781	324	41.5
Noosa (S) Bal	309105758	867	358	41.3
Tiaro (S)	315106850	251	103	41.0
Clarence Valley (A) Bal	125051738	262	107	40.8
Woodridge	305304656	2415	986	40.8
Nanango (S)	315105650	576	235	40.8
Wollongong (C) Bal	115058454	4254	1731	40.7
Sorell (M)—Pt A	605054811	615	250	40.7

TABLE C.10

Fifty SLAs with the Highest Percentage of Total Renter Households in Housing Stress

SLA Name	SLA ID	Estimated No. of Households		
		Total Renter	HS	%
Stuart-Roseneath	345057068	34	20	58.8
Pinjarra Hills	305071465	9	5	55.6
Hervey Bay (C)—Pt B	315103754	262	124	47.3
Byron (A)	120101350	3,203	1468	45.8
West Tamar (M)—Pt B	615105812	60	27	45.0
Cooloola (S) (excl. Gympie)	315102532	1,251	558	44.6
Redland (S) Bal	305506283	829	368	44.4
Clarence Valley (A)—Coast	125051736	1,746	762	43.6
Great Lakes (A)	110103400	3,048	1330	43.6
Bellingen (A)	125050600	1,125	490	43.6
Lismore (C)—Pt B	120104854	766	332	43.3
Tweed (A)—Tweed-Heads	120057554	5,323	2285	42.9
Hall	805403689	21	9	42.9
Maroochy (S) Bal	309104918	1,325	556	42.0
Alexandrina (DC)—Coastal	410200221	912	382	41.9
Caboolture (S)—Hinterland	305202014	288	120	41.7
Hepburn (S)—East	220102911	514	214	41.6
Clarence Valley (A) Bal	125051738	268	111	41.4
Tweed (A)—Tweed Coast	120057556	1,104	457	41.4
Shoalhaven (C)—Pt B	115106952	4,310	1784	41.4
Hastings (A)—Pt B	125103754	2,069	850	41.1
Tiaro (S)	315106850	251	103	41.0
Richmond Valley (A) Bal	120106612	855	350	40.9
Nambucca (A)	125055700	1,732	709	40.9
Eurobodalla (A)	145152750	3,290	1344	40.9
Croydon (S)	350102600	27	11	40.7
Coffs Harbour (C)—Pt B	125051804	1,371	558	40.7
Yarra Ranges (S)—Central	205607451	806	326	40.4
Tweed (A)—Pt B	120107558	1,527	616	40.3
Fairfield (C)—East	105252851	11,865	4759	40.1
Chandler-Capalaba West	305111123	35	14	40.0
Rochedale	305111495	88	35	39.8
Noosa (S) Bal	309105758	924	367	39.7
Caloundra (C)—Hinterland	309102136	559	222	39.7
Kempsey (A)	125104350	2,516	998	39.7

(Continued)

TABLE C.10 (*Continued*)

Fifty SLAs with the Highest Percentage of Total Renter Households in Housing Stress

SLA Name	SLA ID	Estimated No. of Households		
		Total Renter	HS	%
Mount Alexander (S) Bal	235105434	465	183	39.4
Bribie Island	305202002	2,043	803	39.3
Coombabah	307153531	1,393	546	39.2
Coffs Harbour (C)—Pt A	125011801	5,939	2321	39.1
Hastings (A)—Pt A	125033751	4,617	1799	39.0
Wyong (A)—North-East	105708551	6,509	2520	38.7
Bilinga-Tugun	307103511	862	333	38.6
Yankalilla (DC)	410208750	272	105	38.6
Moorabool (S)—West	220105155	128	49	38.3
Greater Taree (C)	125103350	4,118	1573	38.2
Nanango (S)	315105650	632	241	38.1
Caboolture (S)—East	305202013	1,154	435	37.7
Mount Morgan (S)	330155350	268	101	37.7
Gr. Bendigo (C)—Pt B	235102628	385	145	37.7
Bankstown (C)—North-East	105200351	5,798	2178	37.6

Appendix D: Distribution of SLAs, Households, and Housing Stress by SSDs in Eight Major Capital Cities

TABLE D.1

Distributions of SLAs, Households, and Housing Stress by SSDs in Sydney

SSD Name	No. of SLA	No. of Households		
		Total	HS	%
St George–Sutherland	5	150,204	14,748	9.82
Central Northern Sydney	6	133,654	8,815	6.60
Inner Sydney	7	121,059	14,589	12.05
Lower Northern Sydney	6	111,980	9,140	8.16
Gosford–Wyong	4	110,908	14,365	12.95
Outer Western Sydney	4	104,182	11,640	11.17
Fairfield–Liverpool	4	103,162	17,464	16.93
Central Western Sydney	6	101,045	15,352	15.19
Canterbury-Bankstown	4	99,122	15,935	16.08
Eastern Suburbs	3	87,214	8,568	9.82
Blacktown	3	85,609	11,322	13.23
Northern Beaches	3	81,130	6,149	7.58
Outer South Western Sydney	4	74,356	8,837	11.88
Inner Western Sydney	5	59,915	6,731	11.23
The SD of Sydney	*64*	*1,423,540*	*163,655*	*11.50*

TABLE D.2

Distributions of SLAs, Households, and Housing Stress by SSDs in Melbourne

SSD Name	No. of SLAs	No. of Households		
		Total	HS	%
Western Melbourne	7	149,021	17,098	11.47
Eastern Middle Melbourne	8	148,237	12,316	8.31
Southern Melbourne	7	147,154	13,338	9.06
Inner Melbourne	8	116,387	14,264	12.26
Northern Middle Melbourne	4	90,689	9,199	10.14
South Eastern Outer Melbourne	7	87,886	10,446	11.89
Eastern Outer Melbourne	5	86,439	7,826	9.05
Melton–Wyndham	5	61,891	7,331	11.85
Northern Outer Melbourne	6	58,131	5,710	9.82
Boroondara City	4	55,774	4,103	7.36
Moreland City	3	52,151	5,992	11.49
Mornington Peninsula Shire	3	49,870	5,109	10.24
Yarra Ranges Shire Part A	5	47,744	4,649	9.74
Hume City	3	45,678	6,453	14.13
Frankston City	2	43,551	5,484	12.59
Greater Dandenong City	2	42,725	6,384	14.94
The SD of Melbourne	*79*	*1,283,328*	*135,702*	*10.57*

TABLE D.3

Distributions of SLAs, Households, and Housing Stress by SSDs in Brisbane

SSD Name	No. of SLA[a]	No. of Households		
		Total	HS	%
Northwest Outer Brisbane	53	110,702	9,339	8.44
Southeast Outer Brisbane	39 (1)	79,578	8,345	10.49
Northwest Inner Brisbane	27	69,080	7,154	10.36
Southeast Inner Brisbane	21	59,428	5,451	9.17
Logan City	17	57,268	7,670	13.39
Ipswich City	5	47,559	5,573	11.72
Pine Rivers Shire	10	46,963	4,021	8.56
Caboolture Shire	8	45,874	6,324	13.79
Redland Shire	12	44,691	4,526	10.13
Inner Brisbane	18	34,165	4,227	12.37
Redcliffe City	4	20,619	2,806	13.61
Beaudesert Shire Part A	1	12,263	1,282	10.45
The SD of Brisbane	*215 (1)*	*628,190*	*66,718*	*10.62*

[a] No. of missing SLAs in the parenthesis.

TABLE D.4

Distributions of SLAs, Households, and Housing Stress by SSDs in Perth

| SSD Name | No. of SLAs[a] | No. of Households | | |
		Total	HS	%
North Metropolitan	8	159,758	16,090	10.07
South East Metropolitan	7	121,582	13,417	11.04
South West Metropolitan	7	110,803	11,003	9.93
East Metropolitan	5	88,467	8,934	10.10
Central Metropolitan	10 (1)	47,489	4,322	9.10
The SD of Perth	*37 (1)*	*528,099*	*53,766*	*10.18*

[a] No. of missing SLAs in the parenthesis.

TABLE D.5

Distributions of SLAs, Households, and Housing Stress by SSDs in Adelaide

| SSD Name | No. of SLAs[a] | No. of Households | | |
		Total	HS	%
Northern Adelaide	17	131,597	15,626	11.87
Southern Adelaide	15	126,369	12,689	10.04
Eastern Adelaide	13	88,284	8,634	9.78
Western Adelaide	10 (1)	84,523	9,800	11.59
The SD of Adelaide	*55 (1)*	*430,773*	*46,749*	*10.85*

[a] No. of missing SLAs in the parenthesis.

TABLE D.6

Distributions of SLAs, Households, and Housing Stress by SSDs in Canberra

| SSD Name | No. of SLAs[a] | No. of Households | | |
		Total	HS	%
Belconnen	25 (1)	30,325	2083	6.87
Tuggeranong	19 (1)	29,404	1932	6.57
North Canberra	17 (3)	16,046	1333	8.31
Woden Valley	12 (0)	12,480	690	5.53
Gungahlin-Hall	9 (2)	10,618	693	6.53
South Canberra	16 (5)	9,538	515	5.40
Weston Creek-Stromlo	10 (2)	8,423	448	5.32
ACT—Bal	1 (0)	77	6	7.79
The SD of Canberra	*109 (14)*	*116,911*	*7700*	*6.59*

[a] No. of missing SLAs in the parenthesis.

TABLE D.7

Distributions of SLAs, Households, and Housing Stress by SSD in Hobart

SSD Name	No. of SLAs[a]	No. of Households		
		Total	HS	%
The SD of Greater Hobart	8 (1)	76,361	7856	10.29

[a] No. of missing SLAs in the parenthesis.

TABLE D.8

Distributions of SLAs, Households, and Housing Stress by SSDs in Darwin

SSD Name	No. of SLAs[a]	No. of Households		
		Total	HS	%
Darwin City	30 (3)	22,121	2010	9.09
Palmerston-East Arm	9 (1)	7,396	826	11.17
Litchfield Shire	2 (0)	4,620	335	7.25
The SD of Darwin	*41 (4)*	*34,137*	*3171*	*9.29*

[a] No. of missing SLAs in the parenthesis.

Appendix E: Spatial Analyses by Households Tenure Types for the Eight Capital Cities

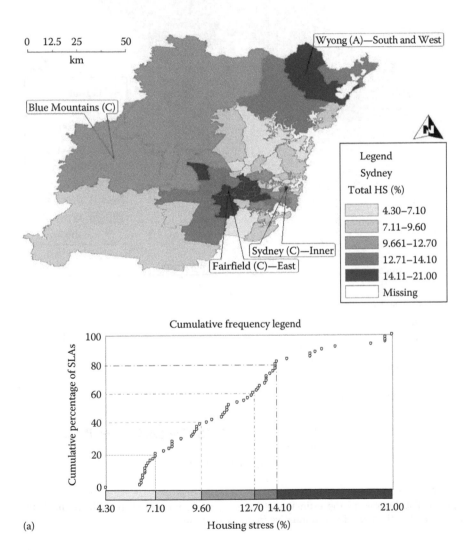

(a)

FIGURE E.1
Spatial distributions of the estimates of housing stress for overall households in Sydney. (a) Percentage estimates with a cumulative frequency legend and   (*Continued*)

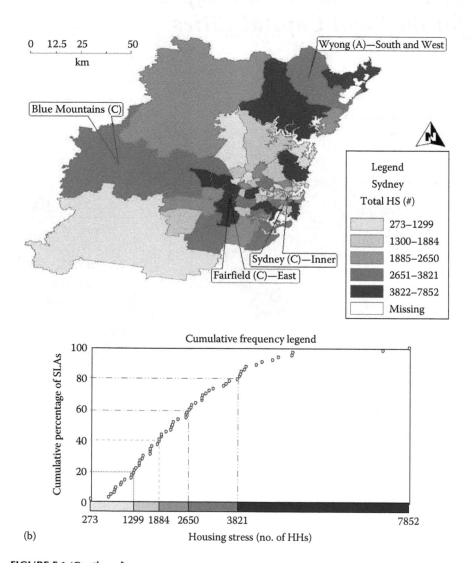

FIGURE E.1 (Continued)
Spatial distributions of the estimates of housing stress for overall households in Sydney.
(b) Number estimates with a cumulative frequency legend.

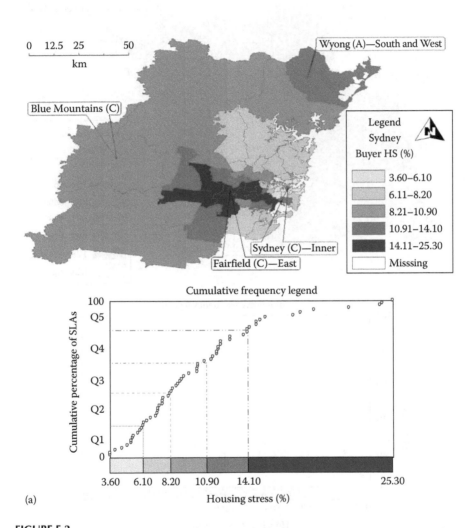

FIGURE E.2
Spatial distributions of the estimates of housing stress for buyer households in Sydney.
(a) Percentage estimates with a cumulative frequency legend and   (*Continued*)

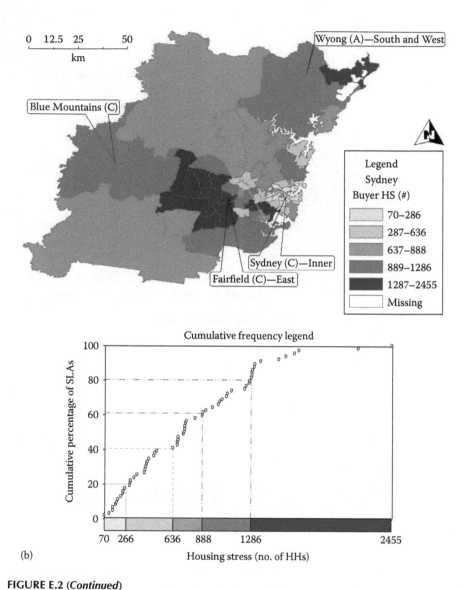

(b)

FIGURE E.2 (*Continued*)
Spatial distributions of the estimates of housing stress for buyer households in Sydney.
(b) Number estimates with a cumulative frequency legend.

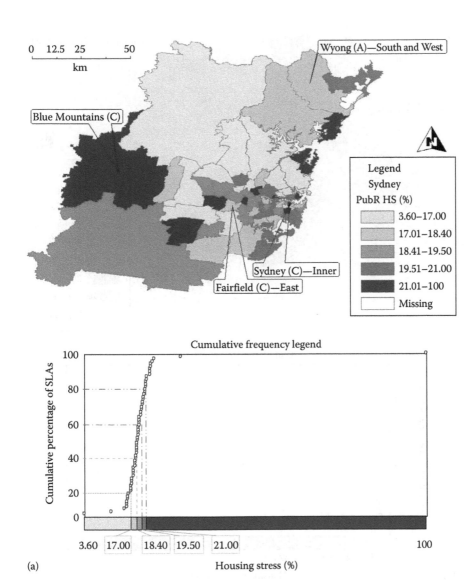

FIGURE E.3
Spatial distributions of the estimates of housing stress for public renter households in Sydney.
(a) Percentage estimates with a cumulative frequency legend and   *(Continued)*

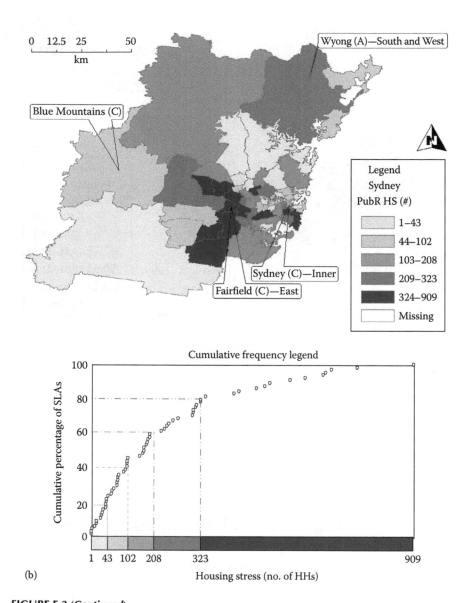

(b)

FIGURE E.3 (*Continued*)
Spatial distributions of the estimates of housing stress for public renter households in Sydney.
(b) Number estimates with a cumulative frequency legend.

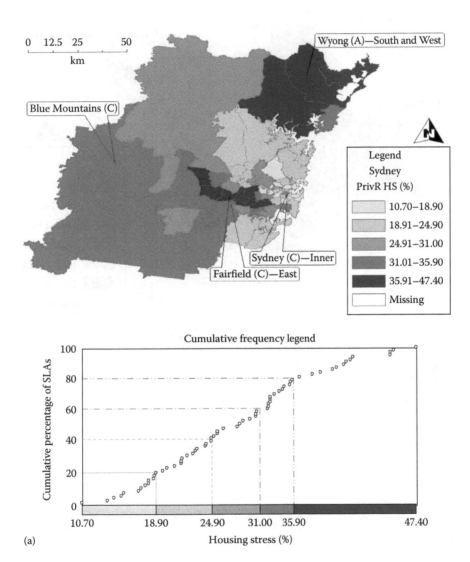

FIGURE E.4

Spatial distributions of the estimates of housing stress for private renter households in Sydney.
(a) Percentage estimates with a cumulative frequency legend and *(Continued)*

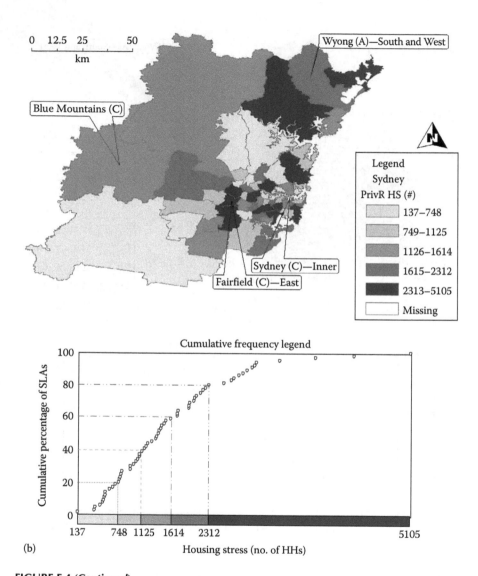

FIGURE E.4 (*Continued*)

Spatial distributions of the estimates of housing stress for private renter households in Sydney.
(b) Number estimates with a cumulative frequency legend.

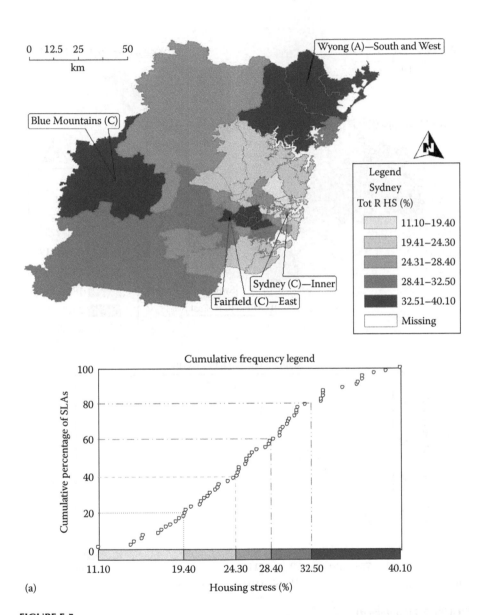

(a)

FIGURE E.5

Spatial distributions of the estimates of housing stress for total renter households in Sydney.
(a) Percentage estimates with a cumulative frequency legend and   (*Continued*)

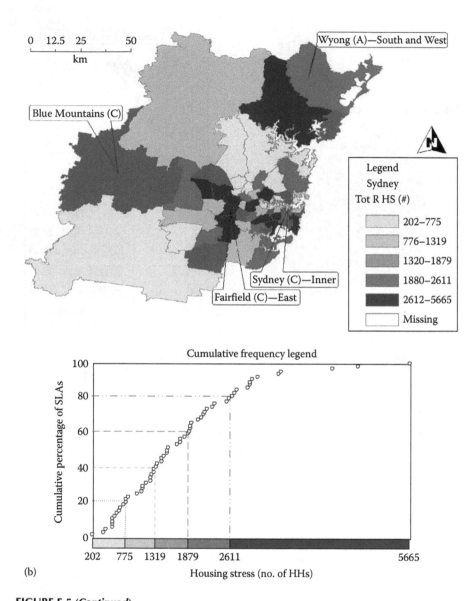

FIGURE E.5 (Continued)
Spatial distributions of the estimates of housing stress for total renter households in Sydney.
(b) Number estimates with a cumulative frequency legend.

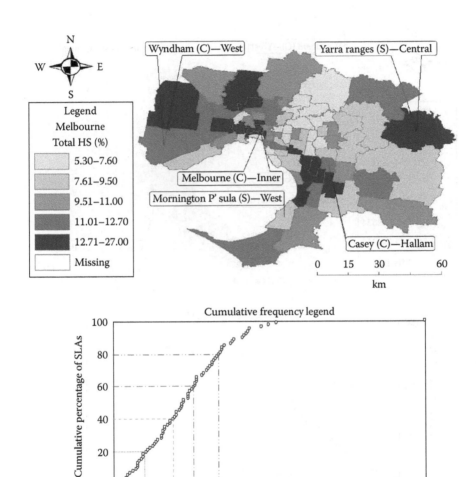

(Continued)

FIGURE E.6
Spatial distributions of the estimates of housing stress for overall households in Melbourne.
(a) Percentage estimates with a cumulative frequency legend and

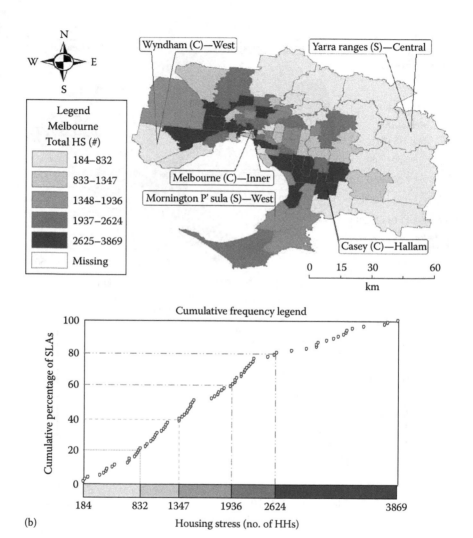

(b)

FIGURE E.6 (Continued)
Spatial distributions of the estimates of housing stress for overall households in Melbourne.
(b) Number estimates with a cumulative frequency legend.

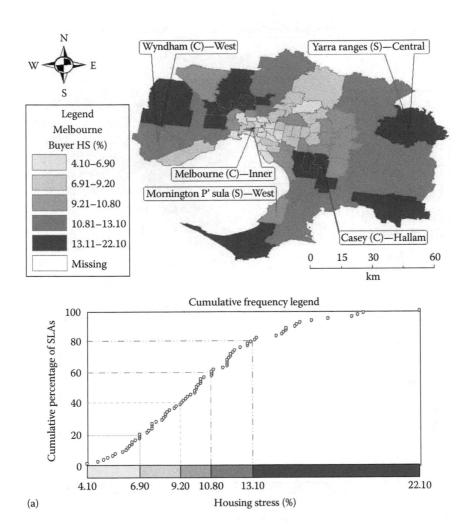

FIGURE E.7
Spatial distributions of the estimates of housing stress for buyer households in Melbourne.
(a) Percentage estimates with a cumulative frequency legend and   (*Continued*)

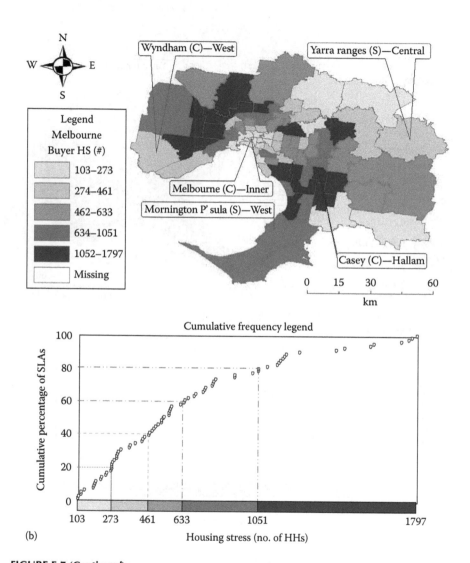

(b)

FIGURE E.7 (*Continued*)
Spatial distributions of the estimates of housing stress for buyer households in Melbourne.
(b) Number estimates with a cumulative frequency legend.

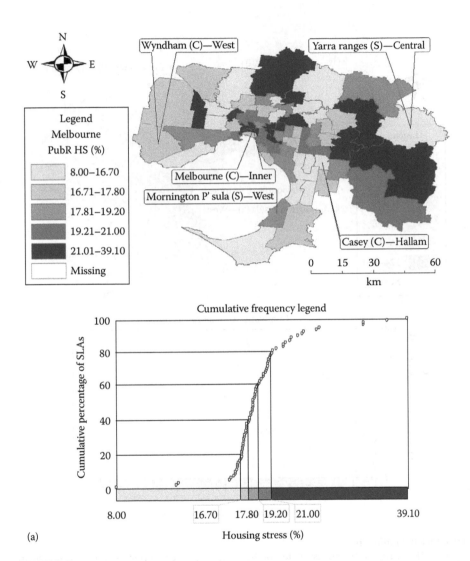

FIGURE E.8
Spatial distributions of the estimates of housing stress for public renter households in Melbourne. (a) Percentage estimates with a cumulative frequency legend and   (*Continued*)

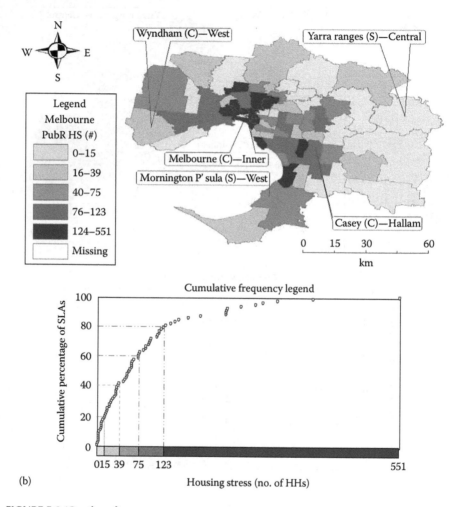

(b)

FIGURE E.8 (*Continued*)
Spatial distributions of the estimates of housing stress for public renter households in
Melbourne. (b) Number estimates with a cumulative frequency legend.

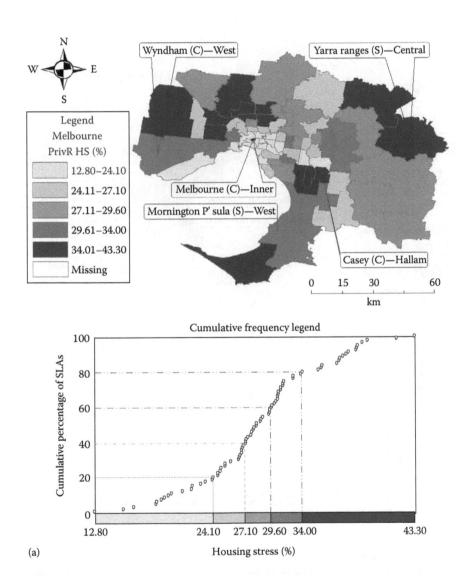

FIGURE E.9
Spatial distributions of the estimates of housing stress for private renter households in Melbourne. (a) Percentage estimates with a cumulative frequency legend and *(Continued)*

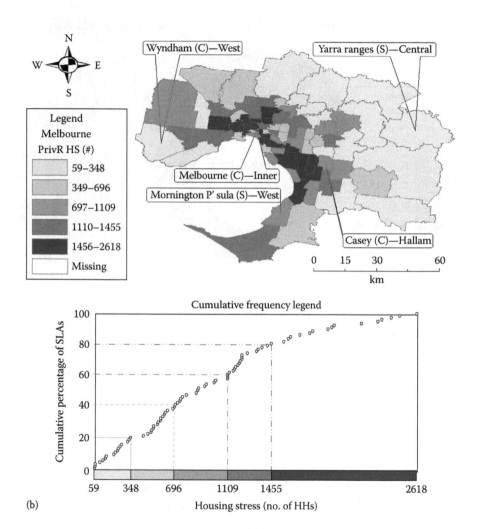

FIGURE E.9 (*Continued*)
Spatial distributions of the estimates of housing stress for private renter households in Melbourne. (b) Number estimates with a cumulative frequency legend.

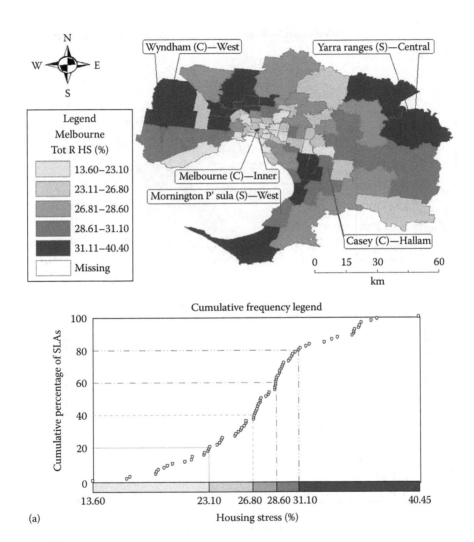

FIGURE E.10
Spatial distributions of the estimates of housing stress for total renter households in Melbourne.
(a) Percentage estimates with a cumulative frequency legend and   (*Continued*)

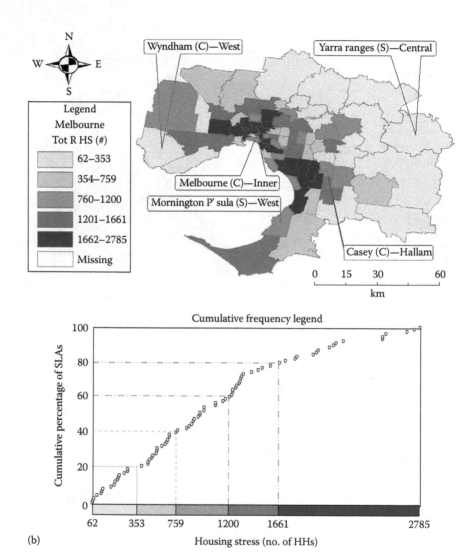

(b)

FIGURE E.10 (*Continued*)
Spatial distributions of the estimates of housing stress for total renter households in Melbourne.
(b) Number estimates with a cumulative frequency legend.

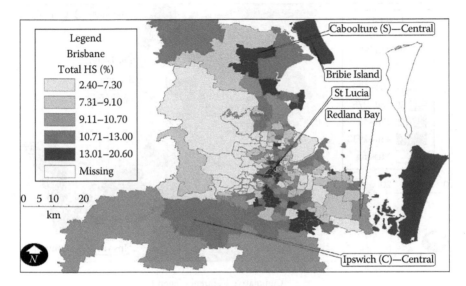

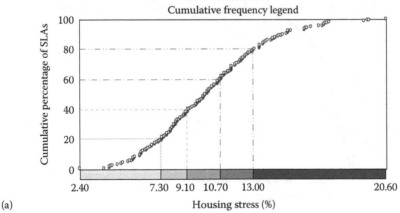

(a)

FIGURE E.11
Spatial distributions of the estimates of housing stress for overall households in Brisbane.
(a) Percentage estimates with a cumulative frequency legend and   *(Continued)*

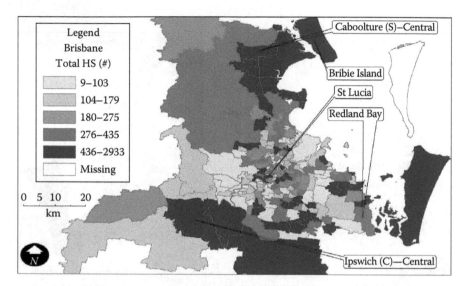

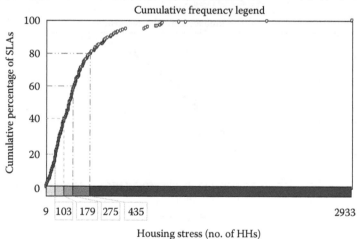

(b)

FIGURE E.11 (*Continued*)
Spatial distributions of the estimates of housing stress for overall households in Brisbane.
(b) Number estimates with a cumulative frequency legend.

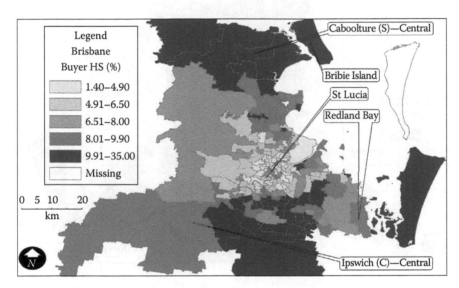

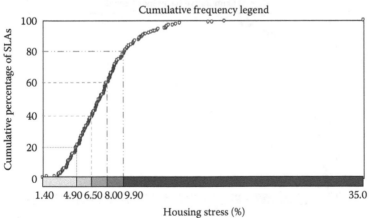

(a)

FIGURE E.12
Spatial distributions of the estimates of housing stress for buyer households in Brisbane.
(a) Percentage estimates with a cumulative frequency legend and   (*Continued*)

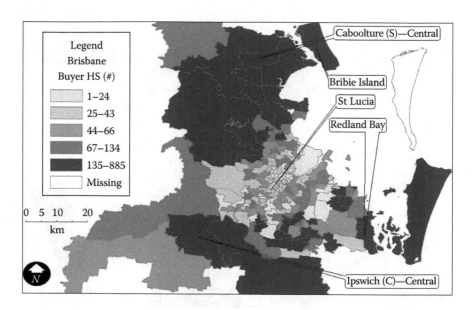

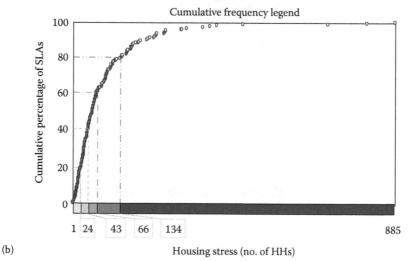

FIGURE E.12 (Continued)
Spatial distributions of the estimates of housing stress for buyer households in Brisbane.
(b) Number estimates with a cumulative frequency legend.

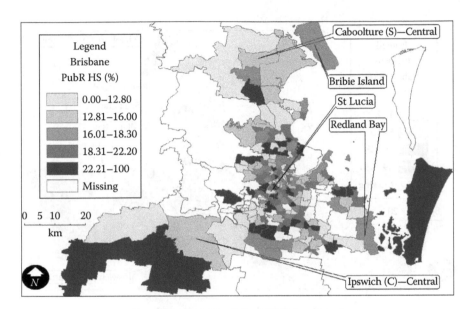

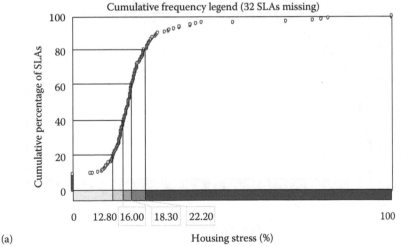

(a)

FIGURE E.13
Spatial distributions of the estimates of housing stress for public renter households in Brisbane.
(a) Percentage estimates with a cumulative frequency legend and   *(Continued)*

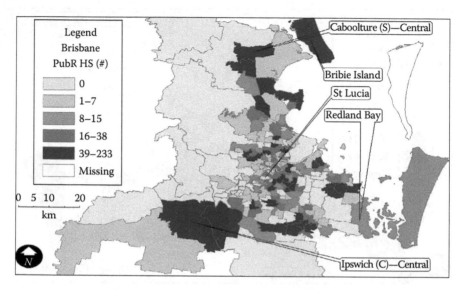

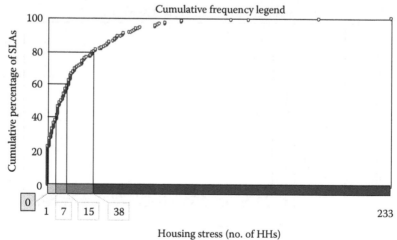

FIGURE E.13 (*Continued*)

Spatial distributions of the estimates of housing stress for public renter households in Brisbane. (b) Number estimates with a cumulative frequency legend.

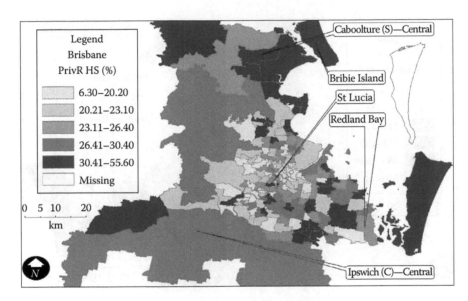

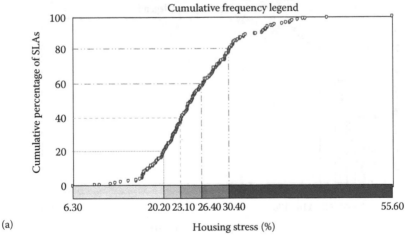

(a)

FIGURE E.14

Spatial distributions of the estimates of housing stress for private renter households in Brisbane. (a) Percentage estimates with a cumulative frequency legend and   (*Continued*)

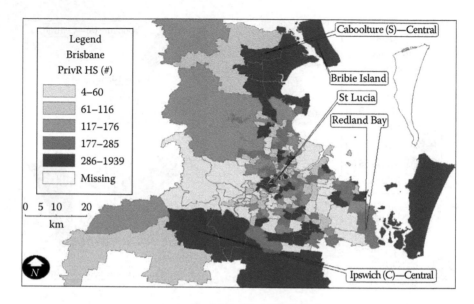

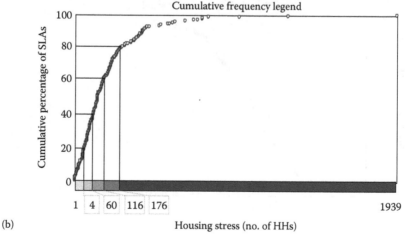

(b)

FIGURE E.14 (*Continued*)
Spatial distributions of the estimates of housing stress for private renter households in Brisbane. (b) Number estimates with a cumulative frequency legend.

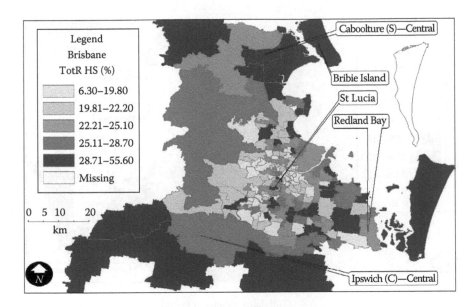

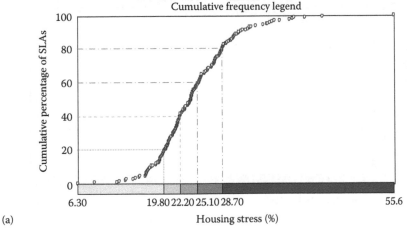

(a)

FIGURE E.15
Spatial distributions of the estimates of housing stress for total renter households in Brisbane.
(a) Percentage estimates with a cumulative frequency legend   (*Continued*)

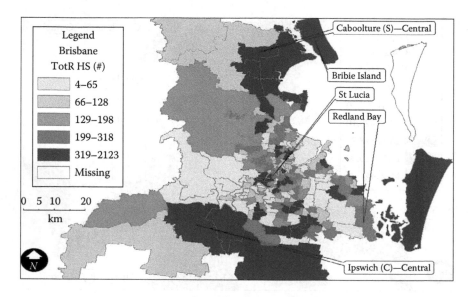

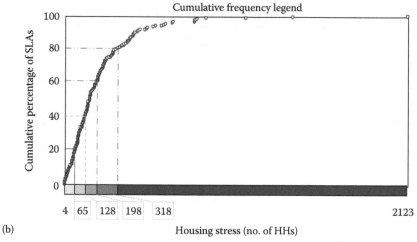

FIGURE E.15 (*Continued*)
Spatial distributions of the estimates of housing stress for total renter households in Brisbane.
(b) Number estimates with a cumulative frequency legend.

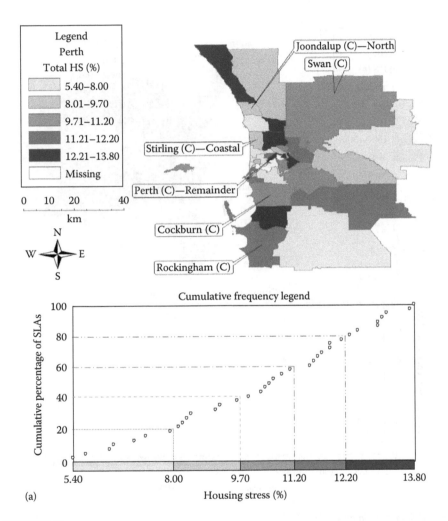

FIGURE E.16
Spatial distributions of the estimates of housing stress for overall households in Perth.
(a) Percentage estimates with a cumulative frequency legend and   *(Continued)*

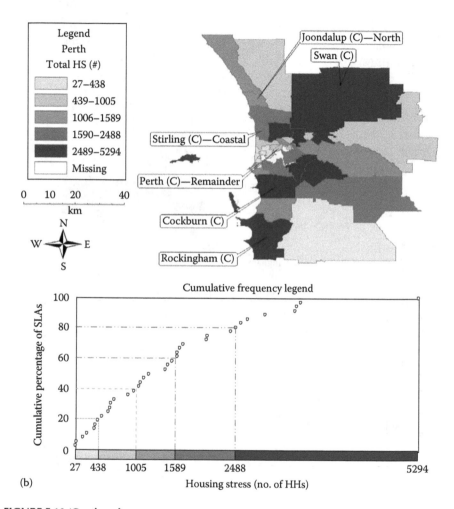

FIGURE E.16 (*Continued*)
Spatial distributions of the estimates of housing stress for overall households in Perth.
(b) Number estimates with a cumulative frequency legend.

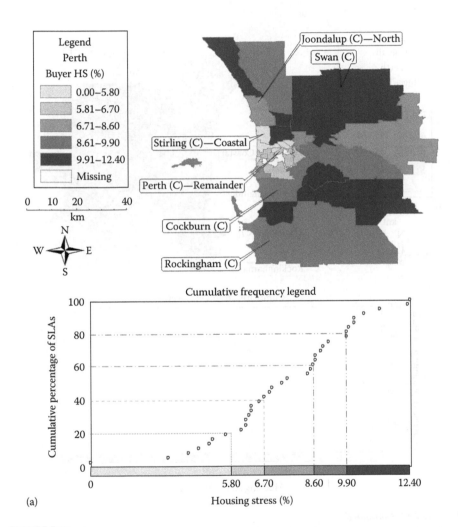

(a)

FIGURE E.17
Spatial distributions of the estimates of housing stress for buyer households in Perth. (a) Percentage estimates with a cumulative frequency legend and *(Continued)*

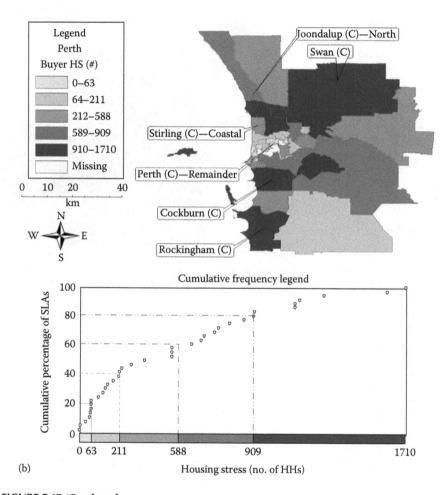

FIGURE E.17 (*Continued*)
Spatial distributions of the estimates of housing stress for buyer households in Perth. (b) Number estimates with a cumulative frequency legend.

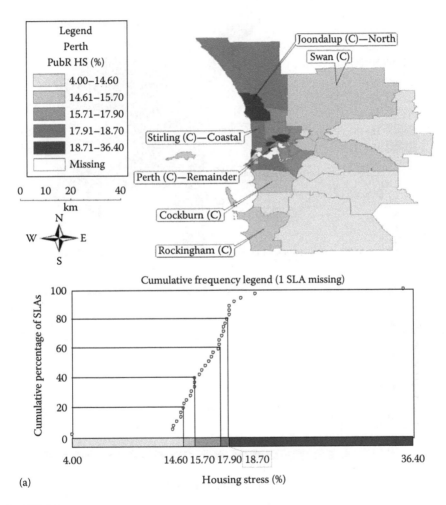

FIGURE E.18

Spatial distributions of the estimates of housing stress for public renter households in Perth.
(a) Percentage estimates with a cumulative frequency legend and *(Continued)*

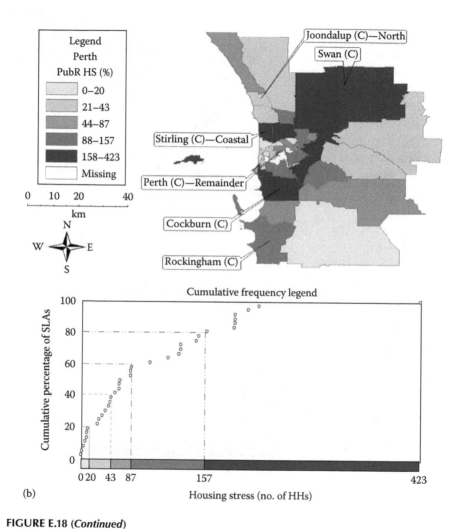

FIGURE E.18 (*Continued*)
Spatial distributions of the estimates of housing stress for public renter households in Perth.
(b) Number estimates with a cumulative frequency legend.

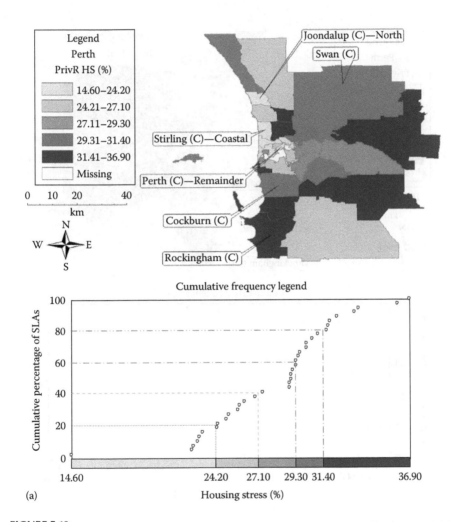

FIGURE E.19
Spatial distributions of the estimates of housing stress for private renter households in Perth.
(a) Percentage estimates with a cumulative frequency legend and　　　　　*(Continued)*

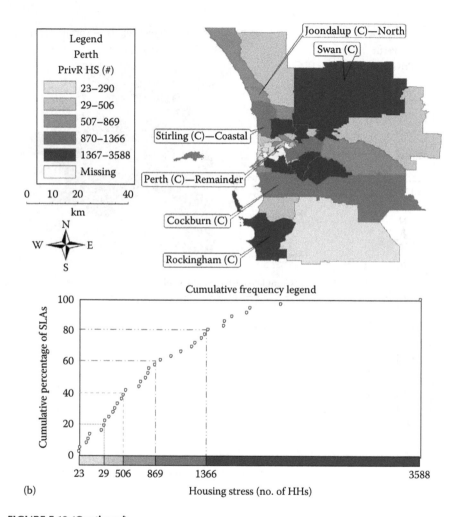

(b)

FIGURE E.19 (*Continued*)
Spatial distributions of the estimates of housing stress for private renter households in Perth.
(b) Number estimates with a cumulative frequency legend.

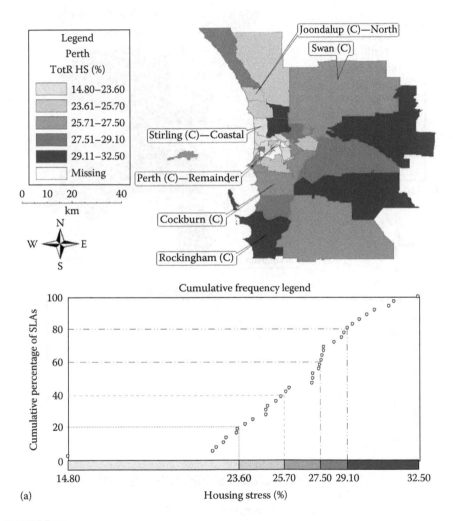

FIGURE E.20
Spatial distributions of the estimates of housing stress for total renter households in Perth.
(a) Percentage estimates with a cumulative frequency legend and   *(Continued)*

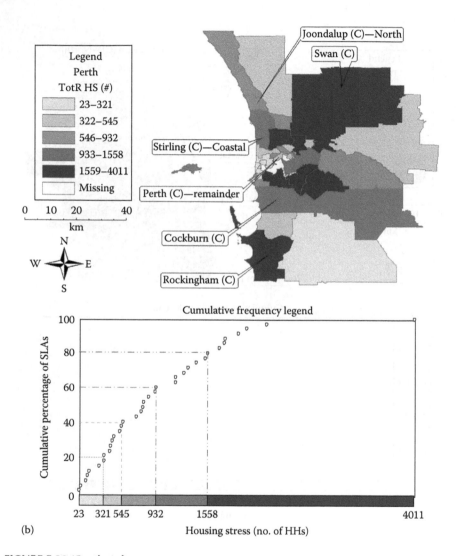

FIGURE E.20 (*Continued*)
Spatial distributions of the estimates of housing stress for total renter households in Perth.
(b) Number estimates with a cumulative frequency legend.

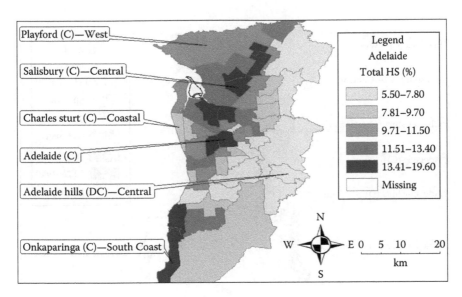

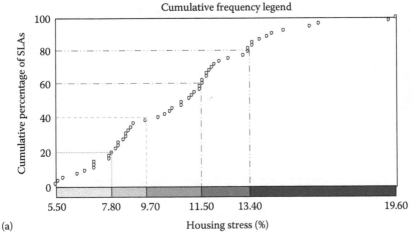

(a)

FIGURE E.21
Spatial distributions of the estimates of housing stress for overall households in Adelaide. (a) Percentage estimates with a cumulative frequency legend and *(Continued)*

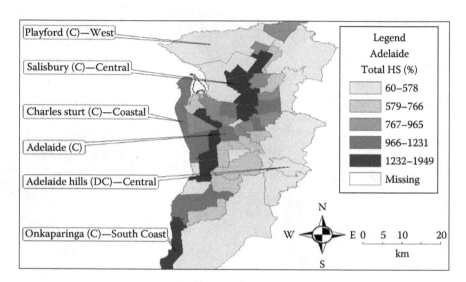

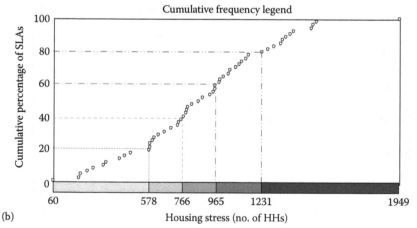

(b)

FIGURE E.21 (*Continued*)
Spatial distributions of the estimates of housing stress for overall households in Adelaide.
(b) Number estimates with a cumulative frequency legend.

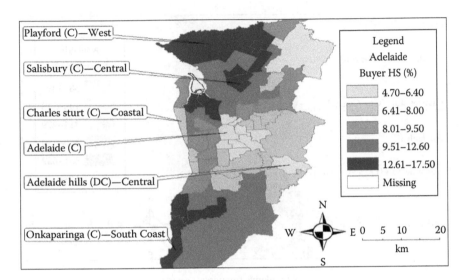

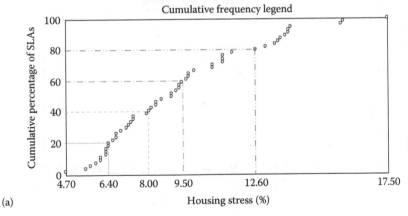

(a)

FIGURE E.22

Spatial distributions of the estimates of housing stress for buyer households in Adelaide. (a) Percentage estimates with a cumulative frequency legend and *(Continued)*

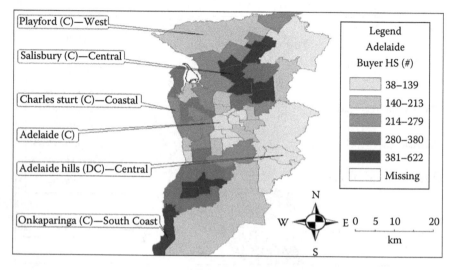

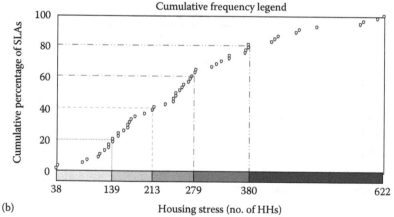

FIGURE E.22 (*Continued*)
Spatial distributions of the estimates of housing stress for buyer households in Adelaide.
(b) Number estimates with a cumulative frequency legend.

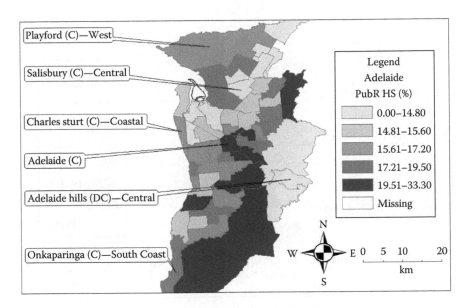

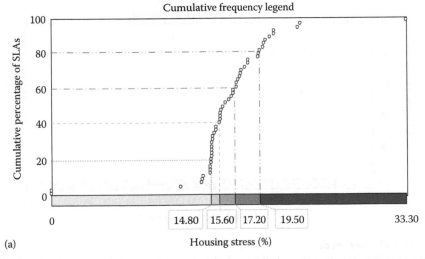

(a)

FIGURE E.23
Spatial distributions of the estimates of housing stress for public renter households in
Adelaide. (a) Percentage estimates with a cumulative frequency legend and   (*Continued*)

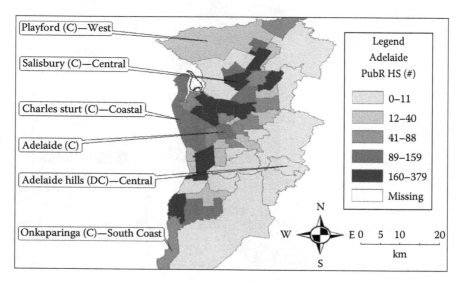

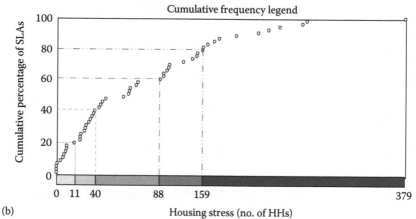

(b)

FIGURE E.23 (Continued)
Spatial distributions of the estimates of housing stress for public renter households in Adelaide. (b) Number estimates with a cumulative frequency legend.

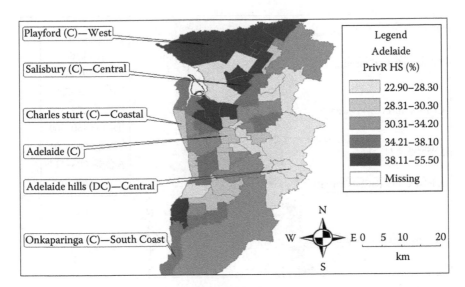

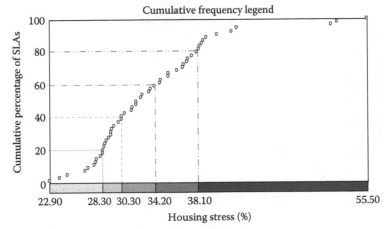

(a)

FIGURE E.24

Spatial distributions of the estimates of housing stress for private renter households in Adelaide. (a) Percentage estimates with a cumulative frequency legend and *(Continued)*

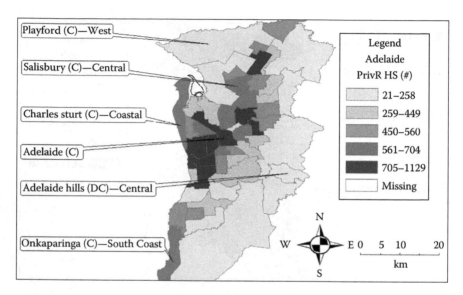

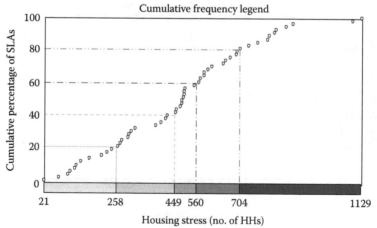

(b)

FIGURE E.24 (Continued)
Spatial distributions of the estimates of housing stress for private renter households in Adelaide. (b) Number estimates with a cumulative frequency legend.

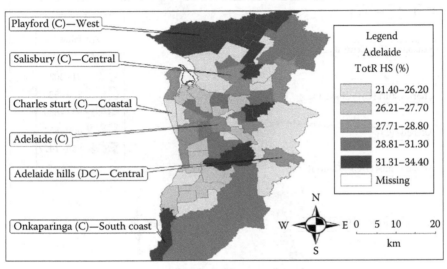

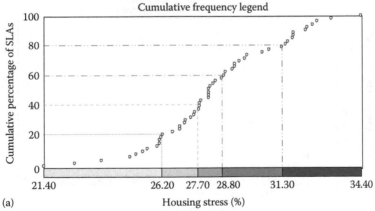

(a)

Housing stress (%)

FIGURE E.25

Spatial distributions of the estimates of housing stress for total renter households in Adelaide.
(a) Percentage estimates with a cumulative frequency legend and   (*Continued*)

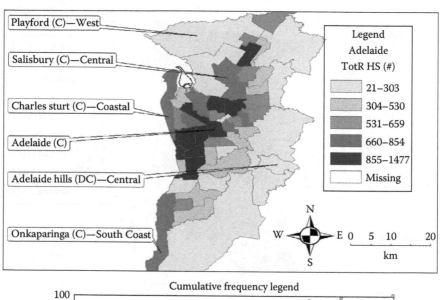

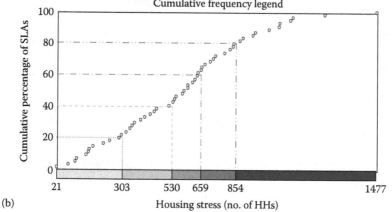

(b)

FIGURE E.25 (Continued)
Spatial distributions of the estimates of housing stress for total renter households in Adelaide.
(b) Number estimates with a cumulative frequency legend.

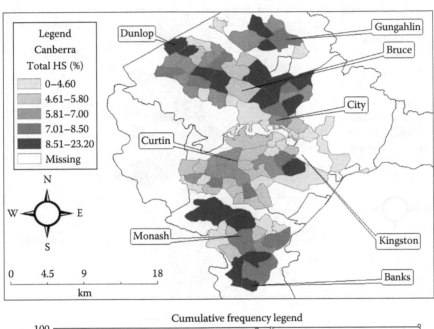

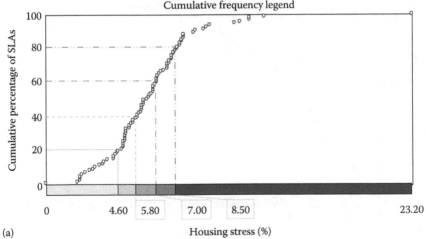

(a)

FIGURE E.26
Spatial distributions of the estimates of housing stress for overall households in Canberra.
(a) Percentage estimates with a cumulative frequency legend and   *(Continued)*

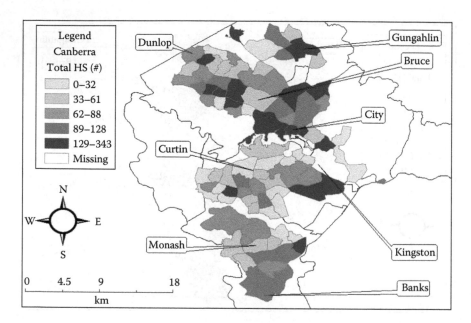

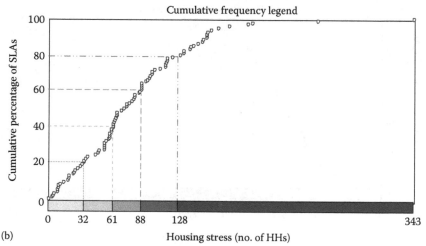

FIGURE E.26 (*Continued*)
Spatial distributions of the estimates of housing stress for overall households in Canberra.
(b) Number estimates with a cumulative frequency legend.

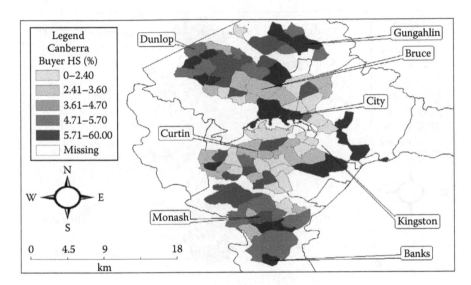

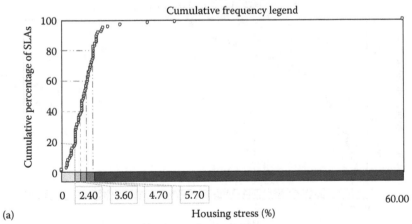

FIGURE E.27
Spatial distributions of the estimates of housing stress for buyer households in Canberra. (a) Percentage estimates with a cumulative frequency legend and   *(Continued)*

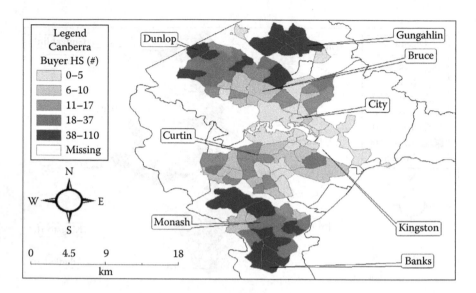

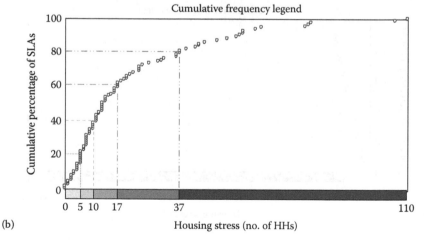

FIGURE E.27 (*Continued*)
Spatial distributions of the estimates of housing stress for buyer households in Canberra.
(b) Number estimates with a cumulative frequency legend.

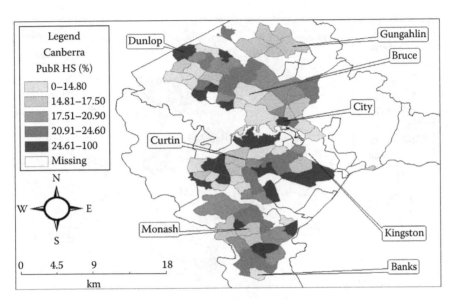

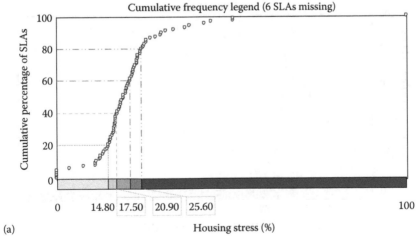

(a)

FIGURE E.28

Spatial distributions of the estimates of housing stress for public renter households in Canberra.
(a) Percentage estimates with a cumulative frequency legend and   (*Continued*)

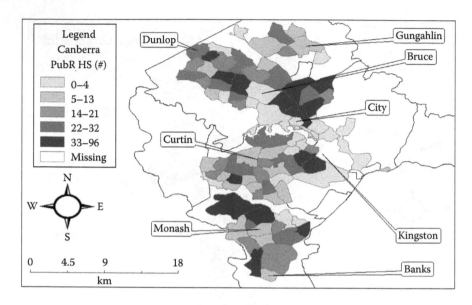

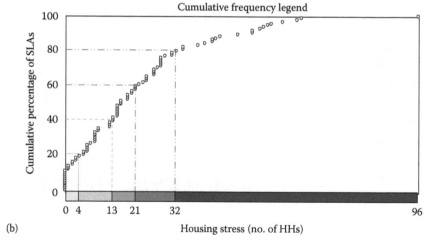

FIGURE E.28 (*Continued*)
Spatial distributions of the estimates of housing stress for public renter households in Canberra.
(b) Number estimates with a cumulative frequency legend.

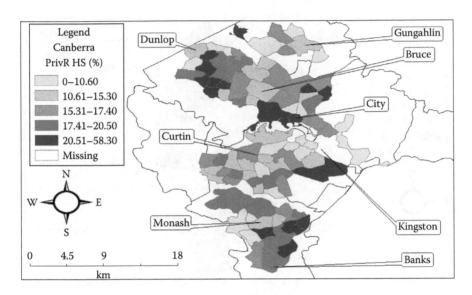

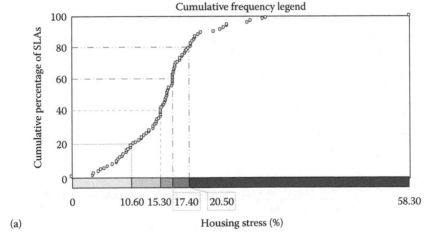

(a)

FIGURE E.29

Spatial distributions of the estimates of housing stress for private renter households in Canberra. (a) Percentage estimates with a cumulative frequency legend and (*Continued*)

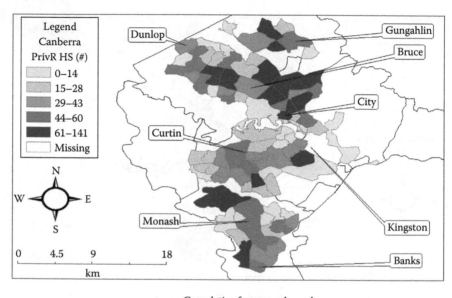

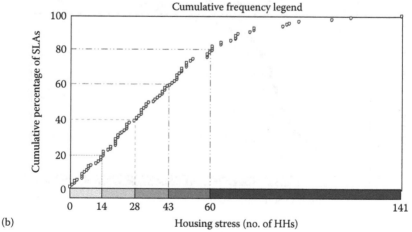

(b)

FIGURE E.29 (*Continued*)
Spatial distributions of the estimates of housing stress for private renter households in Canberra. (b) Number estimates with a cumulative frequency legend.

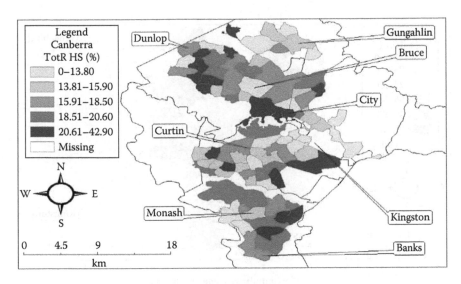

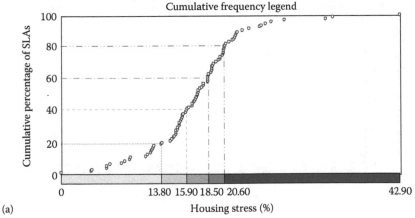

(a)

FIGURE E.30

Spatial distributions of the estimates of housing stress for total renter households in Canberra. (a) Percentage estimates with a cumulative frequency legend and *(Continued)*

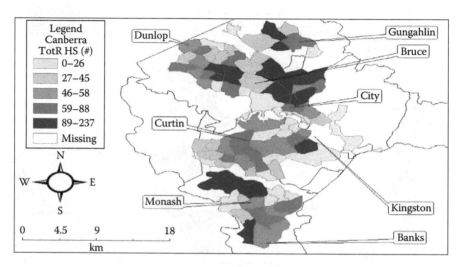

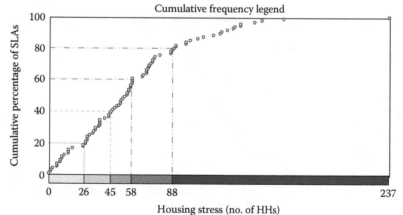

(b)

FIGURE E.30 (*Continued*)
Spatial distributions of the estimates of housing stress for total renter households in Canberra.
(b) Number estimates with a cumulative frequency legend.

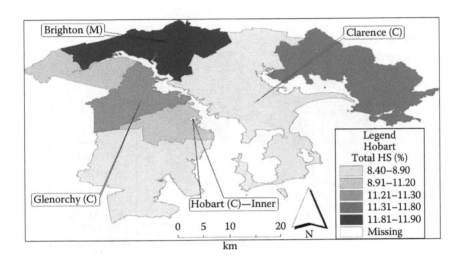

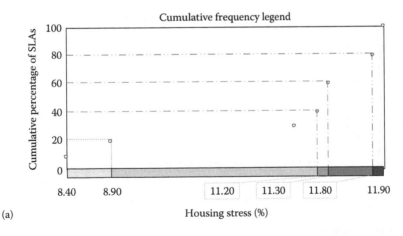

(a)

FIGURE E.31
Spatial distributions of the estimates of housing stress for overall households in Hobart.
(a) Percentage estimates with a cumulative frequency legend and   (*Continued*)

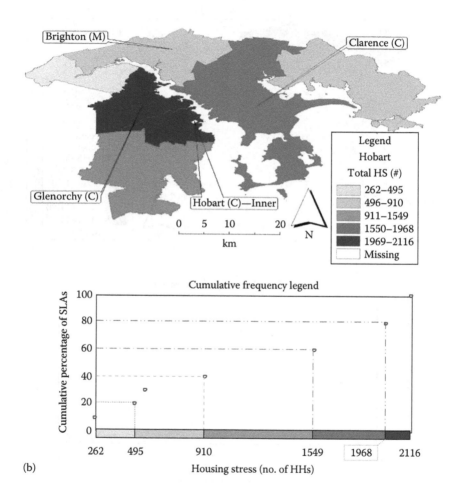

FIGURE E.31 (Continued)
Spatial distributions of the estimates of housing stress for overall households in Hobart. (b) Number estimates with a cumulative frequency legend.

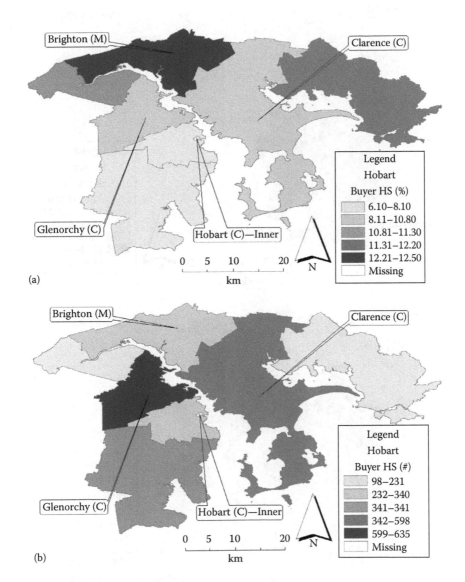

FIGURE E.32
Spatial distributions of the estimates of housing stress for buyer households in Hobart.
(a) Percentage estimates (as there is only 7 SLAs for analysis, the cumulative frequency legend
is ignorable) and (b) number estimates.

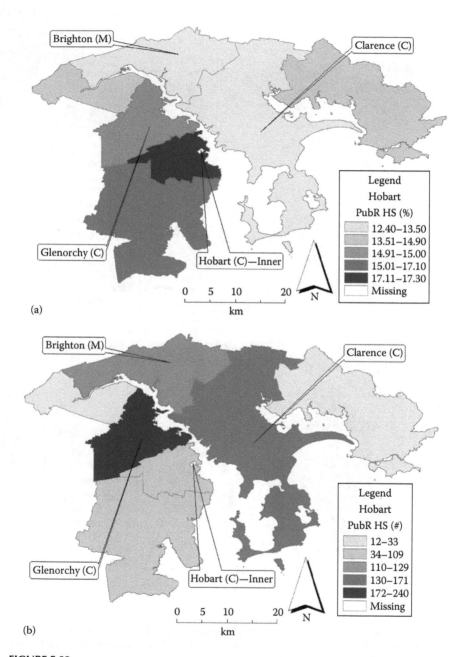

FIGURE E.33
Spatial distributions of the estimates of housing stress for public renter households in Hobart.
(a) Percentage estimates and (b) number estimates.

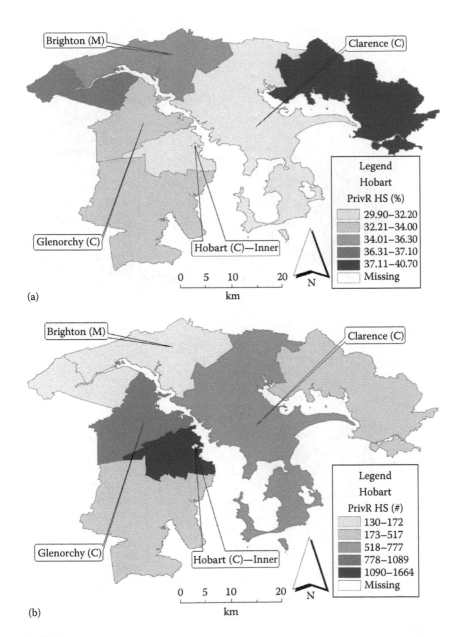

FIGURE E.34
Spatial distributions of the estimates of housing stress for private renter households in Hobart.
(a) Percentage estimates and (b) number estimates.

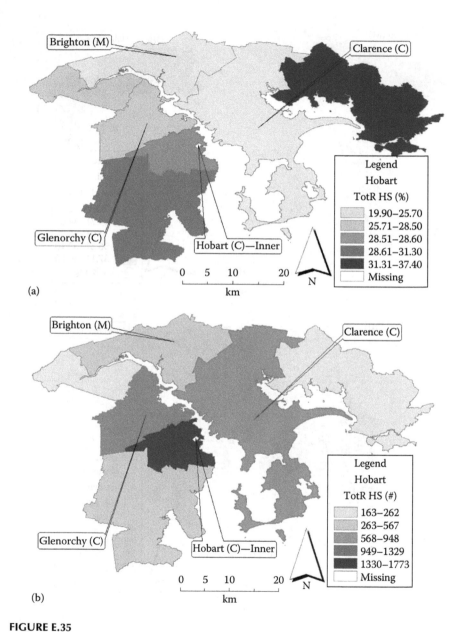

FIGURE E.35
Spatial distributions of the estimates of housing stress for total renter households in Hobart.
(a) Percentage estimates and (b) number estimates.

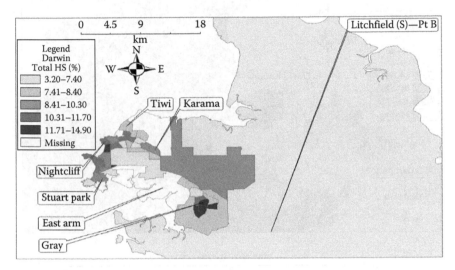

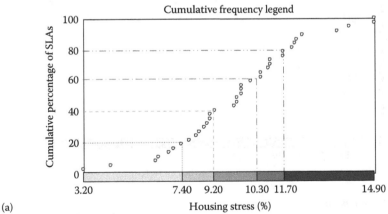

(a)

FIGURE E.36

Spatial distributions of the estimates of housing stress for overall households in Darwin. (a) Percentage estimates with a cumulative frequency legend and   *(Continued)*

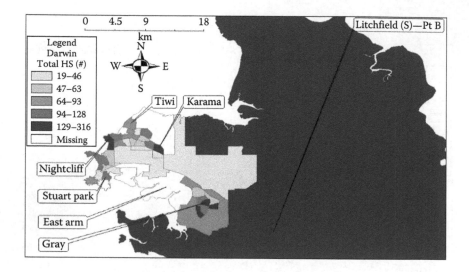

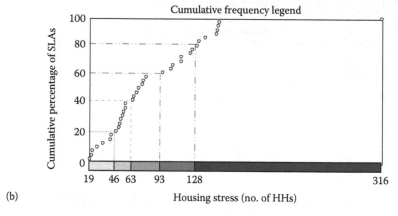

(b)

FIGURE E.36 (*Continued*)
Spatial distributions of the estimates of housing stress for overall households in Darwin.
(b) Number estimates with a cumulative frequency legend.

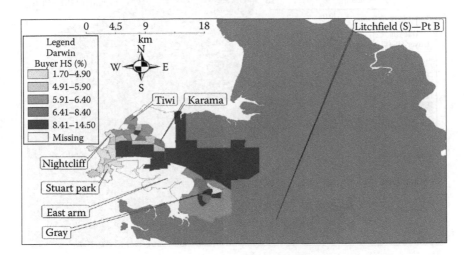

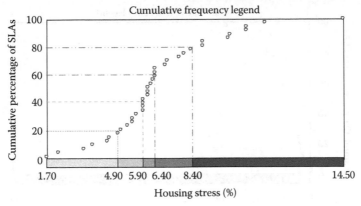

(a)

FIGURE E.37

Spatial distributions of the estimates of housing stress for buyer households in Darwin. (a) Percentage estimates with a cumulative frequency legend and *(Continued)*

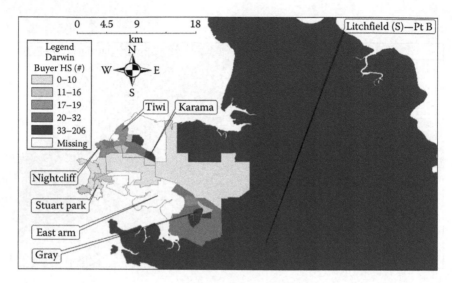

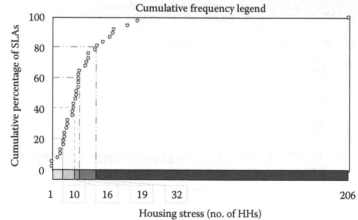

(b)

FIGURE E.37 (*Continued*)
Spatial distributions of the estimates of housing stress for buyer households in Darwin.
(b) Number estimates with a cumulative frequency legend.

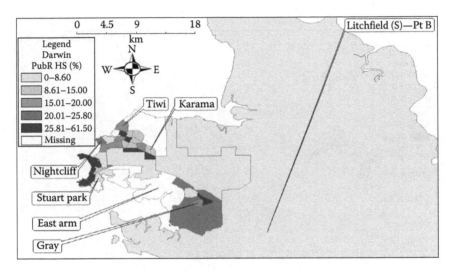

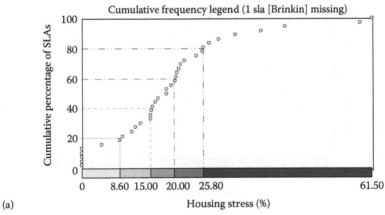

(a)

Housing stress (%)

FIGURE E.38

Spatial distributions of the estimates of housing stress for public renter households in Darwin. (a) Percentage estimates with a cumulative frequency legend and   (*Continued*)

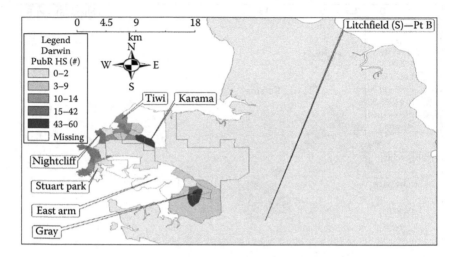

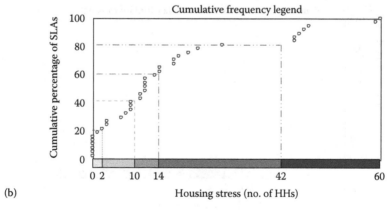

(b)

FIGURE E.38 (*Continued*)
Spatial distributions of the estimates of housing stress for public renter households in Darwin.
(b) Number estimates with a cumulative frequency legend.

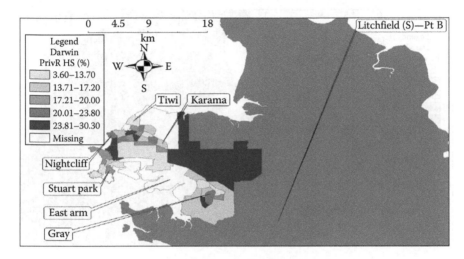

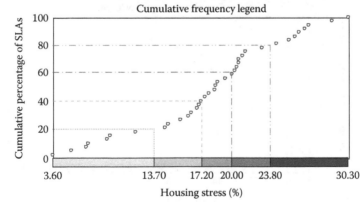

(a)

FIGURE E.39

Spatial distributions of the estimates of housing stress for private renter households in Darwin.
(a) Percentage estimates with a cumulative frequency legend and   (*Continued*)

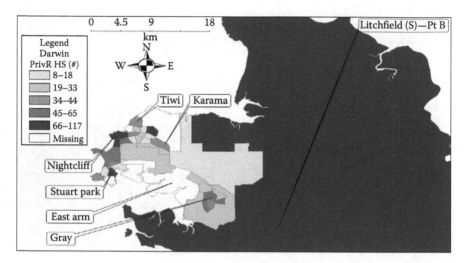

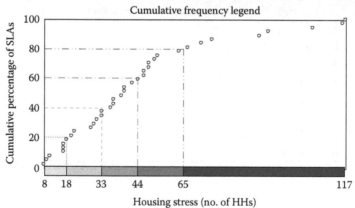

(b)

FIGURE E.39 (*Continued*)
Spatial distributions of the estimates of housing stress for private renter households in Darwin.
(b) Number estimates with a cumulative frequency legend.

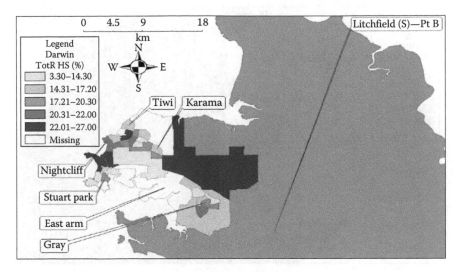

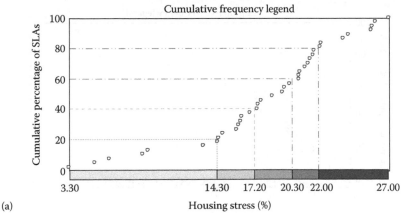

(a)

FIGURE E.40

Spatial distributions of the estimates of housing stress for total renter households in Darwin.
(a) Percentage estimates with a cumulative frequency legend and   *(Continued)*

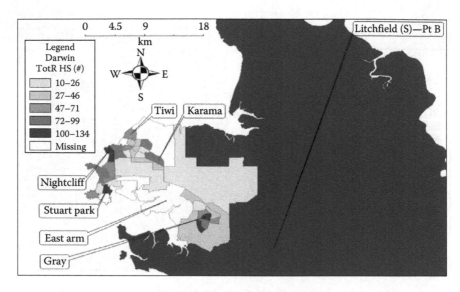

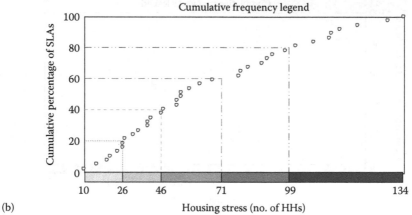

(b)

FIGURE E.40 (*Continued*)
Spatial distributions of the estimates of housing stress for total renter households in Darwin.
(b) Number estimates with a cumulative frequency legend.

Appendix F: SAS Programming for the Reweighting Algorithms from Parts 2 to 10

```
%* PART 2: Set up default values for macro variables *;

%LET GO_END= ;
%LET CK_TRASH= ;

%IF %BQUOTE(&INWEIGHT)= %THEN %LET INWEIGHT=&INWT ;
%IF %BQUOTE(&WEIGHT)= %THEN %LET WEIGHT = &WT ;

%HANDLEBY(BY=&BY &STRATUM &VARGRP &GROUP &UNIT)
%LET UNITCODE=&BYCODE ;
%HANDLEBY(BY=&BY &STRATUM &VARGRP &GROUP)
%LET GRPCODE = &BYCODE ;
%IF %BQUOTE(&GROUP)^= %THEN %LET LASTGRP = &LASTBY ;
%ELSE %LET LASTGRP = 1 ;
%HANDLEBY(BY=&BY &STRATUM &VARGRP)
%LET LASTVG =&LASTBY ;
%HANDLEBY(BY=&BY &STRATUM)
%LET LASTSTRA = &LASTBY ;
%HANDLEBY(BY=&BY) %* produces &BYVARS &BYCODE &PUTBY &LASTBY *;

%***************** defaults ****************************;
%IF (%INDEX(&XOPTIONS,%STR( EXP )))
   OR (%INDEX(&XOPTIONS,%STR( EXPONENTIAL )))
   OR (%INDEX(&XOPTIONS,%STR( EXPA )))
   %THEN %LET EXP = 1 ; %* Exponential *;
   %ELSE %LET EXP = 0 ; %* Linear *;
%IF &EXP %THEN %LET DISTANCE = exp(min(100,cwcutwt{_k_})) ;
   %ELSE %LET DISTANCE = (1 + cwcutwt{_k_}) ;
%IF (%INDEX(&XOPTIONS,%STR( EXP )))
   OR (%INDEX(&XOPTIONS,%STR( EXPONENTIAL )))
   %THEN %LET CWDIST2 = cwcutwt{_k_} ;
   %ELSE %LET CWDIST2 = cwgrpwt{_k_} ;
%LET CWFIRST = %EVAL(%INDEX(&XOPTIONS,%STR( FIRSTWT )) > 0) ;
%IF %BQUOTE(&MAXITER)= %THEN %LET MAXITER = 10 ;
%IF (0&MAXITER<=1) OR ((%BQUOTE(&UPPER&LOWER)=) & ^&EXP)
%THEN %LET MAXITER = 0 ; %* MAXITER=0 is the way to run a
   single iteration *;
%IF (&MAXITER=0) OR (%BQUOTE(&UPPER)=) %THEN %LET UPPER=1E12 ;
%IF (&MAXITER=0) OR (%BQUOTE(&LOWER)=) %THEN %LET LOWER=-1E12 ;
```

```
%IF %BQUOTE(&EPSILON)= %THEN %LET EPSILON=0.001 ;
%IF %BQUOTE(%UPCASE(&ID))=_ALL_ %THEN %DO ;
  %LET ID = ;
  %LET KEEPID = 0 ;
%END ;
%ELSE %LET KEEPID = 1 ;
%IF %BQUOTE(&WEIGHT)= %THEN %LET WEIGHT = &INWEIGHT ;
%IF %BQUOTE(&INWEIGHT)= %THEN %LET INWEIGHT = &WEIGHT ;
%IF %BQUOTE(&INWEIGHT)= %THEN %CHECKERR(1,
 No input weight variable defined) ;
%LET INWEIGHT = %SCAN(&INWEIGHT,1,%STR( -)) ;
%LET WEIGHT = %SCAN(&WEIGHT,1,%STR( -)) ;
%IF (%BQUOTE(&INREPWTS)=) & (%BQUOTE(&REPID)=)
%THEN %LET INREPWTS = &REPWTS ;
%IF %BQUOTE(&REPWTS)= %THEN %LET REPWTS = &INREPWTS ;
%IF %BQUOTE(&WRWEIGHT)= %THEN %LET NEWWT = cwnewwt{0} ;
%ELSE %LET NEWWT = &WRWEIGHT ;

%LOCAL BENPREF BENSUFF ;
%IF %BQUOTE(&BENOUT)= %THEN %LET BENPREF=benout ;
%ELSE %LET BENPREF = %SCAN(&BENOUT,1,%STR(() )) ;
%IF %LENGTH(&BENOUT)>%LENGTH(&BENPREF)
%THEN %LET BENSUFF=%SUBSTR(&BENOUT,%LENGTH(&BENPREF)+1) ;
%ELSE %LET BENSUFF = ;
%LET TEMP = %EVAL(6 + %INDEX(&BENPREF,%STR(.))) ;
%IF %LENGTH(&BENPREF)>&TEMP
%THEN %LET BENPREF=%SUBSTR(&BENPREF,1,&TEMP) ;

%IF %BQUOTE(&BYOUT)= %THEN %LET BYOUT=_byout_ ;
%IF %BQUOTE(&EXTOUT)= %THEN %LET EXTOUT=_extout_ ;
%IF %BQUOTE(&OUTDSN)= %THEN %LET OUTDSN=_outdsn_ ;
%IF %BQUOTE(&REPWTMAX)= %THEN %LET REPWTMAX=5000 ;
%IF %BQUOTE(&NPREDICT)= %THEN %LET NPREDICT = 0 ;

%IF %INDEX(&XOPTIONS,%STR( HMEAN )) %THEN %LET MEANWT=H ;
%ELSE %IF %INDEX(&XOPTIONS,%STR( AMEAN )) %THEN %LET MEANWT=A ;
%ELSE %LET MEANWT= ;
%IF %INDEX(&LOWER,%BQUOTE(%))=0 %THEN %LET LOCODE = &LOWER ;
%ELSE %DO ;
  %LET LOCODE = %SCAN(&LOWER,1,%BQUOTE(%)) ;
  %LET LOCODE = min(0.95,0.01*(&LOCODE))*cwgrpwt{_k_} ;
%END ;
%IF %INDEX(&UPPER,%BQUOTE(%))=0 %THEN %LET UPCODE = &UPPER ;
%ELSE %DO ;
  %LET UPCODE = %SCAN(&UPPER,1,%BQUOTE(%)) ;
  %LET UPCODE = max(1.05,0.01*(&UPPER))*cwgrpwt{_k_} ;
%END ;

%LET EXTRA = %EVAL(%INDEX(&XOPTIONS,%STR( NOLOG ))=0) ;
%LET PRINTREP = (%INDEX(&XOPTIONS,%STR( NOPRINT ))=0) ;
```

```
%LET BADPRINT = (%INDEX(&XOPTIONS,%STR( BADPRINT ))>0) ;
%IF %BQUOTE(&EXTNO)= %THEN
 %IF &BADPRINT %THEN %LET EXTNO = 5 ;
 %ELSE %LET EXTNO = 30 ;
%IF (%BQUOTE(&PENALTY)=_ONE_) OR (%BQUOTE(&PENALTY)=1)
 %THEN %LET PENALTY = ;

%LOCAL I WORD ;
%LET TEMP = ;
%DO I = 1 %TO 999 ;
 %LET WORD = %SCAN(&GROUP,&I) ;
 %IF %BQUOTE(&WORD) = %THEN %LET I = 1000 ;
 %ELSE %LET TEMP = &WORD ;
%END ;
%IF %BQUOTE(&TEMP)^= %THEN %LET TEMP = &TEMP &UNIT ;
%DO B=1 %TO 30 ;
 %IF %BQUOTE(&&B&B.GRP)^=
 %THEN %IF %BQUOTE(&GROUP)= %THEN %CHECKERR(1,
  B&BGRP invalid when GROUP is not specified) ;
 %ELSE %IF %SCAN(&&B&B.GRP,2)>0 %THEN %CHECKERR(1,
  B&BGRP must name a single variable from: &TEMP) ;
 %ELSE %IF %INDEX(
  %UPCASE(XXXXXXXX &TEMP X),
                   %UPCASE(%STR( &&B&B.GRP ))) = 0
 %THEN %CHECKERR(1, B&B.GRP variable &&B&B.GRP not listed in:
   &TEMP) ;
%END ;

%IF %BQUOTE(&OUT)= %THEN %LET OUTGRP = ;
%IF %BQUOTE(&OUTGRP)^=
%THEN %IF %BQUOTE(&GROUP)= %THEN %CHECKERR(1,
 OUTGRP invalid when GROUP is not specified) ;
%ELSE %IF %SCAN(&OUTGRP,2)>0 %THEN %CHECKERR(1,
  OUTGRP must name a single variable from: &TEMP) ;
%ELSE %IF %INDEX(
 %UPCASE(XXXXXXXX &TEMP X),
                  %UPCASE(%STR( &OUTGRP ))) = 0
%THEN %CHECKERR(1, OUTGRP variable &OUTGRP not listed in:
   &TEMP) ;

%IF %BQUOTE(&VAR) = %THEN %LET VAR=_one_ ;
%IF %BQUOTE(&OUT)^= & %BQUOTE(&VARGRP)^= & %BQUOTE(STRATUM)^=
%THEN %LET WTDRES =1 ; %* Calculate weighted residual
  variance *;
%ELSE %LET WTDRES =0 ; %* Default is no calculation *;

%LOCAL CNUM SCNUM CAG1 CAG2 CAG3 CAG4 CAG5 CAG5 CAG7 CAG8 CAG9
CNAM1 CNAM2 CNAM3 CNAM4 CNAM5 CNAM5 CNAM7 CNAM8 CNAM9 ;
%LET TEMP = ;
%LET CNUM = 0 ;
```

```
%DO I = 1 %TO 1000 ;
  %LET WORD = %SCAN(&CLASS,&I,%STR( *+-)) ;
  %IF %BQUOTE(&WORD)= %THEN %LET I = 1001 ;
  %ELSE %DO ;
    %IF %INDEX(&WORD,%STR(#))>0 %THEN %LET CAG&I = 1 ;
    %ELSE %LET CAG&I = ;
    %LET CNAM&I = %SCAN(&WORD,1,%STR(#)) ;
    %LET CNUM = &I ;
    %LET TEMP = &TEMP &&CNAM&I ;
  %END ;
%END ;
%LET CLASS = &TEMP ;
%LET TEMP = ;
%LET SCNUM = &CNUM ;
%DO I = &CNUM+1 %TO 1000 ;
  %LET WORD = %SCAN(&SUBCLASS,&I-&CNUM,%STR( *+-)) ;
  %IF %BQUOTE(&WORD)= %THEN %LET I = 1001 ;
  %ELSE %DO ;
    %IF %INDEX(&WORD,%STR(#))>0 %THEN %LET CAG&I = 1 ;
    %ELSE %LET CAG&I = ;
    %LET CNAM&I = %SCAN(&WORD,1,%STR(#)) ;
    %LET SCNUM = &I ;
    %LET TEMP = &TEMP &&CNAM&I ;
  %END ;
%END ;
%LET SUBCLASS = &TEMP ;

%LOCAL NOUTREPS NUMSTA ;

%IF %BQUOTE(&UNITDSN)^= %THEN %DO ; %* Check that the unit
  dataset exists *;
  %LET NUMSTA = 0 ;
  data ck_dsn ;
   if cwnumsta>0 then do ;
    set &UNITDSN nobs=cwnumsta ;
    call symput('NUMSTA',left(put(cwnumsta,8.))) ;
    _one_ = 1 ;
    cw_temp = 1 ;
    output ;
   end ;
   stop;
  run;

  %CHECKERR((&NUMSTA=0),
    Could not access dataset &UNITDSN) ;
  %IF (&NUMSTA>0) %THEN %DO ;
    %CK_TRASH(ck_dsn)
    %LOCAL NREPWTS NVAR NDEN
            NB1VAR NB2VAR NB3VAR NB4VAR NB5VAR
            NB6VAR NB7VAR NB8VAR NB9VAR NB10VAR
            NB11VAR NB12VAR NB13VAR NB14VAR NB15VAR
```

```
              NB16VAR NB17VAR NB18VAR NB19VAR NB20VAR
              NB21VAR NB22VAR NB23VAR NB24VAR NB25VAR
              NB26VAR NB27VAR NB28VAR NB29VAR NB30VAR
              NB1RVAR NB2RVAR NB3RVAR NB4RVAR NB5RVAR
              NB6RVAR NB7RVAR NB8RVAR NB9RVAR NB10RVAR
              NB11RVAR NB12RVAR NB13RVAR NB14RVAR NB15RVAR
              NB16RVAR NB17RVAR NB18RVAR NB19RVAR NB20RVAR
              NB21RVAR NB22RVAR NB23RVAR NB24RVAR NB25RVAR
              NB26RVAR NB27RVAR NB28RVAR NB29RVAR NB30RVAR ;
    data ck_dsn ;
      set ck_dsn(keep=_one_ &BYVARS &STRATUM &VARGRP &GROUP
        &UNIT
        &INWEIGHT &INREPWTS &REPID &WRWEIGHT
        &CLASS &SUBCLASS &VAR &DENOM
      %DO B=1 %TO 30 ;
        &&B&B.CLASS &&B&B.VAR &&B&B.RVAR
      %END ;
        &ID &PENALTY) ;
      %DO B=1 %TO 30 ;
        array cwvar&B{*} _one_ &&B&B.VAR ;
        call symput("NB&B.VAR",
          left(put(max(1,dim(cwvar&B)-1),6.))) ;
        array cwrvar&B{*} _one_ &&B&B.RVAR ;
        call symput("NB&B.RVAR",left(put(dim(cwrvar&B)-1,6.))) ;
      %END ;
      array cwwt{*} _one_ &INWEIGHT &PENALTY &WRWEIGHT ;
      array cwwts{*} _one_ &INREPWTS ;
      call symput("NREPWTS",left(put(dim(cwwts)-1,6.))) ;
      %IF %BQUOTE(&OUTREPS)^= %THEN %DO ;
       array cwoutrep{*} _one_ &OUTREPS ;
       call symput("NOUTREPS",left(put(dim(cwoutrep)-1,6.))) ;
      %END ;
      array cwvar{*} _one_ &VAR ;
      call symput("NVAR",left(put(dim(cwvar)-1,6.))) ;
      array cwden{*} _one_ &DENOM ;
      call symput("NDEN",left(put(dim(cwden)-1,6.))) ;
      output ;
      stop ;
    run ;
    %LET NREPWTS = &NREPWTS ;
    %CHECKERR((&SYSERR^=0),
      Variable(s) missing or of wrong type on dataset
        &UNITDSN) ;
    %IF (&SYSERR=0) %THEN %DO ;
      %CHECKERR((&NDEN > &NVAR),
      DENOM lists more variables than VAR)
      %IF &NREPWTS > 0 %THEN %LET REPID = ;
    %END ;
  %END ;
%END ;
```

```
%ELSE %CHECKERR(1, No unit dataset given in macro call) ;

%IF %BQUOTE(&OUTREPS)^= %THEN %IF &NOUTREPS < &NREPWTS %THEN
 %CHECKERR(1, Not enough OUTREPS variables for &NREPWTS
   replicates) ;

%IF &GO_END=1 %THEN %GOTO SKIPEND ;

%* PART 3: Initial creation of unit and BY datasets *;

data cw_unit
 %IF &KEEPID %THEN %DO ;
  (keep=_one_ cwgrpflg &BYVARS &INWEIGHT &INREPWTS &REPID
    &WRWEIGHT
  %DO B=1 %TO 30 ;
   &&B&B.CLASS &&B&B.VAR &&B&B.RVAR
  %END ;
  &CLASS &SUBCLASS &VAR &DENOM
  &STRATUM &VARGRP &GROUP &UNIT &ID &PENALTY)
 %END ;
 %ELSE %DO ;
  (drop=cw_beg cw_end cw_nb cw_nrb cwnrep)
 %END ;
 cw_by(keep=_one_ &BYVARS cw_beg cw_end cw_nb cw_nrb) ;

 set &UNITDSN end=_last_ ;
 length _one_ cwgrpflg 3 ;
       %* cwgrpflg = 0 if not last in group
             = 1 if last in group, not last in psu
             = 2 if last in VARGRP, not last in stratum
             = 3 if last in STRATUM, not last in BY
             = 4 if last in BY *;

 &UNITCODE
 retain cw_beg 1 cw_end 0 cw_nb cw_nrb 0 cwnrep 0 ;
*  cw_order = _n_ ; %* Input sort order *;
 _one_ = 1 ;
 cw_end + 1 ;
%IF %BQUOTE(&REPID)^= %THEN %DO ;
  if &REPID = int(&REPID) and 0 <= &REPID <= &REPWTMAX
  then cwnrep = max(cwnrep,&REPID) ;
  else cwnrep = 99999 ;
  if _last_ then call symput("NREPWTS",left(put(cwnrep,6.))) ;
%END ;
 cwgrpflg = 0 ;
 if &LASTGRP then cwgrpflg + 1 ;
 if &LASTVG then cwgrpflg + 1 ;
 if &LASTSTRA then cwgrpflg + 1 ;
 if &LASTBY then cwgrpflg + 1 ;
 output cw_unit ;
```

```
  if &LASTBY then do ;
   output cw_by ;
   cw_beg = cw_end + 1 ;
  end ;
run ;
%CK_TRASH(cw_unit)
%CK_TRASH(cw_by)
%CHECKERR((&SYSERR^=0),
Check unit data is sorted by &BY &STRATUM &VARGRP &GROUP
  &UNIT)
%CHECKERR((&NREPWTS>99998),
 Invalid value for replicate identifier &REPID)

%IF &GO_END=1 %THEN %GOTO SKIPEND ;

%IF &NREPWTS = 0 %THEN %LET OUTREPS= ;
%ELSE %IF (%BQUOTE(&OUTREPS) =) & (%INDEX(&XOPTIONS,%STR
  ( REPS ))>0)
%THEN %DO ;
 %LET OUTREPS = est_1-est_&NREPWTS ;
 %LET NOUTREPS = &NREPWTS ;
%END ;

%*********** Deal with each benchmark dataset ************;
%LOCAL MAXBEN BENVBLS NBENVBLS NBENREPS NNVBLS MAXPARS MAXRPARS
        NBTOT NBREP BNUM CW_MAXNB CWMAXNRB
        BNAM1 BNAM2 BNAM3 BNAM4 BNAM5 BNAM6 BNAM7 BNAM8 BNAM9
        BTYP1 BTYP2 BTYP3 BTYP4 BTYP5 BTYP6 BTYP7 BTYP8 BTYP9
        TNAM1 TNAM2 TNAM3 TNAM4 TNAM5 TNAM6 TNAM7 TNAM8 TNAM9 ;
%LET MAXBEN = 0 ;
%LET BENVBLS = ;
%LET NBENVBLS = ;
%LET NBENREPS = ;
%LET NNVBLS = 0 ;
%LET MAXPARS = %EVAL(&MAXSPACE*1000 / (50 + 16*(&NREPWTS + 1))) ;
%LET MAXRPARS = %EVAL(&MAXPARS*(&NREPWTS+1)) ;
%DO B = 1 %TO 30 ;
 %IF %LENGTH(&&B&B.DSN)=0 %THEN %LET B=1000 ; %* Exit loop *;
 %ELSE %DO ;
  %LET MAXBEN = &B ;
  %IF %BQUOTE(&&B&B.TOT) =
  %THEN %CHECKERR(1,No value for B&B.TOT given) ;
  %IF %BQUOTE(&&B&B.VAR)= %THEN %LET B&B.VAR = _one_ ;
  %IF %BQUOTE(&GO_END)= %THEN %DO ;
  %LET NUMSTA = 0 ;
  data ck_dsn ;
    if cwnumsta>0 then do ;
     set &&B&B.DSN nobs=cwnumsta ;
     call symput('NUMSTA',left(put(cwnumsta,8.))) ;
     _one_ = 1 ;
```

```
    output ;
  end ;
  stop;
run;

%CHECKERR((&NUMSTA=0),
    Could not access dataset &&B&B.DSN) ;
  %IF (&NUMSTA>0) %THEN %DO ;
    %CK_TRASH(ck_dsn) ;
    data ck_dsn ;
      set ck_dsn(keep=_one_ &BYVARS
        &&B&B.CLASS &&B&B.TOT &&B&B.REPS) ;
      array cwtemp{*} _one_ &&B&B.TOT ;
      array cwreps{*} _one_ &&B&B.REPS ;

      output ;
      cwtemp1 = max(1,dim(cwtemp)-1) ;
      call symput("NBTOT",left(put(cwtemp1,6.))) ;
      cwtemp2 = dim(cwreps) - 1 ;
      if cwtemp2 = 0 then call symput("NBREP","0") ;
      else if cwtemp2=cwtemp1*&NREPWTS
      then call symput("NBREP","&NREPWTS") ;
      else call symput("NBREP",".") ;
      stop ;
    run ;
    %CHECKERR((&SYSERR^=0),
Variable(s) missing or of wrong type on dataset &&B&B.DSN)
    %IF (&NBREP = .) %THEN %CHECKERR(1,
B&B.REPS does not match B&B.TOT and INREPWTS) ;
    %ELSE %IF (&NBREP>0) & (&&NB&B.RVAR>0)
    %THEN %CHECKERR((&&NB&B.RVAR^=(&NBTOT*&NBREP)),
B&B.RVAR does not match B&B.TOT and INREPWTS) ;
  %END ;
  %IF (&NUMSTA>0) & (&SYSERR=0) %THEN %DO ; %* dataset looks
    ok? *;
    %IF %BQUOTE(&&B&B.CLASS)^= %THEN %DO ;
      proc contents data=ck_dsn(keep=&&B&B.CLASS)
        out=cwnames(keep=name type length nobs) noprint;
      run ;
      %CK_TRASH(cwnames)
      data _null_ ;
       set cwnames end=_last_;
       call symput("BNAM"!!left(put(_n_,3.)),name) ;
       if type = 1
        then call symput("BTYP"!!left(put(_n_,3.)),
         left(put(length,3.))) ;
        else call symput("BTYP"!!left(put(_n_,3.)),
         '$ '!! left(put(length,3.))) ;
       if _last_ then call symput("BNUM",left(put(_n_,3.))) ;
      run ;
    %END ;
```

```
 %ELSE %LET BNUM=0 ;
%END ;


%IF %BQUOTE(&GO_END)= %THEN %DO ; %* Benchmark dataset ok? *;
 data ck_dsn ;
  set &&B&B.DSN ;
 run ;
 proc contents data=ck_dsn(keep=&&B&B.TOT)
      out=cwnames(keep=name) noprint;
 run ;
 %CK_TRASH(cwnames)
 data _null_ ;
  set cwnames end=_last_ ;
  call symput("TNAM"!!left(put(_n_,3.)),name) ;
 run ;


 data cw_by(keep=_one_ &BYVARS cw_beg cw_end cw_nb cw_nrb
                  cwncat1-cwncat&B)
      cw_unit(drop=cw_beg cw_end cwtbeg cwtend cw_nb
        cw_nrb
               cw_nblo cw_nrblo _count_
               %IF &NBREP > 0 %THEN cwrep ;
               cw_maxnb cwmaxnrb cwi cwj cwtemp cwlo cwhi
               cwdone _name_ _value_ _i_ cwbentot
               cwncat1-cwncat&B &&B&B.TOT &&B&B.REPS)
      cwrep&B(keep=&BYVARS &&B&B.CLASS _i_ _name_ _count_
        _value_)
      cwben&B(keep=cwbentot) ;
   set cw_by(in=incw_by)
      ck_dsn(in=inb keep=&BYVARS
              &&B&B.CLASS &&B&B.TOT &&B&B.REPS)
      cw_unit end=_last_ ;
   &BYCODE
   %* Set up arrays for each category variable.
      The ith element of each array gives the value of that
      category variable for the ith category on B&B.DSN.
      The categories are kept sorted. *;
   %DO I = 1 %TO &BNUM ;
    array cw_&I {0:&MAXPARS} &&BTYP&I _temporary_ ;
   %END ;
   array cwtots{0:&NBTOT} _one_ &&B&B.TOT ;
   %IF &NBREP > 0 %THEN %DO ;
    array cwrtots{1:&NBTOT,1:&NBREP} &&B&B.REPS ;
   %END ;
   array cwagben{0:&MAXRPARS} _temporary_ ;
   array cwcount{0:&MAXPARS} _temporary_ ;
   array cwncats{&B} cwncat1-cwncat&B ;
   array temncats{&B} _temporary_ ;
   length cwcat&B 4 ;
```

```
retain cwncat&B cw_maxnb cwmaxnrb cw_nblo cw_nrblo
 cwtbeg cwtend 0 ;

if incw_by then do ;
 cwncat&B = 0 ;
 cw_nblo = cw_nb ;
 cw_nrblo = cw_nrb ;
 do cwi = 1 to &B-1 ;
  temncats{cwi} = cwncats{cwi} ;
 end ;
 cwtbeg = cw_beg ;
 cwtend = cw_end ;
end ;
else do ;
 cwlo = 0 ; %* 1 below lowest category *;
 cwhi = cwncat&B + 1 ; %* id 1 above highest category *;
 cwdone = cwhi <= (cwlo + 1) ; %* 1 if no categories *;
 do while(^cwdone) ;
  cwi = int(0.5*(cwlo+cwhi)) ; %* Check this category *;
  %* If category cwi is too small set cwlo to cwi, if
    category cwi is too big set
    cwhi to cwi, and if category matches then set
      cwdone=2 *;
  %DO I = 1 %TO &BNUM ;
   if cw_&I{cwi} < &&BNAM&I then cwlo = cwi ;
   else if &&BNAM&I < cw_&I{cwi} then cwhi = cwi ;
   else
  %END ;
  do ;
   cwdone = 2 ;
   if inb then do ;
    cwtemp = &NBTOT*(&NBREP+1)*(cwi-1) ;
    do cwj = 1 to &NBTOT ;
     cwtemp + 1 ;
     cwagben{cwtemp} + cwtots{cwj} ;
     %IF &NBREP > 0 %THEN %DO ;
      do cwrep = 1 to &NBREP ;
       cwtemp + 1 ;
       cwagben{cwtemp} + cwrtots{cwj,cwrep} ;
      end ;
     %END ;
    end ;
   end ;
   else cwcat&B = cwi ;
  end ;
  if ^cwdone then cwdone = cwhi <= (cwlo + 1) ;
 end ;
 if cwdone = 1 then do ;
  if inb
   then if (cw_nblo + (cwncat&B+1)*&NBTOT) > &MAXPARS
```

```
then do ;
 call symput("GO_END","1") ;
 stop ;
end ;
else do ;
 do cwi = cwncat&B to cwhi by -1 ;
  cwtemp = &NBTOT*(&NBREP+1)*(cwi-1) ;
  cwcount{cwi+1} = 0 ;
  do cwj = 1 to &NBTOT*(&NBREP+1) ;
   cwagben{cwtemp+cwj+&NBTOT*(&NBREP+1)}
    = cwagben{cwtemp+cwj} ;
  end ;
  %DO I = 1 %TO &BNUM ;
   cw_&I{cwi+1} = cw_&I{cwi} ;
  %END ;
 end ;
 cwtemp = &NBTOT*(&NBREP+1)*(cwhi-1) ;
 cwcount{cwhi} = 1 ;
 do cwj = 1 to &NBTOT ;
  cwtemp + 1 ;
  cwagben{cwtemp} = cwtots{cwj} ;
  %IF &NBREP > 0 %THEN %DO ;
   do cwrep = 1 to &NBREP ;
    cwtemp + 1 ;
    cwagben{cwtemp} = cwrtots{cwj,cwrep} ;
   end ;
  %END ;
 end ;

 %DO I = 1 %TO &BNUM ;
  cw_&I{cwhi} = &&BNAM&I ;
 %END ;
 cwncat&B = cwncat&B + 1 ;
 end ;
 else if ^inb then cwcat&B = 0 ;
end ;
if ^inb then do ;
 output cw_unit ;
 cwcount{cwcat&B} + 1 ;
end ;
end ;
if &LASTBY then do ; %* Output reports dataset *;
_i_ = cw_nblo ;
do cwhi = 1 to cwncat&B ;
 %DO I = 1 %TO &BNUM ;
  &&BNAM&I = cw_&I{1+(cwhi-1)*&NBTOT} ;
 %END ;
 length _name_ $ 32 ;
 _count_ = cwcount{cwhi} ;
 do cwlo = 1 to &NBTOT ;
```

```
      _i_ + 1 ;
      if 0 then ;
      %DO I = 1 %TO &NBTOT ;
       else if cwlo=&I then _name_="&&TNAM&I" ;
      %END ;
      _value_ = cwagben{1 + (_i_-cw_nblo-1)*(&NBREP+1)} ;
       output cwrep&B ;
      end ;
     end ;
     cw_nb = cw_nblo + cwncat&B*&NBTOT ;
     cw_nrb = cw_nrblo + cwncat&B*&NBTOT*(&NBREP+1) ;
     cw_end = cwtend ;
     cw_beg = cwtbeg ;
     do cwi = 1 to cwncat&B*&NBTOT*(&NBREP+1) ;
      cwbentot = cwagben{cwi} ;
      cwagben{cwi} = . ;
      output cwben&B ;
     end ;
     do cwi = 1 to &B-1 ;
      cwncats{cwi} = temncats{cwi} ;
     end ;
     output cw_by ;
     cwncat&B = 0 ;
     cw_maxnb = max(cw_maxnb,cw_nb) ;
     cwmaxnrb = max(cwmaxnrb,cw_nrb) ;
     cwtend = 0 ;
     cwtbeg = 1 ;
     cw_nblo = 0 ;
     cw_nrblo = 0 ;
     do cwi = 1 to &B-1 ;
      temncats{cwi} = 0 ;
     end ;
    end ;
    if _last_ then do ;
     call symput("CW_MAXNB",left(put(cw_maxnb,6.))) ;
     call symput("CWMAXNRB",left(put(cwmaxnrb,6.))) ;
    end ;
   run ;
   %CK_TRASH(cwrep&B)
   %CK_TRASH(cwben&B)
   %CHECKERR(0&GO_END,
 Not enough storage space: increase MAXSPACE parameter) ;

   %IF &&B&B.GRP ^= %THEN %DO ;
    data cw_unit ;
     set cw_unit ;
     &UNITCODE
     if ^first.&&B&B.GRP then do ;
      cwcat&B=0 ;
     end ;
```

```
      run ;
    %END ;

    %LET BENVBLS=&BENVBLS &&B&B.VAR ;
    %IF (&NBTOT ^= &&NB&B.VAR) %THEN
      %CHECKERR(1,
 B&B.TOT and B&B.VAR contain different numbers of variables) ;
    %ELSE %DO ;
     %LET NBENVBLS = &NBENVBLS &&NB&B.VAR ;
     %LET NNVBLS = %EVAL(&NNVBLS + &&NB&B.VAR) ;
     %LET NBENREPS = &NBENREPS &NBREP ;
    %END ;
   %END ;
  %END ;
 %END ;
%END ;

%IF &MAXBEN=0 %THEN %CHECKERR(1,No benchmark datasets
  specified) ;
%ELSE %IF (&CWMAXNRB < &MAXRPARS) %THEN %DO ;
 %LET MAXPARS = &CW_MAXNB ;
 %LET MAXRPARS = &CWMAXNRB ;
%END ;
%IF &GO_END=1 %THEN %GOTO SKIPEND ;
%IF (%BQUOTE(&REPID)^=) & (&NREPWTS > 1) %THEN %DO ;
 %LET INREPWTS = rwt1-rwt&NREPWTS ;
 %IF %BQUOTE(&REPWTS)=
 %THEN %LET REPWTS = rwt1-rwt&NREPWTS ;
 data cw_unit(drop=cw_ix_
 %IF &KEEPID %THEN &REPID ;)  ;
  set cw_unit ;
  array cwwts{1:&NREPWTS} 4 &INREPWTS ;
  do cw_ix_ = 1 to &NREPWTS ;
   if cw_ix_=&REPID then cwwts{cw_ix_} = 0 ;
   else cwwts{cw_ix_} = &INWEIGHT*&NREPWTS/(&NREPWTS-1) ;
  end ;
 run ;
%END ;

%IF &GO_END=1 %THEN %GOTO SKIPEND ;
%IF %BQUOTE(&OUT)^= %THEN %DO ;
 %LOCAL TOTLENG
  VNAM1 VNAM2 VNAM3 VNAM4 VNAM5 VNAM6 VNAM7 VNAM8 VNAM9
  BYTYP1 BYTYP2 BYTYP3 BYTYP4 BYTYP5 BYTYP6 BYTYP7 BYTYP8 BYTYP9
  CTYP1 CTYP2 CTYP3 CTYP4 CTYP5 CTYP6 CTYP7 CTYP8 CTYP9 ;

 %* PART 4: Develop list of variable names for tables *;

 data ttnames ;
  array temp{*} &VAR ;
  output ;
```

```
run ;
%CK_TRASH(ttnames)
proc contents data=ttnames
   out=ttnames(keep=name npos) noprint;
run ;
proc sort data=ttnames out=ttnames(drop=npos) ;
 by npos ;
run ;
data _null_ ;
 set ttnames end=_last_;
 call symput("VNAM"!!(left(put(_n_,3.))),
             translate(trim(name)
             ,"abcdefghijklmnopqrstuvwxyz"
             ,"ABCDEFGHIJKLMNOPQRSTUVWXYZ")) ;
run ;
%LET VARNAMES="&VNAM1" ;
%DO I = 2 %TO &NVAR ;
 %LET VARNAMES = &VARNAMES "&&VNAM&I" ;
%END ;

%IF %BQUOTE(&CLASS &SUBCLASS &BYVARS)^= %THEN %DO ;

 data ttnames(keep=&CLASS &SUBCLASS &BYVARS) ;
  set cw_unit ;
  output ;
  stop ;
 run ;
 proc contents data=ttnames
      out=ttnames(keep=name type length nobs) noprint;
 run ;
 data _null_ ;
  set ttnames end=_last_;
  length ttctype $ 5 ;
  retain totleng 0 ;
  if type=1 then ttctype=left(put(length,3.)) ;
  else ttctype = '$ ' !! left(put(length,3.)) ;
  %DO I = 1 %TO &SCNUM ;
   if upcase(name)=upcase("&&CNAM&I")
   then do ;
    call symput("CTYP&I",ttctype) ;
    if type=1 then totleng + 8 ;
    else totleng + length ;
   end ;
  %END ;
  %DO I = 1 %TO &NBYVARS ;
   if upcase(name)=upcase("&&BYVAR&I")
   then call symput("BYTYP&I",ttctype) ;
  %END ;
  if _last_ then call symput("TOTLENG",left(put(totleng,3.))) ;
 run ;
```

```
%*DO I = 1 %TO &SCNUM ;
  %*PUT CNAM&I=&&CNAM&I CAG&I=&&CAG&I CTYP&I=&&CTYP&I;
%*END ;
%*DO I = 1 %TO &NBYVARS ;
  %*PUT BYVAR&I=&&BYVAR&I BYTYP&I=&&BYTYP&I ;
%*END ;
  %CK_TRASH(ttnames) ;
 %END ;
 %ELSE %LET TOTLENG = 0 ;

 %LET NVAR_1 = %EVAL(&NVAR - 1) ;
 %LET NVD_1 = %EVAL(&NVAR_1 + &NDEN) ;
 %LET NVD = %EVAL(&NVAR + &NDEN) ;
 %LET TOTLENG = %EVAL(&TOTLENG + 8
                + 8*((&NVAR+&NDEN)*(1+&NREPWTS))) ;
 %IF &WTDRES %THEN %LET TOTLENG
   = %EVAL(&TOTLENG + &MAXPARS*&NVD) ;

 %LET MAXCATS = %EVAL(1000*&MAXSPACE / &TOTLENG) ;
 %IF &OUTGRP ^= %THEN %DO ;
   data cw_unit ;
    set cw_unit ;
    &UNITCODE
    cwoutgrp = first.&OUTGRP ;
   run ;
 %END ;
%END ; %* Get ready now for tables *;
%ELSE %LET MAXCATS = 0 ;
%IF &GO_END=1 %THEN %GOTO SKIPEND ;
%LOCAL CWREG ;
%LET CWREG = 0 ;
%IF &MAXBEN >= 1 %THEN %DO ;
 %LOCAL CATKEEP ;
 %IF %BQUOTE(&GROUP)= %THEN
   %LET CATKEEP=cwgrpflg cwcat1-cwcat&MAXBEN ; %* Run code to
     weight the data *;

 %*PR(cw_unit(obs=50),) *; %*PR(cw_by(obs=50),) *;

 %IF &EXTRA %THEN %DO ;
  %PUT ;
  %PUT Main calculations begin: ;
 %END ;

%* PART 5: Start of main calculations portion *;

 %LOCAL NMAXGRPS NCWEST NMAT NMATRIX CWREP NREPORT CWRULER
   CWPSRAT ;
 %IF (&NNVBLS <= 1) & (&MAXBEN = 1) & (%BQUOTE(&GROUP)= )
 %THEN %LET CWPSRAT = 1 ;
```

```
%ELSE %LET CWPSRAT = 0 ;
%LET NMAXGRPS = 20 ;
%LET NCWXEST = %EVAL((&NREPWTS+1)*&MAXPARS) ;
%IF &CWPSRAT %THEN %LET NMAT = %EVAL(&MAXPARS) ;
%ELSE %LET NMAT = %EVAL(&MAXPARS*(&MAXPARS+1)/2) ;
%LET NMATRIX = %EVAL((&NREPWTS+1)*&NMAT) ;
%IF &NMATRIX > 1000000 %THEN %DO ;
 %PUT WARNING: This request demands a lot of RAM memory ;
 %PUT
* It uses &MAXPARS parameters and &NREPWTS replicate weights ;
 %PUT
* As a result, one matrix in the program has &NMATRIX elements ;
 %PUT
* It will be slow if your computer has insufficient RAM ;
 %PUT * Possible ways to improve speed by: ;
 %PUT * . Use of the BY statement, or ;
 %PUT * . Running repeatedly on subsets of the replicates ;
%END ;
%IF &MAXCATS = 0 %THEN %LET NPREDICT = 0 ;
%IF (&WTDRES) OR (&NPREDICT>0)
%THEN %LET NVECTOR = %EVAL(&MAXPARS*(1+&MAXCATS*&NVD)) ;
%ELSE %LET NVECTOR = &MAXPARS ;
%IF &NREPWTS <=30 %THEN %LET NWREPORT = &NREPWTS ;
%ELSE %LET NWREPORT = 30 ; %* Report on first 30 replicates *;


%LOCAL WRFLAG;
%IF (%BQUOTE(&WROUT)^=) & (&WTDRES) & (&MAXCATS>0) %THEN %DO ;
 %LET WRFLAG = 1 ;
 %IF %BQUOTE(&WRLIST)= %THEN %LET WRLIST=wr_1 ;
%END ;
%ELSE %LET WRFLAG = 0 ;


data cw_wtd(keep=_bygrp_ cwunitid &BYVARS &STRATUM &VARGRP
   &GROUP
                   _inwt_ &WEIGHT &REPWTS &WRWEIGHT &CATKEEP
 %IF (&MAXCATS > 0) & (&NPREDICT>0) %THEN hat_1-hat_&NPREDICT ;)
     cw_out(keep=_bygrp_ cwunitid &BYVARS &STRATUM &VARGRP
        &GROUP
                   _inwt_ _regwt_ _regcwt_)
 %IF &WRFLAG %THEN
     cw_wrout(keep=&BYVARS &STRATUM &VARGRP _ncells_
                 &WRLIST) ;
     cwreport(keep=&BYVARS cw_iter _i2_ _i_ _bench_ _best_
               _crit_ _report_
             rename=(_i2_=_b_))
       cwbyrep(keep=&BYVARS _iters_ _result_
               _nilwt_ _negin_ _negwt_ cwfreq
%IF %BQUOTE(&GROUP)^= %THEN _gnilwt_ _gnegin_ _gnegwt_ _gfreq_ ;
             rename=(cwfreq=_freq_))
```

```
%IF &MAXCATS > 0 %THEN %DO ;
 cw_tab(keep=_est_ &BYVARS
 %IF %UPCASE(&VAR) ^= _ONE_ %THEN varname ;
 %DO I = 1 %TO &SCNUM ;
  &&CNAM&I
 %END ;
 %IF (&NREPWTS > 1) OR (&WTDRES) %THEN _var_ _se_ _rse_ ;
 %IF (&NREPWTS > 1) & (&WTDRES) %THEN _typ_ ;
 &OUTREPS )
%END ;
    ;
length _bygrp_ 4 ;
%IF &MAXCATS > 0 %THEN %DO ;
 length
 %DO I = 1 %TO &NBYVARS ;
  &&BYVAR&I &&BYTYP&I
 %END ;
 %DO I = 1 %TO &SCNUM ;
  &&CNAM&I &&CTYP&I
 %END ;
 &VAR 8 ;
%END ;
array cwncat{&MAXBEN} cwncat1-cwncat&MAXBEN ;
array cwnvbl{&MAXBEN} _temporary_ (&NBENVBLS) ;
array cwnrep{&MAXBEN} _temporary_ (&NBENREPS) ;
array cwcat{&MAXBEN} cwcat1-cwcat&MAXBEN ;
array cwvbl{&NNVBLS} &BENVBLS ;
%IF &NREPWTS > 0 %THEN %DO B = 1 %TO &MAXBEN ;
 %IF &&NB&B.RVAR > 0 %THEN %DO ;
  array cwrvbl&B{&&NB&B.VAR,&NREPWTS} &&B&B.RVAR ;
  array cwrvs&B{&NMAXGRPS,&&NB&B.VAR,&NREPWTS} _temporary_ ;
 %END ;
%END ;
array cwcats{1:&NMAXGRPS,&MAXBEN} _temporary_ ;
array cwvbls{1:&NMAXGRPS,&NNVBLS} _temporary_ ;
array cwstatus{0:&NREPWTS} _temporary_ ;
array cwepsi{0:&NREPWTS} _temporary_ ; %* epsilon *;
array cwxben{0:&MAXRPARS} _temporary_ ; %* Benchmarks for X *;
array cwxest{0:&NCWXEST} _temporary_ ; %* Estimates for X *;
array cwadj {0:&NCWXEST} _temporary_ ;
array matrix{0:&NMATRIX} _temporary_ ;
array vector{0:&NVECTOR} _temporary_ ;
%LET TEMP=%EVAL(&NWREPORT + 1) ;
array cwreptxt{0:&MAXPARS} $ &TEMP _temporary_ ;

%IF &MAXCATS > 0 %THEN %DO ;
 %IF &NPREDICT>0 %THEN %DO ;
  array tthat{1:&NPREDICT} hat_1-hat_&NPREDICT ;
  retain hat_1-hat_&NPREDICT ;
 %END ;
```

```
    %DO I = 1 %TO &SCNUM ;
     array ttc&I{0:&MAXCATS} &&CTYP&I _temporary_ ;
     %IF %SUBSTR(&&CTYP&I,1,1) = $
     %THEN %LET CTYP&I = '' ;
     %ELSE %LET CTYP&I = . ;
    %END ;
    array ttposs{0:&MAXCATS} _temporary_ ;
    %IF &SCNUM > &CNUM %THEN %DO ;
     array ttdposs{0:&MAXCATS} _temporary_ ;
    %END ;
    array ttsums{0:&NREPWTS,0:&NVD_1,0:&MAXCATS} _temporary_ ;
    array ttvds{0:&NVD_1} &VAR &DENOM ;
    array ttvnam{0:&NVAR_1} $ 32 _temporary_ (&VARNAMES) ;
    array ttunpos{1:&NMAXGRPS} _temporary_ ;
    array ttunvds{1:&NMAXGRPS,0:&NVD_1} _temporary_ ;

    %IF &WTDRES %THEN %DO ;
     array ttgest {0:&NVD_1,0:&MAXCATS} _temporary_ ;
     array ttgpred{0:&NVD_1,0:&MAXCATS} _temporary_ ;
     array ttrsum {0:&NVD_1,0:&MAXCATS} _temporary_ ;
     array ttrssq {0:&NVD_1,0:&MAXCATS} _temporary_ ;
     array ttrprod{0:&NVD_1,0:&MAXCATS} _temporary_ ;
     array ttsvar {0:&NVD_1,0:&MAXCATS} _temporary_ ;
     array ttscov {0:&NVD_1,0:&MAXCATS} _temporary_ ;
    %END ;
    %END ;
    retain _cwreg_ 0 ;
    retain ttmaxcat ttnpsu cwgrpfst 0 ;
do cwbygrp = 1 to cwnumby ;
 set cw_by point=cwbygrp nobs=cwnumby ;
 _bygrp_=cwbygrp ;
 _nilwt_ = 0 ;
 _negin_ = 0 ;
 _negwt_ = 0 ;
 cwfreq  = 0 ;
%IF %BQUOTE(&GROUP)^= %THEN %DO ;
 _gnilwt_ = 0 ;
 _gnegin_ = 0 ;
 _gnegwt_ = 0 ;
 _gfreq_  = 0 ;
%END ;
 _k_ = 0 ;
 %DO B = 1 %TO &MAXBEN ;
  do _i_ = 1 to cwnvbl{&B}*cwncat{&B}*(cwnrep{&B}+1) ;
   _k_ + 1 ;
   cwbenp&B + 1 ;
   set cwben&B point=cwbenp&B ;
   cwxben{_k_} = cwbentot ;
  end ;
 %END ;
```

```
do _k_ = 0 to &NREPWTS ;
 cwstatus{_k_} = -1 ;
end ;
cwdone = 0 ; %* Not all weights have converged *;
do _i_ = 0 to &NCWXEST ;
 cwadj{_i_} = 0 ;
end ;

do cw_iter = 0 to (&MAXITER+1) ;

 if ^cwdone or cw_iter > &MAXITER then do ;
 *put cw_iter= ; %* Clear the arrays into which values will
   be aggregated *;
  do _i_ = 0 to &NCWXEST ;
   cwxest{_i_} = 0 ;
  end ;
  do _i_ = 0 to &NMATRIX ;
   matrix{_i_} = 0 ;
  end ;
  _unix_ = 0 ;
  cwgrpfst = 1 ;
  do cwunitid = cw_beg to cw_end ;
   cwuptr = cwunitid ;
   set cw_unit point=cwuptr ;
   array cwwt{0:&NREPWTS} &INWEIGHT &INREPWTS ;
   array cwgrpwt{0:&NREPWTS} _temporary_ ;
   array cwcutwt{0:&NREPWTS} _temporary_ ;
   array cwnewwt{0:&NREPWTS} &WEIGHT &REPWTS ;
   if _unix_ < &NMAXGRPS then do ;
    _unix_ + 1 ;
    do _ix_ = 1 to &MAXBEN ;
     cwcats{_unix_,_ix_} = cwcat{_ix_} ;
    end ;
    do _ix_ = 1 to &NNVBLS ;
     cwvbls{_unix_,_ix_} = cwvbl{_ix_} ;
    end ;
    %IF &NREPWTS > 0 %THEN %DO B = 1 %TO &MAXBEN ;
     %IF &&NB&B.RVAR > 0 %THEN %DO ;
       do cwj = 1 to &&NB&B.VAR ;
        do _ix_ = 1 to &NREPWTS ;
         cwrvs&B{_unix_,cwj,_ix_} = cwrvbl&B{cwj,_ix_} ;
        end ;
       end ;
     %END ;
    %END ;

    %IF &MAXCATS > 0 %THEN %DO ;
     if cw_iter > &MAXITER
     %IF (&WTDRES) OR (&NPREDICT>0) %THEN or cw_iter = 0 ;
     then do ;
```

```
do ttj = 0 to &NVD_1 ;
 %IF %BQUOTE(&OUTGRP)^= %THEN %DO ;
  if ^cwoutgrp then ttunvds{_unix_,ttj} = 0 ;
  else
 %END ;
 ttunvds{_unix_,ttj} = ttvds{ttj} ;
end ;
ttcode = 0 ; %* Indicates nway cell not an aggregate *;
%DO I = 1 %TO &SCNUM ;
 do ttix&I = 0
 %IF (&&CAG&I=1) OR (&I > &CNUM) %THEN to 1 ;;
  if ttix&I = 0 then ttc&I{0} = &&CNAM&I ;
  else ttc&I{0} = &&CTYP&I ; %* Missing *;
  if (ttix&I = 0) or (&&CNAM&I ^= &&CTYP&I) then
%END ;
 if ttcode >= 0 then do ;
  ttlo = 0 ;
  tthi = ttmaxcat + 1 ;
  ttdone = tthi <= (ttlo + 1) ;
  %* 0 = continue search for this category
     1 = category not in arrays
     2 = category found in arrays *;
  do while(^ttdone) ;
   tti = int(0.5*(ttlo+tthi)) ;
   %DO I = 1 %TO &SCNUM ;
    if ttc&I{tti} < ttc&I{0} then ttlo = tti ;
    else if ttc&I{0} < ttc&I{tti} then tthi = tti ;
    else
   %END ;
   do ;
    ttdone = 2 ; %* Found it *;
   end ;
   if ^ttdone then ttdone = tthi <= (ttlo + 1) ;
  end ;
  if ttdone = 2 and ttcode = 0 then do ;
   ttcode = -1 ;
   ttunpos{_unix_} = ttposs{tti} ;
  end ;
  else if ttdone = 2 then do ;
   ttcode = 1 ;
  end ;
  else if ttmaxcat < &MAXCATS then do ;
   do tti = ttmaxcat to tthi by -1 ;
    %DO I = 1 %TO &SCNUM ;
     ttc&I{tti+1} = ttc&I{tti} ;
    %END ;
    ttposs{tti+1} = ttposs{tti} ;
   end ;
   ttmaxcat = ttmaxcat + 1 ;
   if ttcode = 0 then ttunpos{_unix_} = ttmaxcat ;
```

```
          %DO I = 1 %TO &SCNUM ;
           ttc&I{tthi} = ttc&I{0} ;
          %END ;
          ttposs{tthi} = ttmaxcat ;
          if ttcode = 0 then ttnway = ttmaxcat ;
          do ttw = 0 to &NREPWTS ;
           do ttj = 0 to &NVD_1 ;
            ttsums{ttw,ttj,ttmaxcat} = 0 ;
           end ;
          end ;
          %IF &WTDRES %THEN %DO ; %* Clear information for the
            new category *;
%*TESTPR(ttmaxcat)   ;
           do ttj = 0 to &NVD_1 ;
            do ttw = 1 to &MAXPARS ;
             ttsub = (&NVD*(ttmaxcat-1)+ttj+1)*&MAXPARS + ttw ;
%*TESTPR(ttsub)   ;
              vector{(&NVD*(ttmaxcat-1)+ttj+1)*&MAXPARS + ttw}
               = 0 ;
            end ;
            ttgest{ttj,ttmaxcat} = 0 ;
            ttgpred{ttj,ttmaxcat} = 0 ;
            ttrsum{ttj,ttmaxcat} = 0 ;
            ttrssq{ttj,ttmaxcat} = 0 ;
            ttrprod{ttj,ttmaxcat} = 0 ;
            ttsvar{ttj,ttmaxcat} = 0 ;
            ttscov{ttj,ttmaxcat} = 0 ;
           end ;
%*TESTPR(ttj)   ;
          %END ;
          %ELSE %IF &NPREDICT>0 %THEN %DO ;
           do ttj = 0 to &NVD_1 ;
            do ttw = 1 to &MAXPARS ;
             vector{(&NVD*(ttmaxcat-1)+ttj+1)*&MAXPARS + ttw}
              = 0 ;
            end ;
           end ;
          %END ;


          ttcode = 1 ;
          end ;
          else do ;
           call symput("GO_END","1") ;
           stop ;
          end ;
         end ;
        %DO I = 1 %TO &SCNUM ;
         end ;
        %END ;
```

```
     %IF &NPREDICT>0 %THEN %DO ;
       if cw_iter > &MAXITER & cwgrpfst
       then do ttw = 1 to &NPREDICT ;
        tthat{ttw} = 0 ;
       end ;
      %END ;

     end ;
    %END ; %* IF &MAXCATS > 0 *;
   end ; %* if _unix_ *;

   if cwgrpfst = 1 then do ; %* first in group *;
    do _ix_ = 0 to &NREPWTS ; %* group weight is weight from
      first in group *;
     cwgrpwt{_ix_} = cwwt{_ix_} ;
    end ;
    cwgrpfst = 0 ;
%IF %BQUOTE(&PENALTY)^= %THEN %STR(cwgrppen = &PENALTY ;) ;
   end ; %* Code for average weight or harmonic mean weight *;
   %IF &MEANWT = A %THEN %DO ;
    else do _ix_ = 0 to &NREPWTS ;
     cwgrpwt{_ix_} = ((_unix_-1)*cwgrpwt{_ix_}
                          + cwwt{_ix_}) /_unix_ ;
    end ;
   %END ;
   %ELSE %IF &MEANWT = H %THEN %DO ;
    else do _ix_ = 0 to &NREPWTS ;
      if cwgrpwt{_ix_} > 0 & cwwt{_ix_} > 0 then
      cwgrpwt{_ix_} = _unix_/((_unix_-1)/cwgrpwt{_ix_}
                          + 1/cwwt{_ix_}) ;
    end ;
   %END ;
   if cwgrpflg>=1 then do ; %* last in group *;
    cwnunits = _unix_ ; %* First compute the weight for
      this iteration *;
    if cw_iter = 0 then do ;

      do _k_ = 0 to &NREPWTS ;
       cwnewwt{_k_} = cwgrpwt{_k_} ;
       cwcutwt{_k_} = cwgrpwt{_k_} ;
      end ;
    end ;
    else do ;
      do _k_ = 0 to &NREPWTS ;
       if (cwstatus{_k_}=-1) or (cw_iter>&MAXITER) then
        cwcutwt{_k_} = 0 ; %* Start to compute new weight *;
      end ;
      do _unix_ = 1 to cwnunits ;
       cwbase = 0 ;
       cwvbase = 0 ;
```

```
         do _b_ = 1 to &MAXBEN ;
          if cwcats{_unix_,_b_} > 0
          then do _j_ = 1 to cwnvbl{_b_} ;
           _i_ = cwbase + (cwcats{_unix_,_b_}-1)*cwnvbl{_b_}
                 + _j_ ;
            do _k_ = 0 to &NREPWTS ;
             cwtemp = cwvbls{_unix_,cwvbase+_j_} ;
             %IF &NREPWTS > 0 %THEN %DO B = 1 %TO &MAXBEN ;
              %IF &&NB&B.RVAR > 0 %THEN %DO ;
               if _b_=&B & _k_>0
               then cwtemp = cwrvs&B{_unix_,_j_,_k_} ;
              %END ;
             %END ;
             if (cwstatus{_k_}=-1) or (cw_iter>&MAXITER) then
             cwcutwt{_k_} + cwadj{&MAXPARS*_k_+_i_}*cwtemp
%IF %BQUOTE(&PENALTY)^= %THEN / cwgrppen ;
              ;
            end ;
          end ;
          cwbase + cwnvbl{_b_}*cwncat{_b_} ;
          cwvbase + cwnvbl{_b_} ;
         end ;
        end ;

        if cw_iter=1 & cw_iter<=&MAXITER then do ;
         _k_ = 0 ;
         _inwt_ = cwgrpwt{_k_} ;
         if (cwgrpwt{_k_}=0) or (cwgrpwt{_k_}=.) then do ;
          _regwt_ = 0 ;
          _regcwt_ = 0 ;
         end ;
         else do ;
          _regwt_ = cwgrpwt{_k_}*&DISTANCE ;
          _regcwt_ = max(&LOCODE,min(&UPCODE,_regwt_)) ;
         end ;
         output cw_out ;
         if _cwreg_=0 then do ;
          call symput("CWREG","1") ;
          _cwreg_=1 ;
         end ;
        end ;

        do _k_ = 0 to &NREPWTS ;
         if (cwstatus{_k_}=-1) or (cw_iter>&MAXITER) then do ;
          if (cwgrpwt{_k_}=0) or (cwgrpwt{_k_}=.) then do ;
           cwcutwt{_k_} = 0 ;
           cwnewwt{_k_} = 0 ;
          end ;
          else do ;
           cwcutwt{_k_} = cwgrpwt{_k_}*&DISTANCE;
```

```
      %* New value before imposing upper and lower bounds *;
      cwnewwt{_k_} = max(&LOCODE,min(&UPCODE,cwcutwt{_k_})) ;
      %* New value after imposing bounds *;
      if cwcutwt{_k_} ^= cwnewwt{_k_} then cwcutwt{_k_} = 0 ;
      else cwcutwt{_k_} = &CWDIST2 ;
    end ;
   end ;
  end ;
 end ;

%* PART 6: Run calculations via each benchmarks and then each
  x value for the unit. *;

   do _unix_ = 1 to cwnunits ;
    %IF &MAXCATS > 0 %THEN %DO ;
     if cw_iter > &MAXITER then do ;
      do ttw = 0 to &NREPWTS ;
       do ttj = 0 to &NVD_1 ;
        if ttunvds{_unix_, ttj} > .z %* i.e. not missing *;
        then do ; %* Overall estimates *;
         %IF %BQUOTE(&WRWEIGHT)^= %THEN %DO ;
          if ttw = 0 then ttsums{ttw,ttj,ttunpos{_unix_}}
          + &NEWWT*ttunvds{_unix_,ttj} ;
          else
         %END ;
          ttsums{ttw,ttj,ttunpos{_unix_}}
           + cwnewwt{ttw}*ttunvds{_unix_,ttj} ;
        end ;
       end ;
      end ;
      %IF (&WTDRES) OR (&NPREDICT>0) %THEN %DO ;
       do ttj = 0 to &NVD_1 ;
        if ttunvds{_unix_, ttj} > .z %* i.e. not missing *;
        then do ; %* Weighted totals of y variables (needed
          at group level) *;
         %IF &WTDRES %THEN %DO ;
          ttgest{ttj,ttunpos{_unix_}}
           + &NEWWT*ttunvds{_unix_,ttj} ;
         %END ;
         %IF &NPREDICT>0 %THEN %DO ;
          ttw = (ttunpos{_unix_}-1)*&NVD+ttj+1 ;
         %END ;
        end ;
       end ;
      %END ;
     end ;
    %END ;

    cwbase = 0 ;
    cwvbase = 0 ;
```

```
       do _b_ = 1 to &MAXBEN ;
        if cwcats{_unix_,_b_} > 0 then do _j_ = 1 to
          cwnvbl{_b_} ;
         if cwvbls{_unix_,cwvbase+_j_}>.z then do ;
          _i_ = cwbase + (cwcats{_unix_,_b_}-1)*cwnvbl{_b_}
               + _j_ ; %* Calculate the X estimates *;
          do _k_ = 0 to &NREPWTS ;
           cwtemp = cwvbls{_unix_,cwvbase+_j_} ;
           %IF &NREPWTS > 0 %THEN %DO B = 1 %TO &MAXBEN ;
            %IF &&NB&B.RVAR > 0 %THEN %DO ;
             if _b_=&B & _k_>0
             then cwtemp = cwrvs&B{_unix_,_j_,_k_} ;
            %END ;
           %END ;
           if cwstatus{_k_}=-1 or cw_iter>&MAXITER then
           cwxest{&MAXPARS*_k_+_i_} + cwnewwt{_k_}*cwtemp ;
          end ;
          %IF (&WTDRES) OR (&NPREDICT>0) %THEN %DO ;
           if cw_iter = 0 then do _unix2_ = 1 to cwnunits ;
            do ttj = 0 to &NVD_1 ;
             if ttunvds{_unix2_, ttj} > .z %* i.e. not missing *;
             then
             vector{ (&NVD*(ttunpos{_unix2_}-1)+ttj+1)*&MAXPARS
                    + _i_}
               + &NEWWT*cwvbls{_unix_,cwvbase+_j_}
                 *ttunvds{_unix2_,ttj}
%IF %BQUOTE(&PENALTY)^= %THEN / cwgrppen ;
                  ;
            end ;
           end ;
           if cw_iter > &MAXITER then do tti = 1 to ttmaxcat ;
            do ttj = 0 to &NVD_1 ;
             cwtemp = vector{ (&NVD*(ttposs{tti}-1)+ttj+1)
                *&MAXPARS
                           + _i_} ;
             %IF &WTDRES %THEN %DO ;
              ttgpred{ttj,ttposs{tti}} +
               &NEWWT*cwvbls{_unix_,cwvbase+_j_}*cwtemp ;
             %END ;
             %IF &NPREDICT>0 %THEN %DO ;
              ttw = (tti-1)*&NVD+ttj+1 ;
              if ttw <= &NPREDICT then do ;
               tthat{ttw}
                 + cwvbls{_unix_,cwvbase+_j_}*cwtemp ;
              end ;
             %END ;
            end ;
           end ;
          %END ;
```

```
%IF &CWPSRAT %THEN %DO ;
 do _k_ = 0 to &NREPWTS ;
  cwtemp = cwvbls{_unix_,cwvbase+_j_} ;
  %IF &NREPWTS > 0 %THEN %IF &NB1RVAR > 0 %THEN %DO ;
   if _k_>0 then cwtemp = cwrvs1{_unix_,_j_,_k_} ;
  %END ;
  matrix{&NMAT*_k_+_i_} + cwcutwt{_k_}*cwtemp*cwtemp ;
 end ;
%END ;
%ELSE %DO ; %* Calculate the X`X matrix *;
 do _unix2_ = 1 to cwnunits ;
  cwbase2 = 0 ;
  cwvbase2 = 0 ;
  do _b2_=1 to _b_ ;
   cwcatb = cwcats{_unix2_,_b2_} ;
   if cwcatb > 0
   & ((_b2_<_b_) or (cwcatb <= cwcats{_unix_,_b_}))
   then do ;
    if _b2_=_b_ & cwcatb = cwcats{_unix_,_b_}
    then cwtemp = _j_ ;
    else cwtemp = cwnvbl{_b2_} ;
    do _j2_ = 1 to cwtemp ;
     _i2_ = cwbase2 + (cwcats{_unix2_,_b2_}-1)
                     *cwnvbl{_b2_} + _j2_ ;
     do _k_ = 0 to &NREPWTS ;
      cwtemp = cwvbls{_unix_,cwvbase+_j_} ;
      cwtemp2 = cwvbls{_unix2_,cwvbase2+_j2_} ;
      %IF &NREPWTS > 0 %THEN %DO B = 1 %TO &MAXBEN ;
       %IF &&NB&B.RVAR > 0 %THEN %DO ;
        if _b_=&B & _k_>0
        then cwtemp = cwrvs&B{_unix_,_j_,_k_} ;
        if _b2_=&B & _k_>0
        then cwtemp2 = cwrvs&B{_unix2_,_j2_,_k_} ;
       %END ;
      %END ;
      if cwstatus{_k_}=-1 or cw_iter>&MAXITER then
      matrix{&NMAT*_k_+_i_*(_i_-1)/2 + _i2_}
       + cwcutwt{_k_}*cwtemp*cwtemp2
%IF %BQUOTE(&PENALTY)^= %THEN / cwgrppen ;
       ;
     end ;
    end ;
   end ;
   cwbase2 + cwnvbl{_b2_}*cwncat{_b2_} ;
   cwvbase2 + cwnvbl{_b2_} ;
  end ;
 end ;
%END ;
 end ;
end ;
```

```
 %* Now move to base values for next benchmark *;
   cwbase + cwnvbl{_b_}*cwncat{_b_} ;
   cwvbase + cwnvbl{_b_} ;
  end ;
 end ; %* do _unix_ *;

 _unix_ = 0 ;
 cwgrpfst = 1 ;

 %* Put on the unit data output dataset *;
 if cw_iter>&MAXITER then do ;
  _inwt_ = cwgrpwt{0} ;
  if (_inwt_ = 0) or (_inwt_ = .) then _nilwt_ + cwnunits ;
  else cwfreq + cwnunits ;
  if . < &WEIGHT < 0 then _negwt_ + cwnunits ;
  if . < _inwt_ < 0 then _negin_ + cwnunits ;
  %IF %BQUOTE(&GROUP)^= %THEN %DO ;
   if (_inwt_ = 0) or (_inwt_ = .) then _gnilwt_ + 1 ;
   else _gfreq_ + 1 ;
   if . < &WEIGHT < 0 then _gnegwt_ + 1 ;
   if . < _inwt_ < 0 then _gnegin_ + 1 ;
  %END ;

  output cw_wtd ;
 end ;
end ;

%* PART 7: Aggregates calculations for benchmarks across table
 categories *;

 %IF &MAXCATS > 0 %THEN %DO ;
  if cw_iter > &MAXITER
  %IF &WTDRES %THEN & (cwgrpflg >=2) or (cw_iter=0) ;
  %ELSE %IF &NPREDICT > 0 %THEN & (cwgrpflg >= 4) or
   (cw_iter=0) ;
  & cwgrpflg >= 4
  then do ;
  do ttfrom = 1 to ttmaxcat ;
   ttpos = ttposs{ttfrom} ;

    %IF &SCNUM > &CNUM %THEN %DO ;
    %IF &WTDRES %THEN if cwgrpflg = 4 & cw_iter=0 then ;
    if 1
    %DO I = &CNUM+1 %TO &SCNUM ;
     & ttc&I{ttfrom} = &&CTYP&I %* Missing *;
    %END ;
    then ttdposs{ttpos} = ttpos ;
    else ttdposs{ttpos} = -1 ; %*TESTPR(ttfrom
     ttc1{ttfrom} ttdposs{ttpos}) *;
    %END ;
```

```
ttcode = 0 ;
%DO I = 1 %TO &SCNUM ;
 do ttix&I = 0
 %IF (&&CAG&I=1) OR (&I > &CNUM) %THEN to 1 ;;
 if ttix&I = 0 then ttc&I{0} = ttc&I{ttfrom} ;
 else ttc&I{0} = &&CTYP&I ;
 if (ttix&I = 0) or (ttc&I{ttfrom} ^= &&CTYP&I) then
%END ;
 if ttcode = 0 then ttcode = 1 ;
 else do ;
  ttlo = 0 ;
  tthi = ttfrom ;
  ttdone = tthi <= (ttlo + 1) ;
   %* 0 = continue search for this category
      1 = category not in arrays
      2 = category found in arrays *;
  do while(^ttdone) ;
   tti = int(0.5*(ttlo+tthi)) ;
   %DO I = 1 %TO &SCNUM ;
    if ttc&I{tti} < ttc&I{0} then ttlo = tti ;
    else if ttc&I{0} < ttc&I{tti} then tthi = tti ;
    else
   %END ;
   do ;
    ttdone = 2 ;
    ttcpos = ttposs{tti} ;
    %IF &SCNUM > &CNUM %THEN %DO ;
     if ttdposs{ttpos}=-1
     %DO I = &CNUM+1 %TO &SCNUM ;
      & ttc&I{tti} = &&CTYP&I %* Missing *;
     %END ;
     then ttdposs{ttpos} = ttposs{tti} ;
    %END ;

    %* Aggregate into this cell *;
    if cw_iter > &MAXITER & cwgrpflg >=4
    then do ttw = 0 to &NREPWTS ;
     do ttj = 0 to &NVD_1 ;
      ttsums{ttw,ttj,ttcpos} + ttsums{ttw,ttj,ttpos} ;
     end ;
    end ;
    %IF (&WTDRES) OR (&NPREDICT >0) %THEN %DO ;
     if cw_iter = 0 then do ttj = 0 to &NVD_1 ;
     do ttw = 1 to &MAXPARS ;
       vector{ (&NVD*(ttcpos-1)+ttj+1)*&MAXPARS + ttw}
        + vector{ (&NVD*(ttpos-1)+ttj+1)*&MAXPARS + ttw};
      end ;
     end ;
     %IF &WTDRES %THEN %DO ;
      if cw_iter>&MAXITER then do ttj = 0 to &NVD_1 ;
```

```
                    ttgest{ttj,ttcpos} + ttgest{ttj,ttpos} ;
                  end ;
                %END ;
              %END ;
            end ;
            if ^ttdone then ttdone = tthi <= (ttlo + 1) ;
          end ;
          if ttdone ^= 2 then do ;
            call symput("GO_END","1") ;
            stop ;
          end ;
        end ;
      %DO I = 1 %TO &SCNUM ;
        end ;
      %END ;
    end ;

    %LOCAL DORAT ;
    %IF (%BQUOTE(&DENOM)^=) OR (&SCNUM > &CNUM)
    %THEN %LET DORAT = 1 ;
    %ELSE %LET DORAT = 0 ;
    %IF &WTDRES %THEN %DO ;
%* Now for each aggregation level *;
      if cw_iter>&MAXITER & cwgrpflg >= 2 then do ;
        ttnpsu + 1 ;
        %IF &WRFLAG %THEN %DO ;
          length _ncells_ &WRLIST 4 ;
          array ttwr{*} &WRLIST ;
          _ncells_ = min(dim(ttwr),ttmaxcat*(&NVD)) ;
          ttk = 0 ;
        %END ;
        do tti = 1 to ttmaxcat ;
          do ttj = 0 to &NVD_1 ;
            cwtemp = ttgest{ttj,tti} - ttgpred{ttj,tti} ;
            %IF &WRFLAG %THEN %DO ;
              ttk + 1 ;
              ttcpos = ttposs{tti} ;
              if ttk<=dim(ttwr) then
                ttwr{ttk} = ttgest{ttj,ttcpos} - ttgpred{ttj,
                  ttcpos} ;
            %END ;
%*TESTPR(ttj tti (ttgest{ttj,tti}) (ttgpred{ttj,tti})) ;
            ttrsum{ttj,tti} + cwtemp ;
            ttrssq{ttj,tti} + cwtemp*cwtemp ;
            %IF &DORAT %THEN %DO ;
              %* Product with denominator residual *;
              if ttj < &NVAR then do ;
                %IF &SCNUM>&CNUM %THEN tttemp = ttdposs{tti} ;
                %ELSE tttemp = tti ;;
                ttw = min(&NVAR+ttj,&NVD_1) ;
```

```
%*TESTPR(ttj ttw tttemp)   ;
                ttrprod{ttj,tti}
                + cwtemp*(ttgest{ttw,tttemp} -
                  ttgpred{ttw,tttemp})  ;
                end ;
              %END ;
            end ;
          end ;
          %IF &WRFLAG %THEN %STR(output cw_wrout ;) ;
          do tti = 1 to ttmaxcat ;
           do ttj = 0 to &NVD_1 ;
            ttgest{ttj,tti} = 0 ;
            ttgpred{ttj,tti} = 0 ;
           end ;
          end ;
          if cwgrpflg >=3 then do ;
           ttnpsu = max(2,ttnpsu) ;
           do tti = 1 to ttmaxcat ;
            do ttj = 0 to &NVD_1 ;
              ttsvar{ttj,tti} + ttnpsu/(ttnpsu-1)
                *(ttrssq{ttj,tti} - (ttrsum{ttj,tti}**2)/
                  ttnpsu) ;
              %IF &DORAT %THEN %DO ;
               ttw = min(&NVAR+ttj,&NVD_1) ;
               %IF &SCNUM>&CNUM %THEN tttemp = ttdposs{tti} ;
               %ELSE tttemp = tti ;;
               ttscov{ttj,tti} + ttnpsu/(ttnpsu-1)
                    *(ttrprod{ttj,tti}
                      - ttrsum{ttj,tti}*ttrsum{ttw,tttemp}/
                        ttnpsu) ;
              %END ;
             end ;
            end ;
            ttnpsu = 0 ;
            do tti = 1 to ttmaxcat ;
             do ttj = 0 to &NVD_1 ;
              ttrssq{ttj,tti} = 0 ;
              ttrsum{ttj,tti} = 0 ;
              %IF &DORAT %THEN %STR(ttrprod{ttj,tti} = 0 ;) ;
             end ;
            end ;
            if cwgrpflg >=4 then do ;
            end ;
           end ;
          end ;
         %END ;
       end ;
     %END ;
    end ;
```

```
      cwdimen=cw_nb ;
      do _b_=0 to &NREPWTS ;
       if cwstatus{_b_}=-1 or cw_iter>&MAXITER then do ; %*
         Decompose this X`X matrix *;
        cwmbase = _b_*&NMAT ;
        link decomp ;
       end ;
      end ;

      %IF &EXTRA %THEN %DO ;
       if cw_iter=0 then put "BY group " &PUTBY ":" @ ;
      %END ;
      cwreport = (cw_iter>&MAXITER) or (cw_iter<=1) ;
      _result_ = 'C' ;
      if cwreport then do _i_ = 1 to &MAXPARS ;
       cwreptxt{_i_}=repeat(".",&NWREPORT) ;
      end ;
      do _b_ = 0 to &NREPWTS ;
       cwstat = cwstatus{_b_} ;
       if cwstat=-1 or cw_iter>&MAXITER then do ;
        %IF &EXTRA %THEN %DO ;
         if cw_iter>&MAXITER>0 & _b_=0 then
          if cwstat=-1 then put
" did not converge in &MAXITER iterations"  ;
         else if cwstat > 1 then put
" converged in" cwstat 3. " iterations" ;
         else if cwstat = 0 then put
" no iterations required" ;
         else put " used 1 iteration" ;
         else if cw_iter>&MAXITER & &MAXITER=0 & _b_=0 then
           if cwstat = 0 then put " no iterations required" ;
          else put " used 1 iteration" ;
        %END ;
        if cw_iter>&MAXITER>0 & _b_=0 then
         if cwstat=-1 then _iters_ = . ;
         else if (cwstat > 1) or (cwstat = 0) then _iters_ =
           cwstat ;
         else _iters_ = 1 ;
        else if cw_iter>&MAXITER & &MAXITER=0 & _b_=0 then
         if cwstat = 0 then _iters_ = 0 ;
         else _iters_ = 1 ;
        cwepsi{_b_} = 0 ;
        cwbase = 0 ;
        _i_ = 0 ;
        do _i2_ = 1 to &MAXBEN ;
         if cwnrep{_i2_} = 0 then _ir_ = cwbase+1 ;
         else _ir_ = cwbase + _b_ + 1 ;
         do _i3_ = 1 to cwncat{_i2_}*cwnvbl{_i2_} ;
          _i_ + 1 ;
          vector{_i_} = cwxben{_ir_} - cwxest{&MAXPARS*_b_+_i_} ;
```

```
     _ir_ + (cwnrep{_i2_} + 1) ;
   end ;
   cwbase + cwncat{_i2_}*cwnvbl{_i2_}*(1+cwnrep{_i2_}) ;
 end ;

 cwmbase = _b_*&NMAT ;
 ttvbase = 0 ;
 link solve ;
 cwbase = 0 ;
 _i_ = 0 ;
 do _i2_ = 1 to &MAXBEN ;
   if cwnrep{_i2_} = 0 then _ir_ = cwbase+1 ;
   else _ir_ = cwbase + _b_+1 ;
   do _i3_ = 1 to cwncat{_i2_}*cwnvbl{_i2_} ;
     _i_ + 1 ;
     if 1 or (cwxben{_ir_} > &EPSILON) then do ;
       cwtemp = abs(cwxben{_ir_} - cwxest{&MAXPARS*_b_+_i_})
                / max(1,abs(cwxben{_ir_})) ;
       if abs(vector{_i_}) > 1E-9  %* Adjustment worth
         doing *;
       then do ;
        cwepsi{_b_} = max(cwepsi{_b_},cwtemp) ;
         if cwreport then if cwtemp > &EPSILON then do ;
           if _b_ = 0 then _result_='N' ;
           else if _result_ = 'C' then _result_='R' ;
           if _b_<=&NWREPORT
           then substr(cwreptxt{_i_},_b_+1,1) = "N" ;
          end ;
         end ;
         else if cwreport
         then if cwtemp > &EPSILON then do ;
           if _b_ = 0 & _result_^='N' then _result_='I' ;
            else if _result_='C' then _result_='R' ;
           if _b_<=&NWREPORT
           then substr(cwreptxt{_i_},_b_+1,1) = "I" ;
          end ;
        end ;
       _ir_ + (cwnrep{_i2_} + 1) ;
      end ;
     cwbase + cwncat{_i2_}*cwnvbl{_i2_}*(1+cwnrep{_i2_}) ;
   end ;

  if cwstatus{_b_} = -1 and cwepsi{_b_} < &EPSILON
  then cwstatus{_b_} = cw_iter ;
  do _i_ = 1 to cw_nb ;
   cwadj{&MAXPARS*_b_+_i_} + vector{_i_} ;
  end ;
 end ;
end ;
```

```
if cwreport then do ;
 cwbase = 0 ;
 _b_=0 ;
 _i_ = 0 ;

 do _i2_ = 1 to &MAXBEN ;
  if cwnrep{_i2_} = 0 then _ir_ = cwbase+1 ;
  else _ir_ = cwbase + _b_+1 ;
  do _i3_ = 1 to cwncat{_i2_}*cwnvbl{_i2_} ;
   _i_ + 1 ;
   _bench_=cwxben{_ir_} ;
   _best_=cwxest{&MAXPARS*_b_+_i_} ;
   if cwxben{_ir_} > &EPSILON then
    _crit_ = abs(cwxben{_ir_}
      - cwxest{&MAXPARS*_b_+_i_})
           / cwxben{_ir_} ;
   else _crit_=. ;
   _report_=cwreptxt{_i_} ;
   output cwreport ;
   _ir_ + (cwnrep{_i2_} + 1) ;
  end ;
  cwbase + cwncat{_i2_}*cwnvbl{_i2_}*(1+cwnrep{_i2_}) ;
 end ;
end ;

cwdone = 1 ;
do _b_ = 0 to &NREPWTS ;
 cwdone = cwdone and cwstatus{_b_}>-1 ;
end ;

%IF (&WTDRES) OR (&NPREDICT>0) %THEN %DO ;
if cw_iter = 0 then do ; %* Calculate beta parameters for
  each nway table cell *;
 cwmbase = 0 ; %* Use X`X matrix for the 0th weight *;
 do tti = 1 to ttmaxcat ;
  do ttj = 0 to &NVD_1 ;
   ttvbase = (&NVD*(tti-1)+ttj+1)*&MAXPARS ;
   %*TESTPR(ttvbase);
   %*TESTPR(vector{ttvbase+1} vector{ttvbase+2} ) ;
   link solve ;
   %*TESTPR(vector{ttvbase+1} vector{ttvbase+2}) ;
  end ;
 end ;
end ;
%END ;

%IF &MAXCATS > 0 %THEN %DO ;
 if cw_iter > &MAXITER then do ;
```

```
%* PART 8: Copy results into variables for output *;
    do ttcat =1 to ttmaxcat ;
     %DO I = 1 %TO &SCNUM ;
      &&CNAM&I = ttc&I{ttcat} ;
     %END ;

     %DO I = 1 %TO &SCNUM ;
      %IF &&CAG&I^=1
      %THEN %STR(if ttc&I{ttcat} ^= &&CTYP&I then ) ;
     %END ;
     do ;
      ttcpos = ttposs{ttcat} ;
      %IF &SCNUM > &CNUM %THEN %DO ; %* find some subclasses *;
       ttpos = ttdposs{ttcpos} ;
      %END ;
      %ELSE %DO ;
       ttpos = ttcpos ;
      %END ;
      do ttj = 0 to &NVAR_1 ;
       varname = ttvnam{ttj} ;
       length _est_ 8 ;
       %IF &NREPWTS > 1 %THEN %DO ;
        length _var_ _se_ _rse_ 8 ;
        _var_ = 0 ;
       %END ;
       do ttw = 0 to &NREPWTS ;
        %IF %BQUOTE(&DENOM) =
        %THEN %IF &SCNUM > &CNUM %THEN %DO ;
         tttemp = ttsums{ttw,ttj,ttpos} ;
         if (tttemp = 0) or (tttemp <= .z) then tttemp = . ;
         else tttemp = ttsums{ttw,ttj,ttcpos} / tttemp ;
        %END ;
        %ELSE %DO ;
         tttemp = ttsums{ttw,ttj,ttcpos} ;
        %END ;
        %ELSE %DO ;
         tttemp = ttsums{ttw,min(&NVAR+ttj,&NVD_1),ttpos} ;
         if (tttemp = 0) or (tttemp <= .z) then tttemp = . ;
         else tttemp = ttsums{ttw,ttj,ttcpos} / tttemp ;
        %END ;
        if ttw = 0 then _est_ = tttemp ;
        %IF &NREPWTS > 1 %THEN %DO ;
         else if tttemp> .z then _var_ + (tttemp-_est_)**2 ;
        %END ;
        %IF %BQUOTE(&OUTREPS)^= %THEN %DO ;
         array ttrest{0:&NOUTREPS} _est_ &OUTREPS ;
         ttrest{ttw} = tttemp ;
        %END ;
       end ;
```

```
        %IF &NREPWTS > 1 %THEN %DO ;
         _var_ = _var_ * (&NREPWTS-1)/&NREPWTS ;
         _se_  = sqrt(max(0,_var_)) ;
         if _est_ <= 1E-12 then _rse_ = . ;
         else _rse_ = _se_ / _est_ ;
         _typ_ = 'J' ;
         %IF &WTDRES %THEN %STR(output cw_tab;) ;
        %END ;

        %IF &WTDRES %THEN %DO ;
%*TESTPR(ttj ttpos ttcpos) ;
           _var_=ttsvar{ttj,ttcpos} ;
           %IF &DORAT %THEN %DO ;
            %IF %BQUOTE(&DENOM)= %THEN %DO ;
              _var_ = (_var_ - 2*_est_*ttscov{ttj,ttcpos}
              + _est_*_est_*ttsvar{ttj,ttpos})
                 / ttsums{0,ttj,ttpos}**2 ; ;
            %END ;
            %ELSE %DO ;
              _var_ = (_var_ - 2*_est_*ttscov{ttj,ttcpos}
              + _est_*_est_*ttsvar{min(&NVAR+ttj,&NVD_1),ttpos})
                 / ttsums{0,min(&NVAR+ttj,&NVD_1),ttpos}**2 ; ;
            %END ;
           %END ;
           _se_  = sqrt(max(0,_var_)) ;
           if _est_ <= 1E-12 then _rse_ = . ;
           else _rse_ = _se_ / _est_ ;
           _typ_ = 'W' ;
          %END ;
          output cw_tab;
         end ;
%*put ttcat= tti= ttsums{0,ttcat}= ttsums{0,tti}= ttvds{0}= ;
        end ; %* if output required *;
       end ; %* do ttcat *;
%* Clean up for next BY group *;
      %IF &WTDRES %THEN %DO ;
       do ttj = 0 to &NVD_1 ;
        do tti = 1 to ttmaxcat ;
          ttsvar{ttj,tti} = 0 ;
          ttscov{ttj,tti} = 0 ;
         end ;
        end ;
      %END ;
      ttmaxcat = 0 ;
    end ;
   %END ;
  end ;
 end ;
 output cwbyrep ;
```

```
end ;
stop ;
return ;

%* linked code: decomp, solve: inputs are arrays matrix and
   vector and macro variables &DIMEN, &MBASE, &VBASE, &DECEPSI;
   outputs are changed matrix and vector i.e. temporary
   variables used are _i_ _j_ _k_ _i2_ _j2_ *;

%LOCAL DIMEN MBASE VBASE DECEPSI ;
%LET DIMEN=cwdimen ;
%LET MBASE=cwmbase ;
%LET VBASE=ttvbase ;
decomp:
%*   Triangular decomposition:              j
     Symmetric matrix assumed.       1  2  4  7
     Numbering of cells is        i  2  3  5  8
       shown at right for            4  5  6  9
     &MBASE = 0, &DIMEN = 4          7  8  9 10
     That is, only the upper (or lower) half of the symmetric
     matrix is entered.  Result is an upper (or lower)
       triangular
     matrix U such that U`U = A, the original symmetric matrix.
*;
%LET DECEPSI = %SCAN(&DECEPSI?1e-8,1,?) ;
%IF &CWPSRAT %THEN ; %* No need to decompose if matrix is
   diagonal *;
%ELSE %DO ;
 _i2_ = &MBASE ;
 do _i_ = 1 to &DIMEN ;
  do _k_ = 1 to _i_-1 ;
   cwtemp = -matrix{_i2_+_k_} ;
   _j2_ = _i2_ ;
   if fuzz(cwtemp) ^= 0 then
   do _j_ = _i_ to &DIMEN ;
    matrix{_j2_+_i_} + cwtemp*matrix{_j2_+_k_} ;
    _j2_ + _j_ ;
   end ;
  end ;
  _j2_ = _i2_ ;
  do _j_ = _i_ to &DIMEN ;
   if _i_ = _j_ then if matrix{_i2_+_i_} < &DECEPSI
    then matrix{_i2_+_i_} = 0 ;
    else matrix{_i2_+_i_} = sqrt(matrix{_i2_+_i_}) ;
   else if matrix{_i2_+_i_} = 0 then matrix{_j2_+_i_} = 0 ;
   else matrix{_j2_+_i_} = matrix{_j2_+_i_}/matrix{_i2_+_i_} ;
   _j2_ + _j_ ;
  end ;
  _i2_ + _i_ ;
```

```
 end ;
%END ;
return ;

%* Given that matrix contains the triangular decomposition of
   A, and vector contains the elements of a vector b solve
   replaces vector with the elements of x such that Ax = b *;

%IF &CWPSRAT %THEN %DO ;
 do _i_ = 1 to &DIMEN ;
  _k_ = &VBASE + _i_ ;
  _i2_ = matrix{&MBASE + _i_} ;
  if _i2_ > &DECEPSI then vector{_k_} = vector{_k_} / _i2_ ;
 end ;
%END ;
%ELSE %DO ;
 _i2_ = &MBASE ;
 do _i_ = 1 to &DIMEN ;
  _k_ = &VBASE+_i_ ;
  if matrix{_i2_+_i_} > &DECEPSI then do ;
   do _j_ = 1 to _i_-1 ;
    vector{_k_} + -matrix{_i2_+_j_}*vector{&VBASE+_j_} ;
   end ;
   vector{_k_} = vector{_k_} / matrix{_i2_+_i_} ;
  end ;
  _i2_ + _i_ ;
 end ;
 do _i_ = &DIMEN to 1 by -1 ;
  _j2_ = _i2_ ;
  _i2_ + -_i_ ;
  _k_ = &VBASE+_i_ ;
  if matrix{_i2_+_i_} > &DECEPSI then do ;
   do _j_ = _i_+1 to &DIMEN ;
    vector{_k_} + -matrix{_j2_+_i_}*vector{&VBASE+_j_} ;
    _j2_ + _j_ ;
   end ;
   vector{_k_} = vector{_k_} / matrix{_i2_+_i_} ;
  end ;
  else vector{_k_} = 0 ;
 end ;
%END ;
return ;
run ; %* end of main calculations portion *;

%CK_TRASH(cw_wtd)
%CK_TRASH(cw_out)
%CK_TRASH(cwreport)
%CK_TRASH(cwbyrep)
 %IF &GO_END=1 %THEN %DO ;
```

```
 %CHECKERR(1,Not enough space for calcs - increase MAXSPACE)
 %GOTO SKIPEND ;
%END ;
%ELSE %IF &EXTRA %THEN %DO ;
 %PUT Main calculations finished ; %PUT ;
%END ;

%IF &WRFLAG %THEN %DO ;
 options notes ;
 %PUT ;
 %PUT Output weighted residuals dataset: ;
 data &WROUT ;
  set cw_wrout ;
 run ;
 &NOTESOFF
 %CK_TRASH(cw_wrout)
%END ;
%END ;
%LOCAL _REGWT_ _REGRAT_ CWCHANGE ;
%LET CWCHANGE = 0 ;
%LET CWFIRST = %EVAL(&CWFIRST & &CWREG) ;
data cw_wtd(drop=cwunitid cwchange) ;
 %IF &CWREG
 %THEN merge cw_wtd cw_out(keep=cwunitid _regwt_ _regcwt_) ;
 %ELSE set cw_wtd ;
  end=_last_ ;
 by cwunitid ;
 retain cwchange 0 ;
 %IF &CWREG %THEN %DO ;
   if _regwt_ & _inwt_ then _regrat_ = _regwt_/_inwt_ ;
   %LET _REGWT_=_regwt_ ;
   %LET _REGRAT_=_regrat_ ;
 %END ;
 if _inwt_ then do ;
   _wtrat_ = &WEIGHT/_inwt_ ;
   if abs(_wtrat_ - 1)>=0.05 then cwchange = 1 ;
 end ;
 output cw_wtd ;
 if _last_ then call symput("CWCHANGE",put(cwchange,1.)) ;
run ;
%LET CWREG = %EVAL(&CWREG & &CWCHANGE) ;
%IF %BQUOTE(&GROUP)^= %THEN %DO ;
 data cw_wtd ;
   merge cw_unit(keep=&BYVARS &STRATUM &VARGRP &GROUP
     %IF &MAXBEN >=1 %THEN cwgrpflg cwcat1-cwcat&MAXBEN ;
          )
         cw_wtd ;
   &GRPCODE
 run ;
%END ;
```

```
%* PART 9: Diagnostic reports on small area weights *;

%IF &CWCHANGE %THEN %DO ;
 proc univariate data=cw_wtd(where=(&WEIGHT>0)) plot
  %IF %INDEX(&XOPTIONS,%STR( UNIV ))=0 %THEN noprint ;;
   label _inwt_="input weight"
         &WEIGHT="final weight"
         _wtrat_="final weight/input weight" ;
   var _inwt_ &WEIGHT _wtrat_ ;
   &CWTIT "Weights and weight changes, positive weights only" ;
   output out=cwuniv p1=i_p1  w_p1  wr_p1
                  median=i_m    w_m    wr_m
                     p99=i_p99 w_p99 wr_p99
      ;
 run ;
%CK_TRASH(cwuniv)
%END ;
%LOCAL CWEXT ;
data cwext(drop=_one_ w_m cw_keep rename=(&WEIGHT=_finwt_)) ;
%IF &CWCHANGE %THEN %DO ;
  if _n_=1 then set cwuniv ;
   drop
    i_p1 w_p1 wr_p1
    i_m wr_m
    i_p99 w_p99 wr_p99
%END ;
%ELSE w_m = 0 ;;
  _one_ = 1 ;
  set cw_unit(keep=&BYVARS
      %DO B=1 %TO 30 ;
         &&B&B.CLASS &&B&B.VAR
      %END ;
     &STRATUM &VARGRP &GROUP &UNIT &ID &PENALTY) ;
  set cw_wtd(keep=&WEIGHT _inwt_ _wtrat_
    %IF &CWREG %THEN _regwt_ _regcwt_ _regrat_ ;
               ) ;
  length severity 3 ;
  severity = 0 ;
  if . < &WEIGHT < 0 then severity = 8 ;
%IF &CWCHANGE %THEN %DO ;
  else if &WEIGHT > 0 then do ;
   if ^(w_p1 < &WEIGHT < w_p99) then severity + 4 ;
   if ^(wr_p1 < _wtrat_ < wr_p99) then severity + 2 ;
   %IF (&MAXITER > 0) & &CWREG %THEN %DO ;
    if _regwt_ ^= _regcwt_ then severity + 1 ;
   %END ;
  end ;
%END ;
   %IF (&MAXITER > 0) & &CWREG %THEN %DO ;
    drop _regcwt_ _regrat_ ;
   %END ;
```

```
     retain cw_keep 0 ;
     if severity > 0 then do ;
       if cw_keep = 0 then do ;
         if cw_keep = 0 then call symput("CWEXT","1") ;
         cw_keep = 1 ;
       end ;
       if severity = 2 then do ;
         if _wtrat_<= wr_p1 then do ;
           severity = -1 ;
           _order_ = _wtrat_ ;
         end ;
         else do ;
           severity = 1 ;
           _order_ = -_wtrat_ ;
         end ;
       end ;
       else do ;
         if severity = 1 then severity = 2 ;
         if &WEIGHT < w_m then do ;
           severity = -severity ;
           _order_ = &WEIGHT ;
         end ;
         else _order_=-&WEIGHT ;
       end ;
       output cwext ;
     end ;
run ;
proc sort data=cwext ;
  by severity _order_ ;
run ;
options notes ;
%PUT ;
%IF &PRINTREP & (&CWEXT^=1) %THEN
 %PUT Extreme units report (not printed, empty): ;
 %ELSE %PUT Extreme units report: ;
data &EXTOUT ;
 set cwext(drop=_order_) ;
run ;
&NOTESOFF
%LOCAL FULLID ;
%IF (&CWEXT=1) & (&PRINTREP) & (&EXTNO>0) %THEN %DO ;
 %IF %LENGTH(&REPORTID)=0 %THEN %DO ;
  %LET FULLID = &BYVARS ;
  %DO B=1 %TO 30 ;
   %LET FULLID=&FULLID &&B&B.CLASS ;
  %END ;
  %LET FULLID=&FULLID &STRATUM &VARGRP &GROUP &UNIT ;
  %DO B=1 %TO 30 ;
   %LET FULLID=&FULLID &&B&B.VAR ;
  %END ;
```

```
  %LET FULLID = %UPCASE(&FULLID) ;
  %IF &CWFIRST %THEN %LET TEMP = 38 ;
  %ELSE %LET TEMP = 30 ;
  %DO I = 1 %TO 999 ;
   %LET WORD = %SCAN(&FULLID,&I) ;
   %IF %LENGTH(&WORD)=0 %THEN %LET I = 1000 ;
   %ELSE %DO ;
    %IF %INDEX(@@@@@@@@@@@@@@@@@@@@@@@@@@@@@@@@@@@@@@ _ONE_
      &REPORTID @,
            %STR( &WORD )) = 0 %THEN %DO ;
     %LET TEMP = %EVAL(&TEMP + 2 + %LENGTH(&WORD)) ;
     %IF &TEMP <= &LINESIZE %THEN %LET REPORTID = &REPORTID
       &WORD ;
     %ELSE %LET I = 1000 ;
    %END ;
   %END ;
  %END ;
 %END ;
 %GREGPEXT(DATA=cwext, EXTNO=&EXTNO, ID=&REPORTID,
   FIRST=&CWFIRST
 ,OPTIONS=&OPTIONS, TITLELOC=&TITLELOC)
%END ;


%CK_TRASH(cwext)
options notes ; %* put on output log *;
%IF &PRINTREP & (&CWCHANGE^=1) %THEN
%PUT Benchmark report(s) (not printed, <1% change to any
  weight): ;
%ELSE %PUT Benchmark report(s): ;
&NOTESOFF
%* Diagnostic reports on benchmarks *;
data cwbenout ;
  merge cwreport(where=(cw_iter>&MAXITER))
        cwreport(where=(cw_iter=0)
           keep=&BYVARS _i_ _best_ cw_iter
             rename=(_best_=_iest_))
     %IF &CWREG %THEN
       cwreport(where=(cw_iter=1)
         keep=&BYVARS _i_ _best_ cw_iter rename=(_best_=_
           rest_)) ;
           ;
  by &BYVARS _i_ ;
  drop cw_iter ;
run ;
%CK_TRASH(cwbenout)


%LOCAL CONPROB ;
%DO B = 1 %TO &MAXBEN ;
 %LET CONPROB = 0 ;
```

```
 options notes ;
  data &BENPREF&B&BENSUFF ;
   merge cwrep&B cwbenout(where=(_b_=&B)) end=_last_ ;
   retain cw_prob cw_prob2 0 ;
   by &BYVARS _i_ ;
   if _value_ ^= _bench_ then cw_prob2 = 1 ;
   if _report_^=repeat(".",&NWREPORT) then cw_prob = 1 ;
   if _last_ & cw_prob then call symput("CONPROB","1") ;
   if _last_ & cw_prob2
   then put "WARNING: Report benchmarks mismatched" ;
   drop _i_ _b_ _value_ cw_prob2 ;
  run ;
&NOTESOFF
%IF &PRINTREP %THEN %DO ;
 %IF &CWCHANGE=1 %THEN %DO ;
  %LET REPORTID = &BYVARS &&B&B.CLASS ;
  %IF &&NB&B.VAR>1 %THEN %LET REPORTID = &REPORTID _name_ ;
  %GREGPBEN(DATA=&BENPREF&B,BY=&BY,ID=&REPORTID,OPTIONS=&OP
    TIONS
   ,FIRST=&CWFIRST,TITLELOC=&TITLELOC) ;
  %IF (&NREPWTS>0) & &CONPROB %THEN %DO ;
   %GREGPBEN(DATA=&BENPREF&B,BY=&BY,ID=&REPORTID,OPTIONS=&OP
     TIONS
    ,FIRST=&CWFIRST,TITLELOC=&TITLELOC,TYPE=2) ;
   %END ;
  %END ;
 %END ;
%END ;

options notes ;
%PUT Report on overall convergence for BY groups: ;
data &BYOUT ;
 set cwbyrep ;
run ;
&NOTESOFF
%IF &PRINTREP %THEN %DO ;
 %GREGPBY(DATA=cwbyrep,BYVARS=&BYVARS
  ,OPTIONS=&OPTIONS,GROUP=&GROUP,TITLELOC=&TITLELOC) ;
%END ;
options notes ;
%IF &MAXCATS > 0 %THEN %DO ;
%PUT Output table data set: ;
  data &OUT ;
    set cw_tab ;
  run ;
  %CK_TRASH(cw_tab)
%END ;
%PUT Output weighted unit data set: ;
data &OUTDSN ;
  set cw_unit
```

```
  %IF &KEEPID %THEN %DO ;
    (keep=&BYVARS
      %DO B=1 %TO 30 ;
        &&B&B.CLASS &&B&B.VAR
      %END ;
     &STRATUM &VARGRP &GROUP &UNIT &ID
     &CLASS &SUBCLASS &VAR &DENOM)
  %END ;
  %ELSE %DO ;
    (drop=cwgrpflg
     %IF &MAXBEN >=1 %THEN cwcat1-cwcat&MAXBEN ;
     )
  %END ;
  ;
  set cw_wtd(keep=&WEIGHT &REPWTS
%IF (&MAXCATS > 0) & (&NPREDICT>0) %THEN hat_1-hat_&NPREDICT ;
  ) ;
  _one_=1 ;
  drop _one_ ;
run ;
&NOTESOFF

%SKIPEND: %* Any problems cause a jump to here *;
%LET GO_END = ;
data _null_ ;
  call symput("ELAPTIME",put(datetime() - &ELAPTIME, tod7.)) ;
run ;

%* PART 10: Delete all temporary files *;

%IF %INDEX(&XOPTIONS,%STR( DEBUG )) = 0
%THEN %CK_TRASH() ;
&CWTIT ;
options &ORIGNOTE ;
%PUT
NOTE: The macro &MACID took elapsed time &ELAPTIME (hh:mm:ss) ;
%PUT ;

%MEND GREGWT ;
```

Index

A

ABS, *see* Australian Bureau of Statistics
Absolute standardized residual
       estimate (ASRE) analysis, 257
   decision criterion, 235–236
   definition, 235
   microsimulated data sets, 236
   results, 236–239
   SLA-level housing stress
      estimate, 246
   standard notations, 235
   statistical properties, 257
   steps, 235
   unexplained errors, 236
ACT, *see* Australian Capital Territory
Activity-based microsimulation
      models (AMOS), 64
Adelaide, 9, 138, 158–161
   households' tenure types
      for buyers, 205–206, 407–408
      for private renters, 176,
         207, 411–412
      for public renters,
         206–207, 409–410
      for total renters, 207–208, 413–414
   housing stress estimates for overall
      households, 204–205, 405–406
   SLAs, households, and housing
      stress distributions by
      SSDs, 204, 363
Area-level models, 250
ASRE analysis, *see* Absolute
      standardized residual estimate
      analysis
Australian Bureau of Statistics (ABS),
      15, 32, 253
Australian Capital Territory (ACT),
      123, 157, 208
Australian Population and
      Policy Simulator Model
      (APPSIM), 58
Auxiliary data files, 253
Auxiliary information, 18–19

B

Basic area level model
   expression, 33
   linking model, 34
   matching sampling model, 34
   mathematical expression, 34
   parameter estimation, 34
   standard mixed linear model, 35
Basic Community Profile (BCP) data set
      tables, 2006 CPH, 327–328
Bayesian prediction
   basic steps, 103
   Bayes' posterior distribution,
      unknown parameters, 103
   computational method, 5
   data and unobserved sampling
      units, 102
   linkage model, 109–110
   multivariate multiple
      regression model
      data analysis, 104
      generalized gamma function, 106
      linear model, 104
      matrix notations, 105
      matrix-T distribution, 105–106
      multivariate/matrix-variate
         setup, 103
      predictor variables, 104
      probability density function, 106
      Student-t distribution
         modeling, 104–105
   prior and posterior distributions
      generalized gamma function, 109
      invariance theory, 107
      normalizing constant, 108–109
      posterior density, 107
      regression and scale parameters
         matrix, 106
   spatial microdata generation, 102
   unobserved population units
      convenient quadratic form, 112
      covariance matrix, 115
      location matrix, 115

mathematical operations, 112–114
MCMC simulation method, 116
micropopulation data, 116
prediction distribution,
            111–112, 115
probability density function, 111
Bayes' theorem, 103, 107
Benchmarks files, 253
Best linear unbiased prediction (BLUP)
            approach, 42–43
Brisbane
    households' tenure types
        for buyers, 197, 387–388
        for private renters, 198, 391–392
        for public renters,
            197–198, 389–390
        for total renters, 198–199,
            393–394
    housing stress estimates for overall
        households, 196–197, 385–386
    SLAs, households, and housing
        stress distributions by
        SSDs, 195, 362

C

Calibration equation, 21–22, 88
Canberra
    households' tenure types
        for buyers, 210, 417–418
        for private renters, 211, 421–422
        for public renters, 210–211,
            419–420
        for total renters,
            211–212, 423–424
    housing stress estimates for overall
        households, 209, 415–416
    SLAs, households, and housing
        stress distributions by SSDs,
        208–209, 363
Capital cities
    Adelaide
        households' tenure
            types, 205–208
        housing stress estimates for
            overall households, 204–205
        SLAs, households, and housing
            stress distributions by SSDs,
            204, 363
    Brisbane
        households' tenure types, 197–199
        housing stress estimates for
            overall households, 196–197
        SLAs, households, and housing
            stress distributions by
            SSDs, 195, 362
    Canberra
        households' tenure types, 210–212
        housing stress estimates for
            overall households, 209
        SLAs, households, and housing
            stress distributions by SSDs,
            208–209, 363
    characteristics, 181
    Darwin
        households' tenure types, 217–219
        housing stress estimates for
            overall households, 216
        SLAs, households, and housing
            stress distributions by
            SSDs, 215, 364
    estimates mapping at SLA levels
        rural city areas, 186
        spatial distribution
            patterns, 184–185
        urban city areas, 186
    Hobart
        households' tenure types, 213–215
        housing stress estimates for
            overall households, 212–213
        SLAs, households, and housing
            stress distributions by
            SSDs, 212, 364
    households and housing stress
        scenarios, 182–183
        trends, 183–185
    Melbourne
        households' tenure
            types, 192–195
        housing stress estimates for
            overall households, 191
        SLAs, households, and housing
            stress distributions by
            SSDs, 190–191, 362
    Perth
        households' tenure types, 200–203
        housing stress estimates for
            overall households, 200

SLAs, households, and housing
stress distributions by SSDs,
199–200, 363
Sydney
households' tenure
types, 188–190
housing stress
estimates, 186–188
SLAs, households, and housing
stress distributions by
SSDs, 186, 361
Census data
aggregate output data sets, 121–122
in Australia, 120
BCP data tables, 122, 327–328
population and housing, 120
social science researchers, 121
Topics in the 2006 Census of
Population and Housing,
120–121, 326
Census table data sets, 121
Coefficient of variation (CV)
measure, 140
Combinatorial optimization
(CO) reweighting
approach, 251–252
hypothetical data, 85–87
intelligent searching, 82
iterative process, 82
sequential fitting procedure, 82–83
simulated annealing, 83–85
Community profiles, 122, 327–328
Composite estimator, 29–30, 32, 250
Confidence interval (CI)
estimation, 9–10
ASRE analysis, 257
confidence bounds, 241
critical value, 240–241
indicative CI measure, 240
margin of error (ME), 240
range, 240
results, 242–245
standard error of model
estimate, 240
"Confidential unit record file" (CURF),
123–124, 145, 253
Constraint function, 88, 100, 251
Consumer price index (CPI) files,
141–142, 145, 253

D

Darwin
households' tenure types
for buyers, 217, 433–434
for private renters, 218, 437–438
for public renters, 217–218,
435–436
for total renters,
218–219, 439–440
housing stress estimates for overall
households, 216, 431–432
SLAs, households, and housing
stress distributions by SSDs,
215, 364
Data gathering, 3
Demographic estimation, 250
advantages, 32
Australian Bureau of Statistics, 32
birth and death rates, 30
component method, 31
estimated population, small
area, 31
fitted regression equations, 31
immigration statistics, 31
vital rates approach, 30
Design-based *indirect* estimators, 20
Design-based *semidirect* estimators, 20
Deville–Sarndal distance, 89
Dynamic microsimulation models, 55
APPSIM development, 58
APPSIM model, 58
bottom-up strategy, 57
DYNASIM, 57–58
Guy Orcutt's pioneering work, 57
population and cohort models, 58
Dynamic Microsimulation of Income
Model (DYNASIM), 57–58

E

Empirical Bayes (EB) approach,
43–44, 250
(Empirical) best linear unbiased
prediction (EBLUP) approach,
42–43, 250
EURAREA project, 14–15
Exact matching, *see* Statistical data
matching/fusion

F

Fay–Herriot model, *see* Basic area
    level model

G

Generalized linear mixed models
    (GLMMs)
    Bayesian procedure, 37–38
    binomial logistic mixed model, 38
    classical inferential approach, 38
    defined equation, 36
    first-order autoregressive, 37
    GLSM, 39
    nonnormal hierarchical models, 40
    Poisson model, 39
    poverty rate estimation, 37
    robust estimation, 38
    spatial and temporal random
        effects, 36
    SPREE, 39
    unit-specific auxiliary data, 36–37
    unmatched model, 39–40
Generalized linear structural model
    (GLSM), 39
Generalized regression (GREG)
    estimator, 18
Generalized regression weighting
    (GREGWT) algorithm, 124,
    147–148, 251–253
General linear mixed models, 250
General model file, 253, 260–269
Geographic approaches
    CO reweighting approach
        hypothetical data, 85–87
        simulated annealing, 83–85
    GREGWT approach
        explicit numerical solution,
            hypothetical data, 91–97
        new weight generation, 90–91
        theoretical setting, 88–90
    GREGWT *vs.* CO
        methodologies, 97–99
    microsimulation modeling
        advantages, 68–69
        different characteristics, 54
        dynamic models, 57–59
        IPF, 72–74

        population-based/cohort-based
            models, 54
        process of, 52–54
        repeated weighting
            method, 74–80
        reweighting, 80–81
        spatial microdata, 69–70
        spatial models, 59–68
        static models, 55–57
        statistical data matching/
            fusion, 70–72
Gold standard, 72
GREGWT approach, 251–252
    explicit numerical solution,
        hypothetical data, 91–97
    new weight generation, 90–91
    survey estimates, 87
    theoretical setting, 88–90
    truncated chi-squared distance
        function, 87–88
GREGWT program file, 253
"Group jackknife" approach, 96–97

H

Hierarchical Bayes (HB) approach,
    45–47, 250
Hobart
    households' tenure types
        for buyers, 213, 427
        for private renters, 214, 429
        for public renters, 213–214, 428
        for total renters, 214–215, 430
    housing stress estimates for
        overall households,
        212–213, 425–426
    SLAs, households, and housing
        stress distributions by SSDs,
        212, 364
Horvitz–Thompson (H-T)
    estimator, 17, 78
Housing policy, 2
Housing stress
    comparison of measures, 138–141
    definition, 136
    estimation, 125
    fifty SLAs with highest number of
        buyer households, 343–344
    households, 341–342

private renter households,
347–348
public renter households,
345–346
total renter households, 349–350
fifty SLAs with highest percentage of
buyer households, 353–354
overall households, 351–352
private renter households,
357–358
public renter households,
355–356
total renter households, 359–360
final model outputs, 143–144
measurement, 136–138
model execution process, 142–143
second-stage model, 141
Housing stress estimation
Australian households, tenure types,
154–155
different states, 155–157
Lorenz curve, 151–152
model accuracy report, 147–148
number of households
buyer households, 166–167
choropleth map, 164
cumulative percentage, SLAs, 164
estimation, 165–166
micro-level variability, 164
private renters, 169–170
public renters, 167–169
SLA, 162, 165–166
spatial distribution, 163, 165
three different measures, 148–149
total renters, 170–171
percentage distribution, 150–151
proportional cumulative frequency
graph, 152–154
SAE
buyer households, 174–175
cumulative frequency, 173
percentages, low-income
households, 171–174
private renter households, 176
public renter households,
175–176
SLA, 173–174
spatial distribution, 171–172
total renter households, 177

"30 only rule," 149
"30/(10–40) rule," 149–150
various statistical divisions, 157–160
various statistical subdivisions,
159–162

I

Implicit model–based approach, 3, 250
Integrated Land Use, Transport
and Environment (ILUTE)
model, 66–67
Iteration process, 73, 147
Iterative proportional fitting (IPF)
contingency table analysis, 72
determinism property, 73
iteration process, 73
mathematical properties, 74
MCMC sampling, 72
reliability, 73
reweighting/synthetic
reconstruction techniques, 74
utility improvement, 74

J

James–Stein (J-S) estimator, 30

K

Knowledgeable decision-making, 2

L

Lagrange multipliers vector, 89–90
Linear regression, 18
Linear truncated/restricted modified
chi-squared method,
see Truncated chi-squared
distance function
Logistic regression, 22

M

MapStats data sets, 121
Margin of error (ME), 240
Markov Chain Monte Carlo (MCMC)
method, 250, 252
Mean square error (MSE), 29–30

Melbourne
  households' tenure types
    for buyers, 192, 377–378
    for private renters, 193–194,
        381–382
    for public renters, 192–193,
        379–380
    for total renters, 194–195,
        383–384
  housing stress estimates for overall
      households, 191, 375–376
  SLAs, households, and housing
      stress distributions by SSDs,
      190–191, 362
Metropolis Criterion, 251
Microsimulation modeling technology
      (MMT), 3–4
  data sources and issues
    census data, 120–122
    survey data sets, 122–124
  GREGWT approach, 124
  housing stress
    comparison of measures, 138–141
    definition, 136
    estimation, 125
    final model outputs, 143–144
    measurement, 136–138
    model execution process, 142–143
    second-stage model, 141
  model inputs
    auxiliary data files, 128, 132
    benchmark files, 128–131
    general model file, 126
    GREGWT file, 132
    unit record data files, 126–128
  model outputs, 134–135
  reweighting process, 125
  reweight, survey sampling, 119
  small area synthetic weight
      generation, 132–134
Middle class earners, 138
MMT, *see* Microsimulation modeling
      technology
Model calibration/model-free
      calibration procedures, 19
Modeling and Simulation for e-Social
      Science (MoSeS), 60
Monte Carlo Markov chain (MCMC)
      methods, 45

Multinomial logistic GREG (MLGREG)
      estimator, 20–21
Multivariate multiple regression
      model
    data analysis, 104
    generalized gamma function, 106
    linear model, 104
    matrix notations, 105
    matrix-T distribution, 105–106
    multivariate/matrix-variate
        setup, 103
    predictor variables, 104
    probability density function, 106
    Student-t distribution modeling,
        104–105

N

National Center for Health Statistics
      (1968), 13, 28
Newton–Raphson iteration
      method, 251
    accepted root, 322
    first-order adjustment, 321
    first-order derivative, 321
    graphical representation, 322
    iteration algorithm, 322–323
    Lagrange multipliers vector, 322
    root-finding algorithm, 321
    Taylor series, 321
Newton–Raphson method, 91
Northern Territory (NT), 123

O

Object-oriented programming
      approach, 68
Optimal weight, 30

P

Perth
  households' tenure types
    for buyers, 200–201, 397–398
    for private renters, 202–203,
        401–402
    for public renters, 201–202,
        399–400
    for total renters, 203, 403–404

housing stress estimates for overall
households, 200, 395–396
SLAs, households, and housing
stress distributions by SSDs,
199–200, 363
Probability density curve (PDC),
228–229

Q

QuickStats data sets, 121

R

Ratio-synthetic estimator, 28–29
Regression estimator (RE), 76, 78–79
Regression-synthetic estimator, 29
Repeated weighting method
applications, 79
auxiliary data, 75
classification variable, 77–78
common margin evaluation, 77
data reliability, 75
estimation strategy, 76
GREG estimator, 75
Horvitz–Thompson estimator, table
counts, 78
known population totals, 79
RE calibration properties, 76
SSD, 74–75
table estimation, 80
Reweighting algorithm, 147–148

S

SAE, *see* Small area estimate/estimation
Sample regression method, 31
SAS program file, 253
SAS programming, reweighting
algorithms
aggregates calculations, benchmarks
across table category, 467–473
delete all temporary files, 483
diagnostic reports, small area
weights, 479–483
list of variable names, tables,
453–455
run calculations, each benchmarks
and x value, 464–467

set up default values, macro
variables, 441–446
start of main calculations portion,
455–464
unit creation, datasets, 446–453
variables for output, 474–478
Shrinkage estimator, *see* James–Stein
(J-S) estimator
SIH-CURF files, 123, 125, 145, 253
Simulated annealing, 83–85, 99, 116,
251–252
"Simulated estimation," 54
Simulation Model of the Irish Local
Economy (SMILE), 61
Skewed distribution, 154
SLA, *see* Statistical local area
Small area estimate/estimation
(SAE), 1–2
advantages, 6, 12–13
alternative Bayesian prediction-
based microdata simulation
technique, 252
applications, 6
Australian Bureau of Statistics, 15
in Canada, 14
EURAREA project, 14–15
National Center for Health
Statistics (1968), 13
in United States, 13–14
Australia, 8–9
Bayesian approach, 12
Bayesian prediction theory, 8
buyer households, 174–175
in capital cities, 255
cumulative frequency, 173
direct estimation, 13, 250
assisting models, 19
calibration equation, 22
comparison of, 22–24
flexibility, 21
GREG estimator, 18, 20
H-T estimator, 17
MLGREG estimator, 20–21
model-free calibration
estimators, 20–21
modified direct estimator, 18–19
statistical estimators, 20
explicit models, 7
future research areas, 259–260

geographic approach, 7
history, 12
implicit model–based approaches, 7, 15–16
indirect model-based statistical approaches
  explicit models, 250
  geographic approaches, 250–251
  implicit model-based methods, 250
  reweighting algorithm, 251
  spatial microsimulation model-based geographic approaches, 251–252
knowledgeable and effective decision-making, 250
limitations, 257–259
MCMC simulation, 252
MMT, 8
objectives, 5–6
percentages, low-income households, 171–174
policy analysis, 249–250
private renter households, 176
public renter households, 175–176
reweighting methodologies, 16
sample survey, 13
SAS codes and programs, MMT
  general model file codes, 260–269
  reweighting algorithms, 269–283
  second-stage computation process, 283–293
SLA, 173–174, 253–255
small area, definition, 12
spatial analysis reports, 9
spatial distribution, 171–172
spatial microsimulation model
  commonwealth rent-assisted housing, 256
  housing stress estimation, 253–255
  MMT-based construction, 252–253
  specification, 253
  validation techniques, 256–257
"spatial microsimulations," 15
statistical reliability, 257
statistics estimation, 11–12

summary diagram, different methodologies, 15–16
total renter households, 177
validation techniques, 9–10
Small area statistics, 1
Social statistical database (SSD), 74–75
Spatial distributions, housing stress estimates
Adelaide
  for buyers, 205–206, 407–408
  for overall households, 204–205, 405–406
  for private renters, 207, 411–412
  for public renters, 206–207, 409–410
  for total renters, 207–208, 413–414
Brisbane
  for buyers, 197, 387–388
  for overall households, 196–197, 385–386
  for private renters, 198, 391–392
  for public renters, 197–198, 389–390
  for total renters, 198–199, 393–394
Canberra
  for buyers, 210, 417–418
  housing stress estimates for overall households, 209, 415–416
  for private renters, 211, 421–422
  for public renters, 210–211, 419–420
  for total renters, 211–212, 423–424
Darwin
  for buyers, 217, 433–434
  for overall households, 216, 431–432
  for private renters, 218, 437–438
  for public renters, 217–218, 435–436
  for total renters, 218–219, 439–440
Hobart
  for buyers, 213, 427
  for overall households, 212–213, 425–426
  for private renters, 214, 429
  for public renters, 213–214, 428
  for total renters, 214–215, 430

Melbourne
    for buyers, 192, 377–378
    for overall households, 191,
      375–376
    for private renters, 193–194,
      381–382
    for public renters, 192–193,
      379–380
    for total renters, 194–195, 383–384
Perth
    for buyers, 200–201, 397–398
    for overall households, 200,
      395–396
    for private renters, 202–203,
      401–402
    for public renters, 201–202,
      399–400
    for total renters, 203, 403–404
Sydney
    for buyers, 188, 367–368
    for overall households, 186–188,
      356–366
    for private renters, 189, 371–372
    for public renters, 188–189,
      369–370
    for total renters, 189–190, 373–374
Spatial microdata, 54
Spatial microsimulation, 52, 55
  applications, 60
  countries used, 59–60
  economic development and policy
    changes, 59
  health services provision, 62
  human activity–environment
    interaction, 65
  ILUTE model, 66–67
  Irish SMILE model, 64
  land use–focused models, 66
  LocSim model, 62
  micropopulation data set, 59
  MoSeS, 60
  regional development and planning
    model, 67
  Ryuichi's AMOS model, 64–65
  SALSA model, 65
  SimCrime model, 62
  SimLeeds model, 61
  SMILE, 61
  spatial disaggregation, 63

statistical matching,
    micro–household data, 68
survey-based behavioral simulation,
    Nielsen's study, 64
SVERIGE model, 60–61, 66
Swedish SVERIGE model, 63
SYNAGI, 61
tax–benefit microsimulation, 61
theoretical framework, 63
transport model, 65
UrbanSim modeling framework, 67
Spatial microsimulation for Leeds
    (SimLeeds) model, 61
Spatial microsimulation model-based
    geographic approaches, 251
SSDs, *see* Statistical subdivisions
Standard normal curve (SNC), 226
Standard validation methodology, 4
Static Incomes Model
    (STINMOD), 55–56
Static microsimulation
  decisive role, policy reforms, 55
  policy change effects, 55
  STINMOD, 55–56
  TRIM, 56–57
Statistical approaches
  explicit models
    basic area level model, 33–35
    basic unit level model, 35–36
    comparative table, 41
    EB approach, 43–44
    EBLUP approach, 42–43
    generalized linear mixed models,
      36–40
    HB approach, 45–47
    methodological characteristics,
      47–49
  implicit models
    comparative table, indirect
      estimators, 32–33
    composite estimation, 29–30
    demographic estimation, 30–32
    synthetic estimation, 28–29
Statistical data matching/
    fusion, 70–72
Statistical local area (SLA), 157, 173–174,
    186, 253–255, 258, 361
  fifty SLAs with highest number of
    housing stress

buyer households, 343–344
households, 341–342
private renter households,
      347–348
public renter households, 345–346
total renter households, 349–350
fifty SLAs with highest percentage of
      housing stress
  buyer households, 353–354
  overall households, 351–352
  private renter households,
      357–358
  public renter households, 355–356
  total renter households, 359–360
hypothesis test results, 228–229
Statistical reliability measures, 5
Statistical reliability, MMT estimates
  measure, 239–245 (*see also* Confidence
      interval estimation)
  validation methods
    ASRE analysis, 235–239
    basic comparisons, 222
    comparison against ABS census
        data, 222
    constraint and nonconstraint
        variables, 223
    expert validation, 223
    general-level equal variance
        t-test, 222
    individual-level simulated
        micropopulation data, 223
    reaggregation, 222–223
    regression approach, 222–223
    reweighting process, 222
    standard error about identity, 222
    statistical significance test,
        225–235
    two-tailed *t*-test, 223–224
    Z-score-based chi-squared
        measure, 222, 224
    Z-score measure, 222, 224
Statistical significance test
  basic steps, 225
  mathematical notations, 225
  null and alternative hypotheses, 226
  results
    economic growth, 232
    households by tenure
        types, 232–233

inaccurate microsimulation
    modeling technology, 234
noncapital city, 232
PDC, 228–229
poverty rate, 231
private renter households, 234
at SLA level, 228–229
Z-statistic with
    p-value, 229–231
two-sided statistical test, 226
Z-statistic
  definition, 226
  observed significance level, 226
  p-value, 226–227
  standard normal distribution,
      226–227
Statistical subdivisions (SSDs)
  Adelaide, 204, 363
  Brisbane, 195, 362
  Canberra, 208–209, 363
  Darwin, 215, 364
  Hobart, 212, 364
  Melbourne, 190–191, 362
  Perth, 199–200, 363
  Sydney, 186, 361
Structure preserving estimation
    (SPREE), 39
Survey-based national-level *microdata
    files*, 126
Survey data sets
  ABS reports, 124
  ACT/NT, 123
  CURF, 123
  SIH, 122–123
2005–2006 Survey of Income and
    Housing (SIH), 329–339
Survey of Income and Housing–
    Confidentialised Unit Records
    Files (SIH-CURF), 253
Sydney
  households' tenure types
    for buyers, 188, 367–368
    for private renters, 189, 371–372
    for public renters, 188–189,
        369–370
    for total renters, 189–190, 373–374
  housing stress estimates for
      overall households,
      186–188, 356–366

SLAs, households, and housing
  stress distributions by SSDs,
  186, 361
Synthetic Australian Geo-demographic
  Information (SYNAGI), 61
Synthetic estimator, 24, 28–29, 32, 250
Synthetic reconstruction, 69–70, 81
Synthetic weights file, 253
Systematic hypothesis test, 256
System for Visualizing Economic and
  Regional Influences Governing
  the Environment (SVERIGE)
  model, 60
Systems Analysis for Sustainable
  Agricultural production
  (SALSA) model, 65

T

30 only rule-based measure, 149, 253
30/40 rule-based-based measure, 253
30/(40–10) rule-based measure, 253
30/(10–40) rule, 149–150

Topics index
  Basic Community Profile (BCP)
    data set tables, 2006 CPH,
    327–328
  2006 census of population and
    housing, 326
  2005–2006 SIH—CURFs data,
    329–339
Total absolute distance (TAD)
  function, 84
Transfer Income Model (TRIM), 56–57
Truncated chi-squared distance
  function, 87–88, 90
Truncated linear regression method,
  see Truncated chi-squared
  distance function

U

Ultimate small area housing stress
  estimation, 253
Unit-level models, 250
Unit records data files, 253

9 780367 261269